RIVERS OF THE ASIAN HIGHLANDS

Rivers of the Asian Highlands introduces readers to the intersecting headwaters of Asia's eight largest rivers, focusing on the upper reaches of two river systems: the Brahmaputra's highland tributaries in the eastern Himalayan Mountains and the Dri Chu (upper Yangzi), which descends from the Tibetan Plateau's east through the Hengduan Mountains.

This book guides its readers through these two rivers' physical, environmental, cultural, social, and political histories before providing a multifaceted assessment of their present. It uses general and detailed insights from multiple disciplines, including anthropology, conservation, geography, geomorphology, climate science, ecology, history, hydrology, and religious studies. The rivers' stories explain how the catchments' hazards—earthquakes, landslides, floods, droughts, and erosion—interact with their energetic, hydrological, ecological, cultural, and social abundance.

This book's multiple cultural and disciplinary perspectives on the rivers will interest anyone who wants to understand the rivers of this critically important region as the environment faces climate change and other ecological crises.

Ruth Gamble is a senior lecturer in history and environmental humanities in the Department of Archaeology and History at La Trobe University in Melbourne, Australia. She is an environmental and cultural historian of Tibet and the Himalaya. She has previously worked on the Himalaya and Tibet's sacred geography and is writing a book on the environmental history of the Yarlung Tsangpo River.

Gillian G. Tan is a senior lecturer in anthropology at the School of Humanities and Social Sciences at Deakin University in Geelong, Australia. Her research covers the multiple ways humans relate with their natural environment and focuses on the cultural and environmental changes experienced by nomadic pastoralists of the eastern Tibetan plateau. Her current work explores the intersections between religion and ecology.

Hongzhang Xu is a postdoctoral fellow at the Mathematical Sciences Institute, Australian National University in Canberra, Australia. He specializes in transdisciplinary research and explores societal adaptation to changes through power relations, cultural values, and institutional dynamics. His focus includes Aboriginal and Torres Strait Islander mathematics and sciences, and he has additional expertise in biodiversity, water management, and Indigenous knowledge.

Sara Beavis is a senior lecturer at the Fenner School of Environment and Society, Australian National University in Canberra, Australia. She has a background in environmental geology with a focus on water. Her research explores the impacts of anthropogenic and natural processes, including land use and climate variability and change on streamflows and water quality, and the cascading consequences these have for downstream river systems, water-dependent ecosystems, and human communities.

Petra Maurer is a professor at the Institute of Indology and Tibetology, Ludwig Maximilian University in Munich, Germany. Her research focuses on the cultural history of Tibet, particularly on divination, medicine, and lexicography. She compiles the *Wörterbuch der Tibetischen Schriftsprache* at the Bavarian Academy of Sciences and Humanities.

Jamie Pittock is a professor at the Fenner School of Environment and Society, Australian National University in Canberra, Australia. His background includes research into the integration between the management of freshwater ecosystems and responses to climate change, involving case studies from nine countries. His current research focuses on environmental governance, climate change adaptation, energy, and the sustainable management of water.

John Powers is a fellow of the Australian Academy of Humanities and currently holds a joint position in the School of Historical and Philosophical Studies and the Centre for Contemplative Studies at the University of Melbourne in Melbourne, Australia. His research includes works on Buddhist epistemology and Yogācāra soteriology, Chinese government propaganda on Tibetan Buddhism, gender in Indian Buddhism, and human rights issues.

Robert J. Wasson is an emeritus professor at the Australian National University and an adjunct at James Cook University, Cairns, Australia. He is a geomorphologist with expertise in reconstructing landscape change over long periods, catchment processes and management, sediment and nutrient budgeting, extreme flood hydrology and its geomorphic expression and mitigation, and the impact of climate variability and change on fluvial systems. He has undertaken research in many parts of Australia and Asia.

Routledge Planetary Spaces Series

Series Editor: Professor Kimberley Peters,
Helmholtz Institute for Functional Marine Biodiversity,
University of Oldenburg, Germany

This book series captures the emergent, expanding, and important set of discussions on planetary spaces. It provides a home for cross-disciplinary and cutting-edge work that is attentive to the relations between planetary and extra-planetary processes and human and more-than-human life.

In a world of uncertainty, flux, and transformation, the 'planetary' is becoming an ever-important framework for making sense of socio-cultural, economic, political, and environmental change. As it becomes increasingly challenging to draw neat lines around economic crisis, environmental degradation, the reach of the urban, and geopolitical fall-out, the planetary provides a means of understanding the intersecting role of 'worldly,' 'earthly,' and 'more-than-earthly' matter and processes in relation to human and more-than-human life.

We invite scholars from across the social sciences and humanities to publish their original and innovative work in a series that intends to be the 'go to' place for insights concerning planetary and extraplanetary spaces.

Weather: Spaces, Mobilities and Affects
Edited by Kaya Barry, Maria Borovnik and Tim Edensor

Rivers of the Asian Highlands
From Deep Time to the Climate Crisis
Ruth Gamble, Gillian G. Tan, Hongzhang Xu, Sara Beavis, Petra Maurer, Jamie Pittock, John Powers and Robert J. Wasson

For more information about this series, please visit: https://www.routledge.com/Routledge-Planetary-Spaces-Series/book-series/RPS

RIVERS OF THE ASIAN HIGHLANDS

From Deep Time to the Climate Crisis

Ruth Gamble, Gillian G. Tan, Hongzhang Xu, Sara Beavis, Petra Maurer, Jamie Pittock, John Powers and Robert J. Wasson

Routledge
Taylor & Francis Group
LONDON AND NEW YORK

Designed cover image: Getty Images

First published 2025
by Routledge
4 Park Square, Milton Park, Abingdon, Oxon OX14 4RN

and by Routledge
605 Third Avenue, New York, NY 10158

Routledge is an imprint of the Taylor & Francis Group, an informa business

British Library Cataloguing-in-Publication Data
A catalogue record for this book is available from the British Library

Library of Congress Cataloging-in-Publication Data
Names: Gamble, Ruth, 1972– author. | Tan, Gillian G., author. | Xu, Hongzhang, author. | Beavis, Sara, author. | Maurer, Petra H. author. | Pittock, Jamie, author. | Powers, John, 1957– author. | Wasson, Robert J., author.
Title: Rivers of the Asian highlands : from deep time to the climate crisis / Ruth Gamble, Gillian G. Tan, Hongzhang Xu, Sara Beavis, Petra Maurer, Jamie Pittock, John Powers and Robert J. Wasson.
Description: New York : Routledge, 2025. |
Series: Routledge planetary spaces series | Includes bibliographical references and index.
Identifiers: LCCN 2024026689 (print) | LCCN 2024026690 (ebook) | ISBN 9781032490588 (hbk) | ISBN 9781032490595 (pbk) | ISBN 9781003392033 (ebk)
Subjects: LCSH: Rivers—Asia. | Uplands—Asia. | Tibetan Plateau.
Classification: LCC GB1335 .G36 2025 (print) | LCC GB1335 (ebook) | DDC 950.09693—dc23/eng/20240729
LC record available at https://lccn.loc.gov/2024026689
LC ebook record available at https://lccn.loc.gov/2024026690

ISBN: 9781032490588 (hbk)
ISBN: 9781032490595 (pbk)
ISBN: 9781003392033 (ebk)

DOI: 10.4324/9781003392033

Typeset in Sabon
by codeMantra

To our friend and colleague, Bear McPhail, who played a key role in this project's inception and early development but was unable to see its completion.

CONTENTS

ADDITIONAL CONTRIBUTORS

Alexander E. Davis is a senior lecturer in international relations at the School of Social Sciences at the University of Western Australia in Perth, Australia. His research interests include Indian foreign policy, postcolonial international relations theory, and the political ecology of the Himalaya.

Lhamo Kyab is a historian of Tibet and the Himalaya. He is the author of the Tibetan language book, *Bod mdo dbus mtho sgang gi gtsang po* (Rivers of the Central and Eastern Tibetan Plateau) and is now completing a PhD at the Central Institute of Higher Tibetan Studies, India.

Dechen Palmo is a research fellow at the Environment and Development Desk of the Tibet Policy Institute. Her research focuses on Tibet's transboundary rivers and their geostrategic importance, which was also the subject of her PhD research at the University of Madras, India.

Kelzang T. Tashi is a postdoctoral fellow at the University of Heidelberg in Germany. His research focuses on the intersections of religion, society, kinship and gender, health and Indigenous healing, migration, and the environment, with a regional focus on the Eastern Himalaya.

A NOTE ON KEY TERMS AND TIBETAN SPELLINGS

The region we call the Asian Highlands in this book is also called the Hindu Kush Himalaya (HKH) (Wester et al. 2019) and the Pan-Tibetan Highlands (Liu et al. 2022). It includes the Tibetan Plateau and its surrounding mountain ranges: the Pamir, Tian Shan, Himalaya, and Hengduan.

Throughout this book, wherever possible, we use the local and Indigenous names of rivers and other geographic sites followed by their more commonly known names in brackets. As it would be difficult to list all the local names, we have prioritized those from the Tibetic and Tibeto-Burman–speaking regions. We recognize that multiple other languages are spoken in the Highlands and regret that we cannot include all of them. Including them would make the book very difficult for nonlanguage specialists to penetrate. This is particularly true in the case of Highland river naming conventions, which are complicated. For example, the river we have called the Langchen Tsangpo (Sutlej) is often called the Langchen Chu or Langchen Khabap. We have called the river known internationally as the Mekong by its Tibetan name, Da Chu (Tibetan spelling *zla chu*), following its most common Tibetan spelling. We have used the most common international river names in brackets but recognize these are not used widely in China. This means, for example, we use Yangzi instead of Chang Jiang, the Mekong instead of Lancang Jiang, and the Salween instead of Nu Jiang.

The one significant exception to our use of local names is the geological terms for the Asian Highlands' components. Many of these terms entered general scientific discourse through inconsistent Chinese-language transliterations and do not reflect local language naming conventions. However, as these terms are commonly used within Earth science discourse, changing them to reflect local sensibilities would confuse readers. Therefore, we have used the generally accepted scientific names for these terms with explanations of the term's origin following it in brackets. For example, the Qiangtang terrane takes its name from the Tibetan Plateau's Changthang, spelled *chang thang* in Tibetan, meaning "Northern Plain." We have left its name as Qiangtang terrane.

As Tibetan spelling conventions involve many silent letters that make it difficult to read, we have tried to write all Tibetan words in phonetics. All spellings follow the Padmakara Simplified Phonetic phonemic system through the Tibetan phonetics conversion code of the Tibetan and Himalayan Library (https://www.thlib.org/cgi-bin/thl/lbow/phonetics.pl?sep_join=%20), but do not include diacritic markings. The only exceptions to using this system are the local Latinized spellings of Bhutanese and Sikkimese geographic features, which are widely used in these countries. A list of Tibetan spellings of river names is provided as an appendix. We recognize that using this phonetic system does not reflect the diversity of pronunciation across the Highlands' languages, but we needed to use one system to avoid using multiple spellings for the same words.

We have tried to minimize jargon in the book to make it more accessible. But inevitably, particularly given the subject matter and the number of disciplines from whose perspective we are writing, it is neither possible nor helpful to remove all technical terms from the book. To aid those who are not specialists in any of the multiple research areas we discuss in the book, we have added short, bracketed definitions following technical or discipline-specific terms.

ACKNOWLEDGMENTS

This book was made possible through an Australian Research Council Discovery Project grant (DP190101253) and Australian Research Council Discovery Early Career Researcher Grant (DE210101348). We would like to thank our collaborators at the Rivers Workshop in Cairns, Australia, for their insights: Rohan D'Souza, Kunbzang Tsering, Mirza Zulfiqur Rahman, Scott Relyea, and Navarun Varma. We would also like to recognize the debt this work owes to the intellectual legacy of James C. Scott (1936–2024).

1

INTRODUCTION

River Views

The headwaters of the Dri Chu, "the female yak's water," form at the base of the Dangla (Tanggula) Mountains on the northeastern edge of the Asian Highlands. The Dangla Mountains only rise another thousand meters above the already-high Tibetan Plateau, but as they catch snow from winter Westerlies and summer monsoons, they are visible across large distances. Some of the snow they catch compacts into ice, a frozen reservoir that suspends water for centuries or millennia. Other snow melts with the spring, and the water flows downhill, accumulating at the range's base and creating small streams that water grasslands on which marmots, deer, yak, and sheep graze and snow leopards hunt. Tibetan pastoralists, known as *drokpa*, traverse these plains, moving with seasonal pulses to high summer and low winter pastures. As well as providing a home for the *drokpa* and their animals, the Dangla Mountain slopes also give rise to three of Asia's largest rivers: the western Gyelmo Ngul Chu (Salween) rises to their south, the Da Chu (Mekong) to their east, and the Dri Chu, known downstream as the Yangzi, forms to their north.

After coalescing in the *drokpa*'s pastures, the Dri Chu turns south, descending through a 1,000-kilometer gorge that drops its waters 2,000 meters at rates up to 2.7 meters per kilometer. Large hydropower dams extract energy from the falling water's powerful kinetic energy. At the river's southernmost point, it realigns with the Da Chu and western Gyelmo Ngul Chu as the rivers, separated by 5-kilometer-high mountains, flow through the deep gorges of the Three Parallel Rivers region. The East and South Asian Monsoon rains wet these ravines, adding substantially to the river's flow and creating an array of ecological niches for plant and animal life. This life, in turn, produces biologically rich soils that the river erodes and carries downstream.

DOI: 10.4324/9781003392033-1

After descending through the Three Parallel Rivers region, the Dri Chu turns abruptly again, away from the other two rivers, and flows northeast through the Hengduan Mountains, where two other Highland mountain rivers, the Nyak Chu (lower Yalong) and the eastern Gyelmo Ngul Chu (Dadu) drain into it. In Tibetan, this land is known as Kham, and its people are called Khampa. The Khampa have another name for their homelands, Chuzhi Gangtruk, "Four Rivers, Six Ridges," in recognition of its precipitous topography. After the Dri Chu leaves Chuzhi Gangtruk, it traverses the Naxi and Yi homelands, crosses the Sichuan Basin, plummets through the Three Gorges, and onto the heavily populated central Chinese plains. On these plains, it is known as Chang Jiang, "the long river." Locals only describe it by its internationally known name, Yangzi, near its delta on the East China Sea. In the Lowlands, it runs parallel to yet another river that descends from the northeast of the Plateau, the Ma Chu (Huang He or Yellow River). The Ma Chu descends from the next mountain range north of the Dangla, the Bayan Ha Mountains, and crosses China's Northern Plains.

The Yarlung Tsangpo, "the pure river of Yarlung," accumulates waters in the wetlands between the Khangtise (Gangdese) Range and the Western Himalaya in the Asian Highlands' southwestern corner. Like the Dri Chu, its headwaters are fed by glacial runoff and seasonal snow melt, and it rises with other rivers. The Sengge Tsangpo (Indus) rises to the Khangtise's north, the Langchen Tsangpo (Sutlej) to its southwest, and the Macha Tsangpo (Karnali) rises to the mountains' south. The Macha Tsangpo then transects the Himalaya and flows into the Ganga (Ganges), which rises just south of the Himalayan watershed.

After the Yarlung Tsangpo's waters seep out of its headwater wetlands, it forms a braided river in an otherwise high-altitude desert and continues east along the Indus-Tsangpo suture across the southern Plateau into Central Tibet. Two gorges interrupt the river's journey through Central Tibet, separating it into its three regions: from Tsang, the river flows through the Kampala Gorge into U, and from U, it flows through the Gyatsa Gorge into Kongpo. As the river travels east, the southern monsoon penetrates its catchment, bringing moisture. Willows appear on its banks in Tsang and U, along with centuries-old, irrigated barley fields. Tsang's largest town, Zhigatse, sits near the Myang Chu's confluence with the Yarlung Tsangpo. The Kyi Chu (Lhasa River), which flows through Tibet's capital city, Lhasa, joins the Yarlung Tsangpo in U before it enters the Gyatsa Gorge, where a series of recently installed hydropower dams draw energy from its flow. In Kongpo, on the other side of Gyatsa Gorge, pine forests, fruit trees, and tea plantations grow along the river's banks, and the Nyang Chu joins it from the north after flowing past Central Tibet's fastest-growing city, Nyingtri. At this point, the Yarlung Tsangpo is only 300 kilometers away from the Three Parallel Rivers region and the western Gyelmo Ngul Chu, Da Chu, and Dri Chu rivers. The Ayerarwady River's headwaters sit between them.

FIGURE 1.1 *Left*, The Dri Chu in Tiger Leaping Gorge, Yunnan. Photograph by Ruth Gamble, 2009. *Right*, The Yarlung Tsangpo in western Tibet, Yunnan. Photograph by Ruth Gamble, 2018.

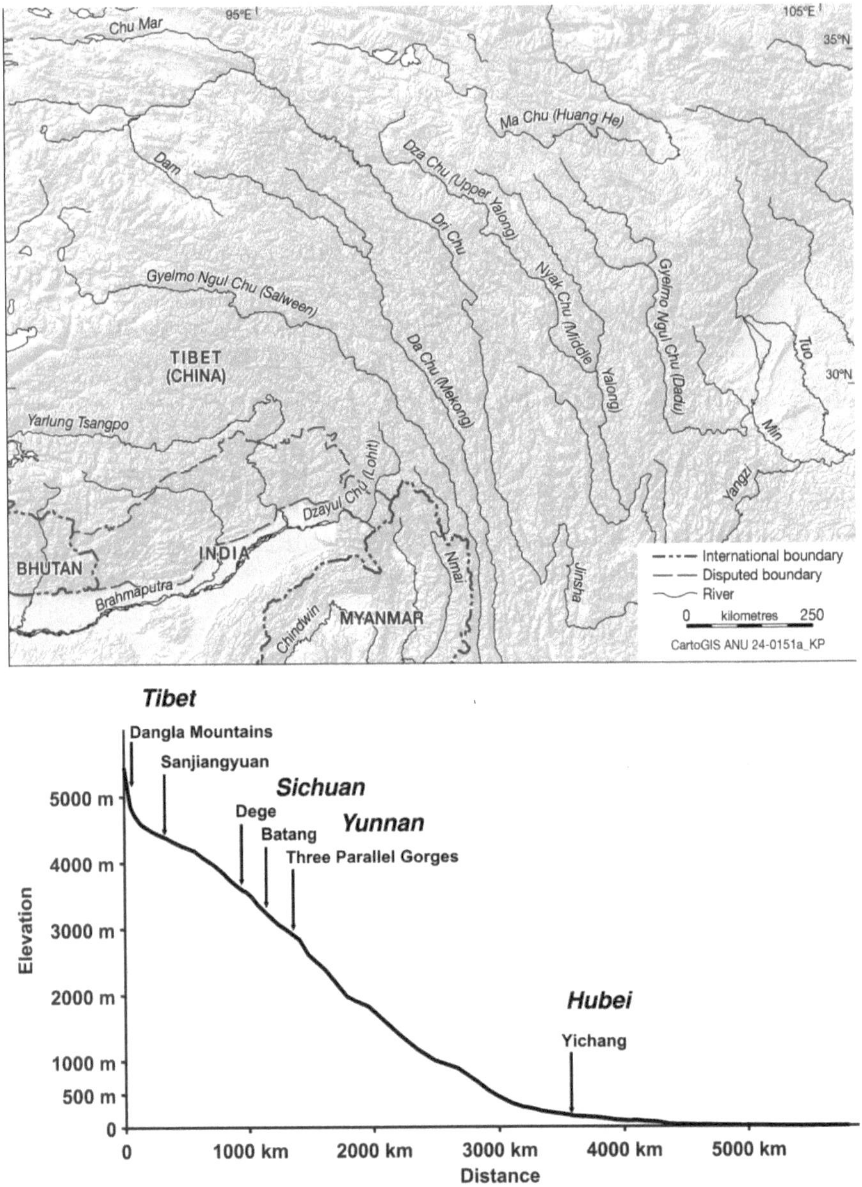

FIGURE 1.2 *Top*, Map of the Dri Chu and other Upper Yangzi Catchment rivers. *Bottom*, Long profile of the Dri Chu.

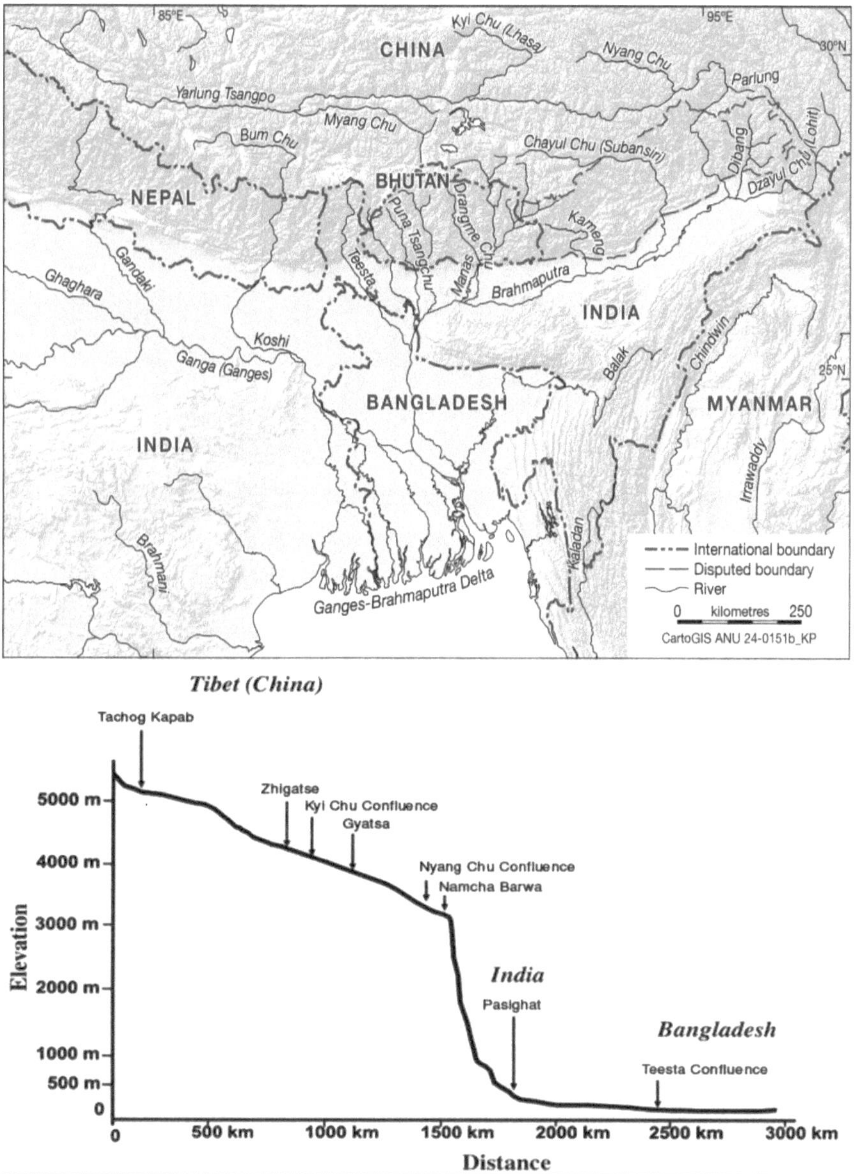

FIGURE 1.3 *Top*, Map of the Yarlung Tsangpo and other upper Brahmaputra tributaries. *Bottom*, Long profile of Yarlung Tsangpo.

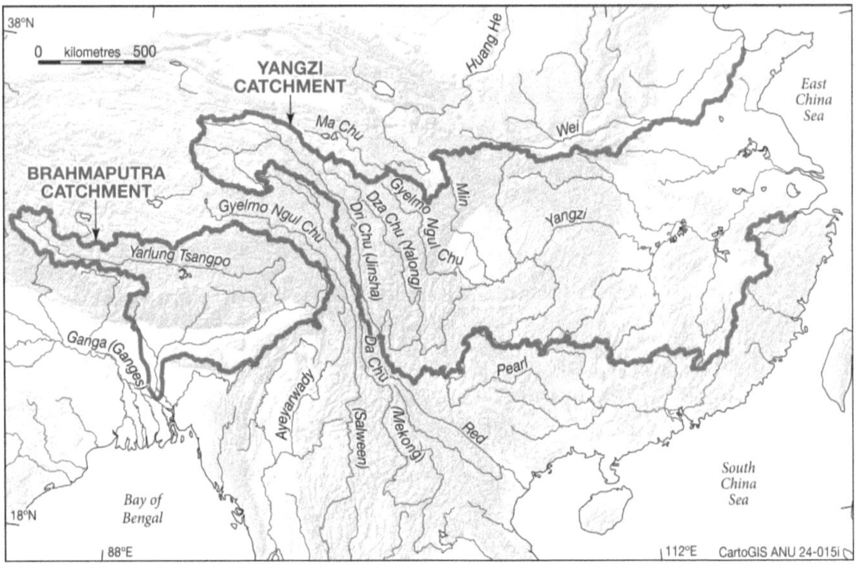

FIGURE 1.4 The upper and lower catchments of the Brahmaputra and Yangzi Rivers.

At the Plateau's southeastern edge, the Yarlung Tsangpo turns sharply southwest. Descending through its—and the terrestrial Earth's—deepest gorge, its waters drop from 3,500 to 500 meters within 250 kilometers. The Chinese government plans to build the world's largest hydropower station on this bend, but for now, its waters flow unimpeded through a mosaic of ecological zones, from alpine forests through broadleaf canopies to tropical rainforests. The gorge's steep descent has also encouraged cultural diversity; Tibetans, Pemakopa, and Adi people—who call the river Siang—all live in and around the gorge. At the gorge's end, the river empties into Assam, where the Dzayul Chu (Lohit) and Dibang rivers join it. Their combined flow, the Brahmaputra River, then moves west toward its delta on the Bay of Bengal. As it flows, more Himalayan rivers spill into it, including the Chayul Chu (Subansiri) from Arunachal Pradesh, Drangme Chu (Manas) from Bhutan, and the Teesta from Sikkim.

1.1 Asian Highland Rivers, a Planetary Space

Together, the rivers that rise below the Dangla and Khangtitse Mountains support around three and a half billion people, nearly half the world's population. What happens in these rivers' catchments has shaped and will shape human history. They are unequivocally a planetary space, and this planetary space is suffering from multiple crises. Climate change, biodiversity loss, water shortages, and pollution are all stressors to the rivers and the populations they support. The causes of this crisis stretch back through history and out around the globe. They are underpinned and intensified by unequal global systems that extract from the Global South to give to the Global North and the inequalities within the catchments between rich and poor, powerful and powerless, and the Highlands and Lowlands.

This book responds to the rivers' multiple crises by taking a different perspective than most studies of rivers. Rather than focusing on the heavily populated river plains, it seeks to reset the Highland–Lowland knowledge imbalance by concentrating on the rivers' upper reaches. Without their Highland headwaters, these rivers would not exist. They are all geologically and ecologically complex catchments, lightly populated by a mosaic of local and Indigenous peoples, and far removed from the centers of regional power and economic activity. Few places on Earth are as little-known and globally consequential as these headwaters.

We consider the many flows of river water, evaporation and precipitation, people, ideas, economies, armies, and governance between the Lowlands and Highlands, particularly those of our two case study rivers: the Yangzi and Brahmaputra. Our view extends the rivers' temporality, positioning them in their deep-time earth-water-climate cycles and the *longue durée* of generational care by local and Indigenous people. We ask how this *longer-than-human* perspective provides insights into the rivers' current crises and possible future. By becoming a geological force in the Anthropocene, humans have stepped into this deep-time story, making decisions with implications that spread across the planet and further into time. From the perspective of the Asian Highlands as a planetary space, we ask what led us into these crises, how much the rivers' legacies have locked in its future or whether the threads that bind the rivers' present can be unraveled.

1.2 Braiding Research

As thinking on this *longer-than-human* scale about a planetary space like the Asian Highlands is so expansive, we have brought together a group of authors to tackle it. The book's principal co-authors have been trained in nine disciplines: anthropology, climatology, conservation ecology, cultural history, environmental history, geography, geology and geomorphology, political ecology, and religious studies. As only one of our co-authors is from the Highlands, Hongzhang Xu is from Yunnan, we have tried to mitigate our Global North perspective by co-writing sections of the book with four other scholars with expertise in anthropology, international relations, and cultural history, all but one of whom are Indigenous Highlanders. We conducted research in the mountains in collaboration with multiple communities and had hoped to co-write much more of the book with them, but the region's complicated politics made the inclusion of these other collaborative researchers' work in this book problematic.

FIGURE 1.5 Satellite image of the braided Yarlung Tsangpo River in Central Tibet. Google Earth.

Mostly, this work is multidisciplinary in that we wrote it by comparing and contrasting our disciplinary observations. Occasionally, however, these disciplinary perspectives overlap entirely, becoming interdisciplinary. Our diverse sources—reflective of our disciplines—include empirical data sets, manuscripts, archives, literature, art, and ethnographies. We present case studies from the Earth, biological and social sciences, and stories from the humanities.

Co-writing with this many people across multiple disciplines has been fraught but productive. We have had to balance the positivist—usually physical science—tendency to present rivers as independent phenomena that experts can enumerate with the constructivist—generally humanist—tendency to insist rivers exist within place-based, language-dependent cultures. To avoid polarity, we have understood positivism and constructivism as a spectrum rather than a choice and approached both the spectrum's edges as problematic. The positivist extreme creates elite "hydrocracies" that fail to reflect human and *more-than-human* lived experiences of rivers (Boelens et al. 2022). The relativist extreme cannot acknowledge the systemic scientific studies at the planetary scale that explain significant shifts in physical phenomena like climate and tends to fetishize local knowledge and activists as feisty underdogs (Baghel and Stepan 2017).

Some have suggested that a compromise or synthesis of positivist and constructivist approaches could be reached (Tvedt 2019). But, as the philosopher of science, Isabelle Stengers (2018) has argued, in practice, such compromises allow powerful positivist scientific hegemonies to co-opt the aspects of cultural understanding that suit their purposes. By playing the hegemonic game, scientists also leave themselves open to the state and commercial appropriation of their findings.

This is why, in this book, we have chosen to place our variant understandings of the rivers alongside each other and in conversation rather than seeking compromise or synthesis. This approach mitigates positivist–constructivist extremes and either/or discourses. It forestalls conceptual hegemonies and allows commonalities and tensions to unfold. We are not looking for a grand synthesis of our views and have not set up a competition between our perspectives. Instead, we allow the book's perspectives to converge and diverge like channels in a braided river.

Presenting multiple river views not only resists hegemonies but also answers Stenger's (2018) call for "slow science," which she argues leads to more deliberate engagement with the rivers and, thus, more precise insights. Just as braided rivers flow more slowly than those in gorges, thinking with scholars from other disciplines has complicated and slowed our river studies. Slowing the research down also allows us to introduce many more voices, including those of Indigenous and local communities whose generational knowledge is often overlooked, and to compare and contrast the multiple temporal scales within which our disciplines operate. In this way, by following the multiple channels and temporal pulses of our research, we are using the river as method to find meaning in our multiple disciplinary meanderings.

1.3 Three River Channels: Planetary, Social, and Regulated

To add coherence to our braided approach, we channeled our insights into three main perspectives on rivers: Planetary, Social, and Regulated. Each of these approaches has specific but not exclusive visions. The Planetary Rivers view examines these waterways' geophysical and biological systems within the temporal framework of deep time. The Social Rivers view looks at human–river relationships that operate within webs of meaning. Within these human–river relationships, the river is an endowment to be cared for and passed to the next generation. Regulated Rivers assesses the ways humans use technologies to control rivers' flow. Human attempts to regulate rivers have a much shorter history than the other perspectives, and this view regards rivers as resources from which water and power can be extracted (D'Souza 2008). These socio-ecologies and socio-hydrologies operate within Earth Systems but regularly try to subvert them, often approaching them with High Modernist technological fixes that then require "a fix to fix the fix" (Harrell 2023, 2) when they are undermined by Earth Systems. The book's analysis of the Regulated Rivers approach from the perspectives of "deep-time Planetary River" and "generational Social River" perspectives provides many of its key insights.

All three perspectives rise from and reflect what Michel Foucault called "knowledge genealogies" (2003, 306). How we see rivers depends not only on our shared need for them and the human ability to cognize them but also on the knowledge about them that we inherit. Multiple genealogies inform our three perspectives. Physical science knowledge lineages shape the Planetary

Rivers perspective. The Social Rivers perspective encounters and engages with the Highlands' Indigenous knowledge lineages and those they imported from India and China. Studies of colonialism and its descendants, globalization and development, along with knowledge about river-management systems, inform the Regulated Rivers perspective.

Each genealogy has a history. They have developed, converged, and diverged over millennia, centuries, and decades, shifting human understandings of rivers. Before we journey further into the river catchments, we need a better understanding of the perspectives that will guide us. We not only need to understand the rivers but also learn how we have learned to look at them.

1.3.1 Planetary Rivers: Hypothesizing River Systems

Like all knowledge lineages, the physical sciences were shaped by historical and social contexts. This construction is often obscured by the hegemonic reproduction of uncontextualized scientific data as an unfolding, progressively discovered truth. Most scientific disciplines developed from the seventeenth-century "Scientific Revolution," whose advocates claimed the world could be understood mathematically and empirically through rational thought. This way of thinking depended on a "hyperseparation" of reasoning men, representing Culture, from an external object they called Nature, which was their object of study (Latour 1993, Plumwood 1993). As this tradition implicitly assigned "rationality" and "empiricism" to European men, their science was imbricated within the colonial project, and reduced rivers to quantifiable resources. For centuries, the scientific process—the refinement of knowledge through the scientific method—encouraged governments and corporations to think human subjects would garner ever greater control over Nature. Furthermore, this teleological march would surpass or at least challenge all other knowledge systems.

In the 1940s, however, new strands of scientific theory began to develop that placed humans back within Nature's "systems." The Macy Conference of Cybernetics in 1946 enabled early proponents of systems thinking, such as Norbert Wiener, a mathematician, and Gregory Bateson, an anthropologist, to exchange ideas and subsequently develop the field of cybernetics and critical ideas such as the double bind. Key to these discussions was a theory of cybernetic information, namely how information is communicated because of a disparity or hierarchy between two parts. For Bateson, information not only regulated the respective learning processes of the parts (what he calls "learning to learn") but also contributed to how the part connected to something larger (Tan 2014). From the start, then, the theory of cybernetic information and early expressions of systems thinking underpinned subjects as varied as stochastic process and schizophrenia. The ideas of cybernetics greatly influenced an important figure of modern ecology, G. E. Hutchison, who would use the ideas of hierarchy, learning to learn through feedback, and interdependence between part and context to develop the foundation for ecology. Specifically, he showed how the conditions under which groups of organisms exist cause the actions of one member of the group on others (Tan 2014). These actions he called "circular causal paths" (Hutchinson 1948) and the self-regulating mechanism of feedback loops, negative when an output inhibits the system and positive when an output amplifies the system, is the basis for ecosystems analysis to this day.

Understanding humans as existing in ecosystems was one of multiple strands of Earth Systems science thinking that became mainstream during the 1970s and 1980s. During this period, Earth Systems science began to describe several interpenetrating "spheres"—the geosphere, atmosphere, hydrosphere, cryosphere, and biosphere—and examined their interactions (Warde et al. 2018). Importantly for the scientific worldview, this placed humans within a system; they were no longer observers of nature but rather actors within it. At the same time, an increasing number of physical scientists began to voice concerns about human impacts on the spheres. Rifts appeared between some lineages—or disciplines—of scientific knowledge and the globally dominant, extraction-driven political economies built on seventeenth-century European imperial empiricism. New fields, such as Hutchinson's ecology, offered direct and scathing critiques of this system (Coen et al. 2022). Since the turn of the millennia, the multiple crises in Earth Systems—climate change, biodiversity loss, and pollution—have intensified the focus on these systems' interconnected functions.

The Earth System paradigm transformed all river studies, from the physical sciences to the humanities. Placing rivers, humans, and other riverine inhabitants, animate and inanimate, within the spheres offered a new way to understand rivers. In Earth Systems, rivers are, to quote Ellen Wohl (2020), "broadly connected systems with bidirectional fluxes of energy and matter between the channels of the river network and the greater environment."

In this cycle, water takes three forms: gas (vapor), liquid (water), and solid (ice and snow). It cycles through the atmosphere as a gas, the hydrosphere as water, and the cryosphere as ice and snow. Solar radiation evaporates water, generally from oceans and lakes. As this water vapor rises (often forced upward by mountains), it cools, condenses, and precipitates. Rain, snow, and ice fall back onto the earth. Some frozen water stays within ice caps, glaciers, or permafrost for millennia.

Liquid water flows across the Earth's surface through river systems and lakes. Rivers, accumulations of flowing waters, are our greatest stores of accessible freshwater. River courses are determined by topography, geology, climate, and human interventions (structures or anthropogenic climate change). If the gradient is steep and straight, they can flow relatively quickly, or their water can be diverted into wetlands, lakes, and constructed reservoirs and may take centuries or millennia to complete the cycle. River courses pulse with floods and low flows and, over time, where gentle gradients occur, their primary channels can move large distances diagonally, traversing flood plains. Some liquid water infiltrates the ground and feeds aquifers. Some water flows down to the ocean, where evaporation produces vapor and the cycle restarts.

The broad Asian Highlands' rivers system is Earth's most significant concentration of the water cycle. The Highlands create and catch precipitation from summer monsoonal systems and winter Westerlies. In summer, vapor rises from the Indian and Pacific Oceans, is drawn up to the Highlands by low-pressure currents, and precipitates when it reaches their high altitudes. In winter, the process reverses: a westerly overland system brings less precipitation and colder temperatures from inner Eurasia. Winter precipitation is primarily snow. Summer precipitation also falls as snow at high altitudes. Much of this snow remains frozen in the Highlands' cryosphere, creating Earth's third largest concentration of ice, a "Third Pole," with permafrost, snow cover, and glaciers. Spring and summer snowmelt and lower-altitude rain combine to create an intense seasonal water cycle, the consequences of which are felt as rains and floods across monsoonal Asia.

The Highlands are also one of the most critical sites in the Earth's geosphere. Their geological and hydrological cycles co-created the rivers. Both processes began when Insular India, a tectonic plate that had broken away from the supercontinent Gondwana 72 million years ago, collided with the Eurasian plate around 50 million years ago. Geologists and geomorphologists see this deep-time story unfolding in the Highlands' rocks. They see sedimentary rocks that formed on ocean floors before being thrust up to mountainous heights as the subduction zone between the two plates grew thicker. They see metamorphic and igneous rocks that formed when the plates buckled, and magma seeped and burst through Earth's crust. They also know that this orogenic (mountain-building) story is not over. The Himalaya and Hengduan mountains remain intensely geologically active, creating earthquakes, landslides, avalanches, and outburst-flood hazards for residents to negotiate. Not only do mountains play a significant role in precipitation creation but they also store gravitational potential energy (GPE), which adds great force to the rivers' descent. This allows the rivers to incise further into the mountains, eroding their soils and carrying them downstream. Some of these soils sink to the ocean floor, where water pressure creates the same sedimentary rocks thrust upward by the Indian-Eurasian plate collision millions of years ago.

Over shorter periods, between the mountains and the sea, waterway sediments and atmospheric dust lay the building blocks for the biosphere. Minerals and biomatter flowing through hydrological and geological systems are life-giving; they co-create aquatic and liminal (river-edge) biodiversity, wetlands, and grasslands. These wet, fertile spaces—especially those on the Highlands' edge where altitude and climatic variability co-create ecological niches—are sites of prolific biodiversity. The Dri Chu, upper Brahmaputra, and the river systems between them, the Gyelmo Ngul Chu (Salween) and

Da Chu (Mekong), house Earth's most biodiverse terrestrial, wetland, and aquatic spaces. From these spaces, flood pulses and irrigation carry biomatter onto floodplains, depositing the fertile soil upon which agriculture, pastoralism, and the planet's greatest concentration of humanity depend.

Along with describing Earth Systems, a Planetary Rivers perspective also outlines the multiple anthropogenic effects on the spheres, particularly humanity's uneven role in environmental crises. It notes how climate change is already disturbing the Highlands' intermingled atmosphere–cryosphere–hydrosphere, and these disturbances will increase with rising temperatures. In addition, this global anthropogenic event is intensified by regional actions. One example that we will explore more in Chapter 4 is black carbon, a type of ash created in the Lowlands that that monsoons carry up to ice fields, depositing a layer of soot that intensifies warming.

A Planetary Rivers perspective explains how the region's gravitational energy production has influenced Highland development by encouraging hydropower constructions and making these buildings vulnerable to earthquakes, landslides, and extreme weather events. It also explains how urbanization and hydropower installations have profoundly impacted biodiversity within and around the rivers. It notes that risk assessments for dams and other infrastructure are based on short records rather than deep-time Earth science geoarchives. This perspective outlines these impacts and speaks to the disconnect between the scientific insights of Earth Systems and governmental and commercial behavior in the catchments, which is further investigated by a Regulatory Rivers perspective in later chapters. It also acknowledges the new data showing Indigenous knowledge systems have conserved biodiversity better than modernist river-management systems. Estrada et al. (2022) have encouraged physical scientists to engage more fully with other knowledge lineages. But as historians and philosophers like Stengers (2018) have suggested, the scientists' engagement with other knowledge genealogies has raised further questions for the scientific project that cannot necessarily be answered by experiments alone, including: How can scientists engage with these other genealogies ethically and honestly? How can they recognize the glaring power differentials between the various views? And how do locals understand rivers? The authors of this book are attempting to grapple with these pressing questions.

1.3.2 Social Rivers: Rivers as Endowments

A Social Rivers perspective seeks to answer these questions by diving into the *longue durée* of human–river relations and explaining how this relationship has affected neighboring people, other animals, and the environment. This perspective is influenced by Jamie Linton and Jessica Budd's (2014) call for the recognition of "hydrosocial" rivers in management regimes and Ruth Morgan's (2017) insistence that river histories need to be understood as hydrosocial phenomena. Humans depend on freshwater, meaning rivers have played an analogous role in all human communities (Strang 2005). Over humans' long history, we have used rivers for drinking, washing, transport, and fishing. We practiced agriculture and pastoralism on the silt-rich, moist floodplains created by the rivers' avulsions and seasonal flood pulses. We dammed, diverted, and channeled their waters, drained their wetlands, and wet their drylands, developing strategies to deal with flood pulses' dual nature as an enlivener and a threat. We used rivers' flows to carry away sewage and other pollution, and as an energy provider, grinding grain and transporting water to our fields.

Rivers have also been a source of political power. Early states arose around rivers, and as the influential political theorist Karl Wittfogel (1957) noted, their rulers often utilized the human need for water to control populations. Water's specific qualities, such as its bulkiness and weight, meant that the infrastructure required to utilize it for agriculture was large-scale and, therefore, needed often-forced, community-wide labor for its construction. Whoever mobilized this labor and controlled water infrastructure held great power. Later scholars nuanced Wittfogel's often orientalist and ahistorical deterministic lens on hydropolitics. By providing details of the myriad hydrosocial configurations—ranging from locally irrigated fields to large, state-sponsored infrastructure—in Asia and elsewhere and highlighting humans' shifting relations with rivers over time. Importantly, for this study, they also noted how these relations included non-material cultural and cosmological understandings of water, which underpin politics and knowledge systems, even if they are naturalized and unacknowledged.

Some of the most influential studies of these hydrosocial systems have examined the lower reaches of Asian Highland rivers in South and East Asia. They tell the story of China's long history with river interventions, framed by the story of the mythical, divine Emperor Da Yu (Yu the Great, reputed c. 2200–2100 BCE). Yu inherited his river-management responsibilities from his scholar-official father, who had unsuccessfully tried to stop flooding by blocking river flows. Rather than blocking them, Yu dug channels that drained floodplains, creating fertile agricultural land and directing water to the sea. His people were so impressed that they made him emperor (Sima Qian 2009, 34.4; Pietz 2015, 8; Mostern 2021, 3, 85). His story narrated the state's responsibility for river management, and the state management of rivers has played a central role in Chinese hydropolitics. However, China's successive empires did not enact river regulation evenly across their major river systems. They were much more invested in controlling the Huang He (Yellow River, known as the Ma Chu in its upper reaches) than the Chang Jiang (Long River or Yangzi).

The Huang He runs through China's most populated area, the semi-arid, wheat-growing Northern Plains. Control of the capricious and sediment-rich river has preoccupied Chinese governments for thousands of years (Zhang 2016; Mostern 2021). By the twentieth century, there were 870 kilometers of levees along the river, and silt deposits had raised much of the river above the surrounding floodplain, making the Northern Plains one of the most anthropogenic spaces in pre-modern history (Pietz 2015). Pre-modern governance in the Chang Jiang catchment was very different. Before it was impounded in the 1990s, the Chang Jiang flowed consistently and powerfully across wetland-dotted, rice-growing floodplains through a long course to the sea. It provided China with surplus food and water that were transported north through the Da Yunhe (Grand Canal) to the Huang He valley and—as early as the thirteenth century—beyond it to Beijing. Huang He Valley-based empires depended on the Yangzi but disparaged it, viewing it as uncontrolled and uncivilized. As historian Yan Gao (2022, 28–63) has shown, the Chinese state was only able to "order" the Yangzi Plain in the late nineteenth century, reclaiming lands, building levees and dams, and straightening rivers so that they could eliminate "disorderly waters" such as wetlands (see also Courtney 2018, 56–89).

India has had a very different riverine history. Unlike China, Indian emperors were not as fixated on water governance. Instead, a decentralized complex of hydrological and hydraulic knowledge lineages and practices arose across the subcontinent, all responding to the region's dominant climate system, the South Asian Monsoon (Amrith 2018). Within this complex, societies that developed around the perennial rivers descending from the Himalaya—the Indus, Ganga, and Brahmaputra—were the most populous and enjoyed the closest relations with Highland residents. The Indus Valley Civilization (3300–1900 BCE) was the earliest plains society to leave a substantial legacy. It depended on perennial western Himalayan rivers for irrigation, drinking water, and sewage disposal before disbanding when the region's climate and hydrology changed.

As population centers moved eastward from the Indus into the Ganga (Ganges) Valley in the first millennium BCE, they set up "adaptive" riverine systems (Shah 2011). These systems used monsoonal rains to plant cereal crops—wheat in the west, rice in the wetter east—and preserved water for drinking, bathing, pastoralism, and limited secondary dry-season crops. The goddess Ganga became the paradigmatic Indian water deity. As with other river deities, she resides in the sky and brings flood-pulse fertility to the earth before descending into undersea worlds. Hindus hold that she can purify negative karma as she travels through the terrestrial sphere (Da Cunha 2018).

Asian Highland communities continuously exchanged ideas, goods, and people with these Lowland population centers but also developed distinct cultures and societies. Non-material cultural and cosmological understandings of water have played and continue to play an important role in the Asian Highlands' complex linguistic and cultural network of human–river relations.

Highlanders' linguistic and cultural diversity can be overwhelming for those approaching them from the Lowlands. Sometimes, these languages and cultures change dramatically between valleys; at other times, linguistic and cultural connections stretch thousands of kilometers. The Highlands' largest cultural-linguistic group is those who speak the multiple Tibetic languages. Smaller, Tibeto-Burman (but not Tibetic) speaking groups live between the Tibetic speakers and the large Lowland Chinese and Indian populations. Some Tibeto-Burman–speaking communities are Buddhists who use Tibetan script; others have different scripts and distinct local knowledge lineages. The communities—or, more appropriately, nations—of this region have been variously minoritized by the region's territorialization over the past 150 years and refer to themselves as "local" or "Indigenous." In recognition of this self-classification, we will refer to them as "local and Indigenous" people throughout this book.

For Tibetic-speaking Highlanders, life depends on *chu* (water), which moves in and out of river channels. They extract drinking and irrigation water from *chugyun* (flowing water), *chuwo* (rivers and streams), and *chulam* (waterways). Rivers can also be known simply as *chu*, "the water," even when this is part of their proper names, as it is for the Dri Chu. Living at high altitudes also means that Highlanders experienced mingling between the hydrosphere, atmosphere, and geosphere. *Khawa* (seasonal snow) and *khang* (permanent snow) adorn their mountains. Clouds pass through their houses, and they live with the constant threat of *chulok* or *churu* (outburst floods), *damru* (mudslides), and *khangru* (avalanches). Water played a significant role in the Highlands' political history. Foreshadowing Wittfogel, *chabsi* (*chab* as the honorific form of *chu* and *si* meaning "governance") was and is the most common word for kingdom, territory, polity, and politics. The histories of imperial-era kings (sixth to ninth centuries CE) are replete with tales of water governance. However, their government did not directly control the Plateau's water or create bureaucracies to control Tibet's rivers. Instead, it extracted taxes from those who controlled water.

Although each mountain valley had its own genealogies of knowledge, other lineages have operated on broader scales across the Highlands. Dominant among these were the (1) Indigenous, (2) Buddhist, and (3) *mentsi* ("medicine and calculation" of astrology, astronomy, geomancy, and other forms of divination) knowledge "accretions" (Mueggler 2011).

Most Highland Indigenous knowledge lineages recognize the presence of supernatural inhabitants, whom they consider the land's primary and original owners. Tibetic speakers classify these beings into three habitat-based groups: (1) *lha* (territorial deities) in the heavens, (2) *sinpo, tsen, and nojin* demons in the intermediate space, and (3) *sadak* (earth lords) or *nepo* (inhabitants) and *lu*, underground and underwater. Highlanders know these beings collectively as *lha-lu*, "(from) *lha* (in the heavens) to *lu* (underground)."

The places where *lha-lu* reside are called *ne*, which some sources say are the *lha-lu*'s homes and others say are their embodiments. The most famous *ne* are associated with mountains and often have partners who live in nearby lakes. But the region's oldest residents are the *lu*, who have been attached to the land since before it rose from the primordial ocean and now live in and around water sources and other sites of moisture, like trees.

The *lha-lu* influence the elements—earth, water, fire, air, and ether—and create disasters when insulted. As the cause of their affront is often obscure, the Highlands' human residents ask experts in ritual and divination to intercede with the *lha-lu* on their behalf. Even then, the *lha-lu*'s reaction is not guaranteed and may be cryptic. The relationship between humans and *lha-lu* is reciprocal and complicated, affected by kinship, rebirth, and spiritual power. In the Highlanders' cosmology, all beings are connected. Deities become humans through rebirth and transformation, and humans become non-humans. Some humans have great spiritual power, and others, especially royalty, have mighty *rugyu* (bone lineage). These semidivine humans' actions affect the environment more than ordinary beings, for good or bad. Traditional histories explain, for example, that when the much-maligned Tibetan emperor Langdarma (r. 838–842) "abandoned the traditions of his forefathers," floods, fire, and sandstorms swept across the Yarlung Tsangpo catchment (Mkhas pa lde'u jo sras 1987, 366).

Lha-lu also have specific relationships with water sources. Large rivers' frozen sources and headwaters are called *khabap* (snow descended) and are associated with *lha*. The rivers that descend from there are called *tsangpo* (the pure one), as their flow carries away pollution. The Yarlung Tsangpo is one such watercourse. Bubbling springs are called *chumig* (water eyes), and local surface runoff is called *chugo* (headwaters). These water sources are less revered than *khabap* but are essential to Highland life and are respected as the home of capricious and ancient *lu*.

The arrival of Buddhism from India between the sixth and tenth centuries CE modified but did not end Highlanders' relationship with the *lha-lu*. As Hildegard Diemberger (2021, 20) has noted, the *lha-lu* did not represent "specific ontological claims"—like those of a universalistic creator god—but rather the relations between humans and their surroundings and could, therefore, be incorporated into Buddhist practice. Rather than decrying the *lha-lu,* Buddhist teachers included their *dulwa* (calming) in their conversion narratives and transformed many already-existing *ne* into Buddhist sites by recognizing their *lha* as manifestations of celestial buddhas (Buffetrille 2013). They also aligned them with Buddhist cosmology's other supernatural entities. They associated the *lu*, for example, with the Indic *naga*, half-human and half-serpent creatures that reside underground, and the stories associated with each creature began to influence each other.

Buddhism did, however, bring Indian cosmology, which was the most expansive in the pre-modern world, to Tibet. Within this vast cosmology, humans are said to reside in the Desire Realm, where six types of beings—gods, demigods, humans, animals, *preta* (hungry spirits), and hell-beings—transmigrate from birth to birth, impelled by desire. Residents of this sphere cycle through births in all six forms, often unknowingly operating in the same space. A common analogy for the experiences of the six classes of beings describes them drinking from a single stream: the gods and demigods taste nectar, humans and animals taste water, *preta* taste pus and blood, and hell-beings consume molten iron. The worlds where the six types of Desire Realm beings live are variously shaped and sized. Our world is described as a flat disk with a large mountain, Sumeru, at its center and four continents stretching out in cardinal directions, each watered by a large river. Hinduism, Buddhism, and the institutionalized form of the Indigenous religion, Bon, all associate Mount Khangtise in the Khangtise Mountains with Sumeru.

Buddhist teachings combine this vastness of geography and temporality with another idea, *tongpanyi* (emptiness). *Tongpanyi* is sometimes described as a "mere negation" and sometimes as a "nonduality." In both readings, phenomena have no inherent existence and arise through *tendrel* (interdependent origination), influenced by *gyu* (direct causes) and *kyen* (general conditions). This association between the ultimate and negation explains how adepts who have realized emptiness can affect the world around them. If nothing exists inherently, all is impermanent, and much is possible. *Tendrel* also allows Buddhist lineages to frame environments ethically. As individual and collective negative and positive actions or *karma* create non-inherently existing reality, inside and out, our actions create our environments. Our world consists of a *no* (a non-sentient container, the outside world) and *chu* (its contents, the beings within it). These two components of reality, *no-chu*, non-sentient and sentient, arise from beings' *karma* and interpenetrate each other ("like butter tea softens a leather cup"). Environments shape beings. Beings shape environments.

The Highlands' other significant knowledge tradition is "medicine and calculation": *mentsi*. Adepts in this system specialized in evaluating bodies (medicine) or the influence of heavenly bodies (astrology) and enlivened environments (geomancy). From these three knowledge lineages, Tibetan geomancy, heavily influenced by the Chinese geomantic tradition of *feng shui* (wind and water), is most directly engaged with rivers. Geomancers read topography—the directions of mountains, boulders, trees, forests, springs, lakes, and rivers—to determine the presence and activities of supernatural forces. They encourage humans to live near freshwater sources, particularly springs, in houses that back onto mountains and front onto flowing water. The lineage also preserves memories of floods by warning against constructions near river confluences (Maurer 2009, 68; Maurer 2019, 11–12).

Before the rise of sovereign states in the mid-twentieth century, Indigenous (including the reformed Bon religion), Buddhist, and geomantic knowledge holders worked with local rulers to govern water. Village chiefs and households sought assistance communicating with *lha-lu* before interfering with hydrological systems. They asked for their advice in dealing with *churab* (floods with too much water), *chulok* (floods of displaced water), *churu* (outburst floods), *serwa* (hail), *thenpa* (droughts), and *muke* (famines). Like other local water governance practices, these knowledge systems continue, but those who maintain them have much less power to exercise their authority than before.

1.3.3 Regulated Rivers: Rivers as Resource

In the pre-modern era, Asian Highlanders and Lowlanders tried to control the river to varying degrees, but generally, their lifeways approached the rivers with a light touch. They did not use the limited technology and energy they had to terraform riverine environments. Even the Huang He plains, one of the world's most anthropogenic pre-modern regions, was only partially transformed, and was constantly being reformed by Earth Systems.

In the last two centuries, however, two combined socio-political-technological events have transformed human–river relations globally and many of the globe's river systems themselves. The first was the Industrial Revolution of the eighteenth and nineteenth centuries, which granted humans access to hitherto unused material and energy sources. The second was the Great Acceleration of the mid to late twentieth century, during which humans used Earth's resources at ever greater rates with consequential impacts on Earth Systems (Steffen et al. 2015). The effects of both these events took decades to climb the mountains, but with their arrival came profound shifts in the Asian Highlands' socioeconomic systems. As technologies became available to them, governments and corporations used these technologies to build hard infrastructure on rivers, regulate their flood pulses, and change their fluvial characteristics. In this book, we call our focus on these river control and extraction mechanisms the Regulated Rivers perspective.

Other conceptual transformations in governance and environmental management accompanied these technological and consumerist transitions. Around the same time as the Industrial Revolution, the rise of modern European nation-states and large institutions such as schools and hospitals led to an increase in what Michel Foucault called "governmentality," or the control of human conduct and the legibility of human populations (Burchell et al. 1991). Governmentality led, in turn, to new forms of soft infrastructure—laws, norms, knowledge frameworks, government programs, resource use patterns—that were dedicated to restricting rivers' flows and encouraging their use. The rectification of the Rhine, for example, which made trade and transport more efficient among European polities following the Congress of Vienna, left the river reduced to a canal where "multiple branches are redundant" (Cioc 2002, 3). This self-sustaining combination of technology and governance underpinned the European colonial project, and the Europeans exported their regulatory frameworks globally. The river transformation this industrial technology enabled made even China's pre-modern riverine projects seem limited.

The European understanding of rivers owed much to the ancient Mediterranean world, most notably the Roman Empire, from whom they inherited the word "river." Derived from the Latin word *ripa* (bank), a river is a defined body of water flowing from a source between banks to the sea, a lake, a wetland, or another river. This confined sense of a river erased the monsoon flood pulses, which is evident in many Asian words for these waterways. By the eighteenth century, when the European took control of large swathes of Asia, their understanding of rivers had combined these ancient Roman models, two millennia of Christian dominion over nature narratives, and, perhaps more profoundly, the scientific knowledge we discussed in Planetary Rivers.

New technologies and engineering methods allowed rivers to be channelized, widened, deepened, straightened, and extended through locks and weirs, canals, barrages, and sluices, turning them into sizeable water-based transport networks. Embankments and dams mitigated flood pulses and controlled water flow into agricultural fields, industries, and cities. Piped drinking water and sewerage developed symbiotically with urban areas. Industries increasingly used the power of falling, fast-flowing water to turn mechanisms in their mills. Eventually, they developed turbines and generators that transformed the rivers' kinetic energy into electricity. Then, they intensified this kinetic energy by using wood and fossil fuels to turn it into steam. (These are still the mechanisms for producing hydro and fossil-fuel electricity today.) These new technologies disconnected rivers from their floodplains and ecological complexity. Those operating them reimagined rivers as resources, reducing them to units of flow and meters of expanse.

These European inventions were aided by and, in turn, aided colonization. Colonizers used the technologies they had developed in European rivers to explore and exploit the Americas, Asia, Africa, and Oceania. While the British, Dutch, and French Empires claimed to spread liberal values, they prioritized their economic system and defended their control of colonies through racial hierarchies.

In addition to hard infrastructure, British India developed the soft infrastructure of knowledge production and bureaucracies to facilitate its hydrological and hydraulic enterprises. They proliferated mapping to govern and define borders, often using watersheds as natural boundaries (Simpson 2021; Gamble and Davis 2024), and militarized expeditions to determine the Brahmaputra's course and established borders accordingly. Their engineering paradigm led to a reductionist, technocratic approach that replaced traditional water management with a less effective system based on a new engineering concept, the "hydrological unity" of river catchments. These hydrocracies enforced already stark social hierarchies.

In combination, the energy revolution, governmentality, and colonialism of the nineteenth and early twentieth centuries encouraged other broader international trends to take hold from the mid-twentieth century, such as the consumerism, territorialization, globalization, and developmentalism of the Great Acceleration. These trends have led to profound changes in human–river relations, assisting increasingly powerful states and corporations in regulating and exploiting river flows.

After gaining independence in the mid-twentieth century at the start of the Great Acceleration, the Chinese and Indian governments became devotees of the postwar doctrine of national development, which influenced the communist, capitalist, and non-aligned worlds alike. The People's Republic of China began building large modern dams under the guidance of the Soviet Union. India (and Pakistan) adapted the United States Tennessee Valley Authority model. Both experienced catastrophic dam collapses and anti-dam protests. During the 1980s and 1990s, protests in India increased until they persuaded the World Bank to pull its funding from large dam projects. It seemed like the era of the big dam was over, but China continued to build the world's largest Three Gorges Dam, and then in the 2010s, hydropower came back into favor as carbon offset technology. Hydropower's rehabilitation coincided with periods of growth in both countries' economies, and this initiated the largest dam rush in world history in the Highland river catchments. This boom has increasingly strained the Highlands' cultures, societies, and ecologies, including its rivers (Pomeranz 2017).

The rivers' colonial and post-colonial histories give us some insight into the layers of exploitation occurring in the rivers' catchments. At the global scale, these rivers are part of the Global South, and their resources and residents' labor continue to be exploited by the Global North, the old colonial powers. At the regional scale, much of the upper Yangzi and Brahmaputra catchments lie within the Global South's two most populous and powerful nations, China and India, and their resources are often redirected to aid these countries' large Lowland populations. Given stark global inequalities, the need for development and economic justice in the Global South, particularly in minoritized areas, is undeniable. But how this development unfolds is important, especially in sites as necessary to Earth Systems as the Asian Highlands. Increasing governmentality in the mountains, which Arun Agrawal (2005) coined "environmentality," has meant that any push from science-informed conservationists to preserve environments reifies their connections to colonial and national sciences and complicates the lives of already-marginalized Highland communities. The book's authors disagree on whether it is proper to offer suggestions to undo the systems that have created the rivers' current predicament (see

Concluding Reflections). We do agree that the rivers' current socio-ecological situation, resulting from increasing regulation, is not sustainable.

1.4 River Guide: Book Overview

The chapters of this book move from a primarily Planetary Rivers perspective (2–5) through a Social Rivers Perspective (6–7) to a Regulated Perspective (8–10). Many of the main chapters are accompanied by boxes that provide an alternate perspective on the materials they present. Chapter 2, "Mountain Rivers," is concerned with river geomorphology and deep history and is accompanied by a box on Highlander cosmological origin stories. Chapter 3, "Climatic Rivers," describes the climate production of rivers and is supplemented by two boxes on GPE and wind action on dust and sand in the Highlands. The fourth chapter, "Frozen Rivers," examines the rivers' cryosphere and includes a box on the Highlands' flood types, including large and outburst floods. The fifth chapter, "Living Rivers," explains the ecologies of the riverine system and its many threats and includes two supplementary boxes. The first continues the themes of the main chapter by defining wetlands from a conservation perspective. The second box on *lu* and water describes Indigenous and local relationships with this class of supernatural being.

Chapter 6, "Human Rivers," traces the *longue durée* of human–river relations from their peopling to the nineteenth century. It is accompanied by a box on fish–human relations. Chapter 7, "Agricultural and Pastoral Rivers," describes the lives of contemporary pastoralists along the Dri Chu and agriculturalists in Bhutan's river valleys. A box on the persistent and widespread use of the "four rivers" motif complements it.

Chapter 8, "Territorializing Rivers," traces the history of the region's political ecology, describing the rivers' pre-colonial societies, their reconfiguration in the colonial period through mapping and commodification, and how this led to the hardening of borders, the creation of international rivers, and Lowland–Highland tensions in the post-colonial era. Chapter 9, "Managing Rivers," examines river-management regimes in the catchments' four sovereign states—China, India, Bhutan, and Bangladesh—and demonstrates the challenges of river conservation. A box on ecological systems services, one highly influential form of conservation practice, accompanies this chapter, along with another box on anti-hydropower protests in Sikkim. Chapter 10, "NGOs and Rivers," examines the often-problematic relationship between international conservation projects, ecological system services, and local knowledge traditions. A boxed sidebar accompanies it that describes a fish-conservation NGO program in Yunnan, China.

As our fields use timescales that range from multi-million-year-long geological cycles through the millennia of human history and much more recent development cycles, we have included "turtle timelines" at the start of each chapter to indicate the respective timeframes within which the chapters are situated. We chose turtles to tell the chapters' time because they repeatedly surface in our combined research. In Indigenous Tibetan origin stories, the world emerges from a turtle, and turtles are the world's first inhabitants. Turtles hold up the world in Buddhist and geomantic cosmologies. Riverbed turtle fossils grant insights into the waterway's geomorphological formations. Turtles are a key species in river bioconservation. Turtles also move quickly in rivers, slowly on land, and dip in and out of water. We hope our readers can take the same approach to our book. Some can dip in and out of the parts of the book that most interest them, and others can take the slow approach, traveling the rivers with us, at least trying to take everything in.

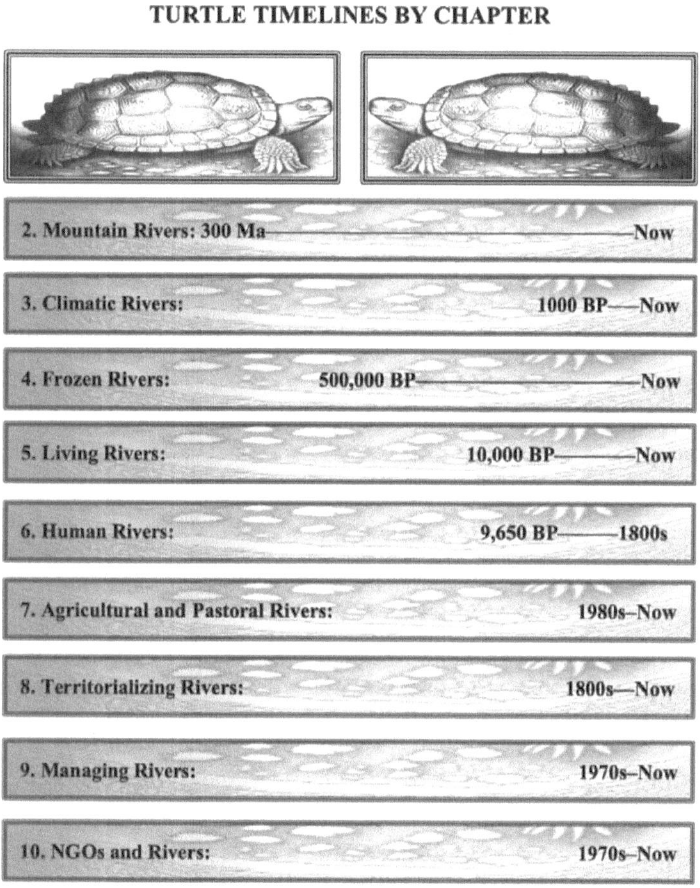

FIGURE 1.6 A comparison of the Turtle Timelines with which we begin each chapter.

2

MOUNTAIN RIVERS

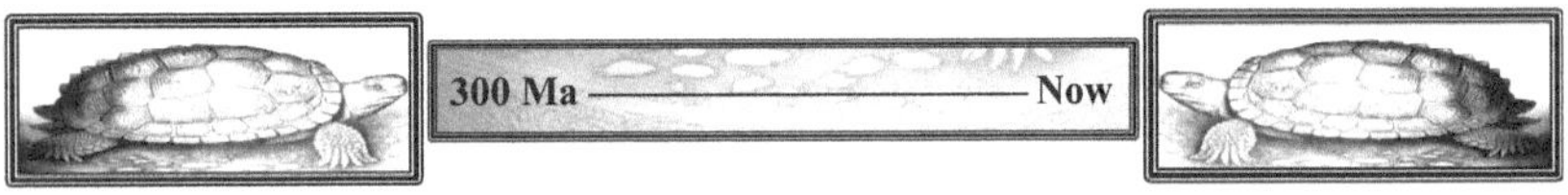

This chapter's most common time markers are Ma (mega annum, or million years ago) and Ka (kilo annum, or thousand years ago).

Humans have lived with origin stories of the Asian Highlands for millennia. Those looking up at the Highlands from the Indian and Chinese river plains associated their formation with the workings of the gods. Those living in them knew this piece of Earth's ever-changing moods and believed capricious *lha-lu* were responsible for them. In the last 150 years, Earth scientists have developed a new way to look at the Highlands, using measurements and models to explain the deep past with increasing sophistication. In some ways, the scientists' evidence tells a similar tale to those that preceded it. Like Highland cosmologies, it claims an ocean once existed between Eurasia and India. It also highlights the key roles of earth and water in land formation. But unlike the Highland origin stories, which we will explore in Box 2a following the main text, Earth scientists do not look to the *lha-lu* to explain the formations. Instead, they look to the deep-time physical processes of the geosphere, reading the evidence these processes leave behind.

According to these scientists' current evidence, the geologic and geomorphic processes that created today's Asian Highlands began tens of millions of years ago, are still unfolding today, and will continue for millions of years. They are slow processes by human standards—some operate at about

DOI: 10.4324/9781003392033-2

the rate of fingernail growth—but fast by geological norms. Insular India, a continental plate that had broken away from the southern supercontinent Gondwana, only began indenting Eurasia around 60 million years ago (Ma), deforming a vast area of Earth's crust (the planet's outer shell). The world's highest mountains, deepest river gorges, most extensive high-elevation plateau, and the sources of Asia's largest rivers formed within this area of deformed crust. Along with this topography, the deformation also caused some of Earth's largest earthquakes, landslides, floods, and highest erosion and rock uplift rates. Unlike the slow, fingernail-scale movement of the continental plates, these events are dramatic on both human and geological scales.

The rising Plateau transformed the Northern Hemisphere's climate, including the monsoon (Kutzbach et al. 1993; Li et al. 2021), and led to the aridification of much of central Asia (Stroeven et al. 2009) (see Chapter 3). Some have even argued that Asian Highland weathering drew down large amounts of atmospheric carbon dioxide (Raymo and Ruddiman 1992; although Rugenstein et al. 2019 dispute this). The Asian Highlands have also transformed their surrounding river plains. As the Highlands grew, water started flowing down them, creating the river systems that formed and molded the plains. In the rainy season, the Ganga's flow is so massive that it suppresses small earthquakes in the Himalaya (Bollinger et al. 2007). The Highlands' creation was a key driver in the threefold increase of sediment transport to the world's oceans between 60 and 30 Ma (million years) (Salles et al. 2023). To the Highlands' south, the Ganga and Brahmaputra river systems flow across the Indo-Gangetic Plain (see Figure 2.1). Such large sediment loads flow through the Ganga-Brahmaputra system that they affect Earth's gravity field (Klemme et al. 2023). Soil eroded from the Tibetan Plateau and the Himalaya creates the world's largest source-to-sink sediment system and a submarine fan that creates much of Bangladesh's land mass and stretches far into the Bay of Bengal (Krishna et al. 2016).

The ongoing collision in the Himalaya is partially responsible for the crustal stress field, topography, and earthquakes that stretch over a vast area of Central, East, and Southeast Asia (Schellert et al. 2019) and as far away as Australia (Braun et al. 2009). The Early/Mid-Miocene (see Figure 2.8 for the geologic timescale) slowdown of the Indian plate's movement led to the deformation of and ongoing earthquakes in the Indian Ocean lithosphere (the crust plus the rigid part of Earth's upper mantle) that stretched well south of India. Scientists have not determined the exact cause of this extended impact, but it is possibly caused by southward-directed compressional force from the thickened Tibetan crust (Iaffaldano 2021).

Many aspects of the story that this chapter tells will be well-known to Earth scientists, but they may not have encountered them pieced together

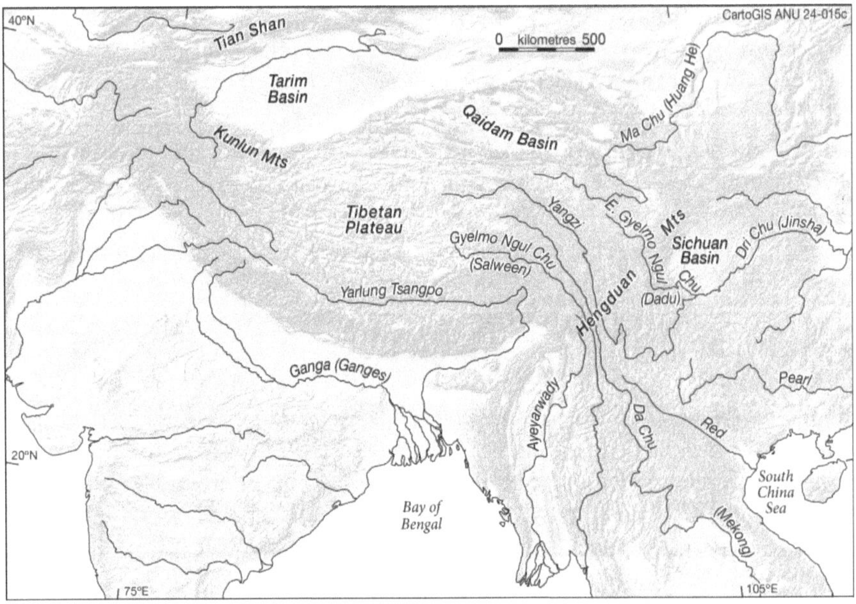

FIGURE 2.1 Topographic map of the region with major rivers. Based on data developed by Dr. Shukla Acharjee (Dibrugarh University) from SRTM data (https://www.earthdata.nasa.gov/sensors/srtm).

in this way. Hopefully, telling the Highlands' evolution as a story will also make it accessible to non-specialists whose lives—whether they are aware of it or not—are directly affected by it. This story, uncovered by generations of Earth scientists, starts with the Highlands' deep history, exploring a timescale inaccessible before their work. It begins by describing the Asian Highlands' pre-collision components before explaining how they came together to form this intensely consequential, high-altitude land mass. After describing this assembling, it explains how the Highlands' presence facilitated the development of river networks from about 113 Ma.

Before continuing, however, we have one caveat. Earth scientists have been fascinated by the Asian Highlands for over two centuries. For them, the Highlands represent a natural laboratory for understanding the interactions of tectonics (processes that produce structures in Earth's crust), climate, and landform evolution. This interest has produced a large and growing scientific literature and many controversies. As this literature proliferates, some findings may differ from the outline of the Highlands' deep history presented here. Even if some of the chapter's detailed data is quickly superseded, it will still provide readers with an overview of this dynamic region's geomorphological and geological processes.

2.1 The Asian Highlands' Components and Assembly

The primary and ongoing process creating the Highlands is the collision of the Indian and Eurasian tectonic plates. This event began when Insular India left Gondwana and drifted north over 130 Ma. It continues today. Insular India's northward drift closed the Tethys Ocean that used to exist between it and the Eurasian plate and caused crustal thickening that created the region's present-day topography, including areas on the Tibetan Plateau with a maximum altitude of 7,439 meters above sea level (ASL) (Jolivet et al. 2010; Zhao et al. 2020). But it is not the only factor in the Highlands' creation, and it was not the earliest (Kapp and DeCelles 2019).

The Highlands consist of successively accreted litho-tectonic terranes (pieces of Earth's crust accreted to one tectonic plate from another plate) (Figure 2.2) associated with subduction (when one tectonic plate moves beneath another). The deformation and uplift caused by this subduction began in the Paleozoic late Mesozoic (Guillot and Replumaz 2013) and also continue to this day.

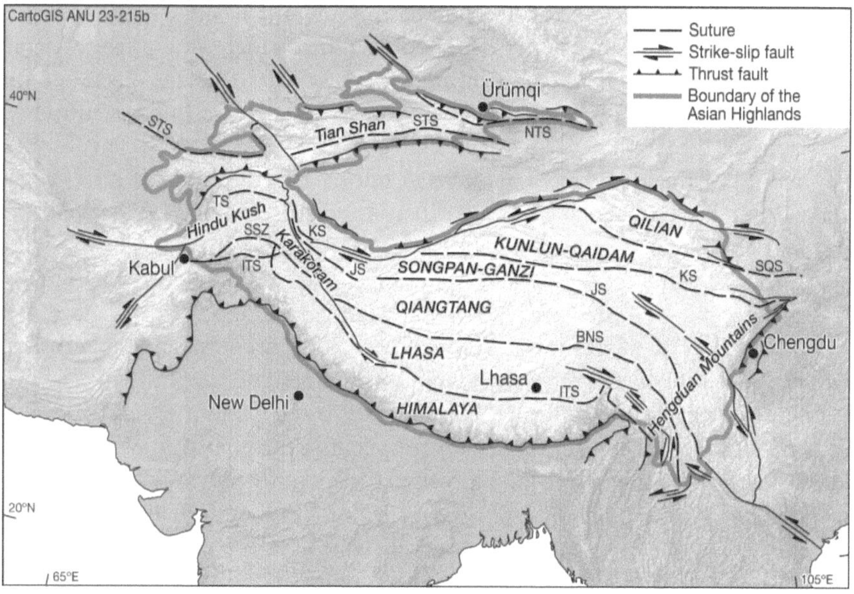

FIGURE 2.2 The major litho-tectonic terranes in the Asian Highlands (based on Liu et al. 2022b). The dashed lines mark the major sutures (faults and intensely deformed rocks), as follows: SQS South Qilian; KS Kunlun; JS Jinsha; BNS Bangong Nujiang; ITS Indus-Tsangpo; NTS North Tian Shan; STS South Tian Shan; TS Tanymas; SSZ Shyok. Note that the names for the sutures sometimes differ between publications.

To understand how these terranes come together, we need to understand the Highlands' tectonic neighborhood. They sit between the Tarim Basin in the north and the Indo-Gangetic Plains in the south. The Tarim Basin contains about 10 kilometers of Cainozoic-age sediment that is still being tectonically deformed (Yang and Liu 2002). Rapid uplift over the last few million years has formed the anomalously high Tian Shan and Pamir ranges to their northwest. This uplift is partly a result of major earthquakes, including a magnitude 7.5 event in 1885 that occurred near the Tian Shan-Pamir boundary (Seong et al. 2009).

The Tibetan Plateau begins south of the Tarim Basin. Its most northerly terrane is the Qilian terrane, a set of northwest–southeast trending mountain ranges separated by fold-thrust systems (folded rocks and thrust faults where one block is moved over another at an angle). This terrane's rocks contain evidence of the Tethys Ocean. The Qilian terrane began to form when this ocean's floor began subducting (when one tectonic plate moves beneath another) about 275 Ma. After a quieter spell for over 100 million years, tectonic activity in the region renewed at 145 Ma, then again at 50 Ma with the India–Eurasia collision.

Within the Qilian terrane is the Tsadam (Qaidam) Basin, a high-altitude, hyper-arid, tectonically active fold-thrust belt that contains around 12 kilometers of Cainozoic sediment (Mischke et al. 2010). This Basin and its surrounding mountains are still uplifted and, along with the Qilian terrane, are the youngest developing parts of the Plateau (Chen et al. 1999). This means they should be much higher than they are, but periods of flowing water (Kapp et al. 2011) and intense wind erosion that lasted since the Late Pliocene have removed anywhere from hundreds of meters to many kilometers of material from the Basin. Indeed, the wind erosion rates in the Basin have reached up to 0.42 millimeters per year, which deforms the underlying sediments isostatically (when land rises as weight is removed). This interaction between wind erosion and tectonics may be the most extreme on the planet.

The next terrane south of the Qilian terrane is the Kunlun-Qaidam, which resulted from the accretion of crustal fragments onto the southeastern margin of the North China Craton. Cratons are ancient tectonically stable pieces of crust. The North China Craton sits to the northeast of the Tibetan Plateau. The Kunlun-Qidam terrane's accretion to this Craton occurred from the end-Neoproterozoic to the Triassic. As it accreted, it collected island arc volcanics (chains of volcanoes above oceanic subduction zones, e.g., Japan) and metamorphic rocks that are now covered by several kilometers of Cainozoic sediment (Sun et al. 2021).

South of the Kunlun-Qaidam terrane are the folded and metamorphosed rocks of the Songpan-Ganzi terrane (Figure 2.2), formed in the late Triassic (Chang 2000). This terrane is seismically active; an earthquake of magnitude

7.4 struck it at Mato (Maduo) in 2021 (Yue et al. 2022). The Songpan-Ganzi terrane's northern margin is formed by the Chimentagh (Qimantag) and Kunlun Mountain belt. To its south is the Qiangtang (Changthang or "Northern Plain") terrane. Although it now sits within the Tibetan Plateau, the Qiangtang terrane used to be attached to Gondwana, a large landmass that sat in the southern hemisphere. In the latest Devonian, the Qiangtang terrane separated from the part of Gondwana that became India (Zhu et al. 2011) and traveled north. After drifting across the Tethys Ocean, it collided with the Songpan-Ganzi terrane in the Triassic (Guan et al. 2021).

Between the Songpan-Ganzi and Qiangtang terrains is the Yidun terrane (not shown in Figure 2.2) in eastern Tibet. The Yidun terrane is a body of sedimentary and granitic rocks with a history similar to that of the Songpan-Ganzi terrane. It collided with the Qiantang terrane in the early-middle Triassic and with the Songpan-Ganzi terrane at the end of the Triassic (Wang et al. 2013a).

The Lhasa terrane sits south of the Qiangtang terrane. It is also a fragment of Gondwana that used to be attached to what is now Western Australia before traveling north as a microcontinent in the late Devonian (Zhu et al. 2011). The Lhasa terrane is bounded to its north by the Bangong Nujiang (Pangong-Gyelmo Ngulchu) suture and to its south by the Indus-Tsangpo suture. The Bangong Nujiang suture closed in the Cretaceous (Zhu et al. 2011). The Indus-Tsangpo suture, which separates the Lhasa terrane from the Himalayan terrane, closed with the India–Eurasia collision (Laskowski et al. 2017).

The Himalayan terrane is formed partly from an island arc—like modern Japan—that used to sit within the Tethys Ocean at the southern margin of Eurasia. Like Japan, this arc had oceanic subduction and explosive volcanoes (Wang et al. 2013a). After the Indian-Eurasian tectonic plates collided, its remnants became the Khangtise (Gangdese) Mountains, which now run parallel to the Himalaya from western to central Tibet. During the collision, island arc volcanic rocks and ophiolites (pieces of oceanic crust thrust onto land) also accreted to the northern part of India, becoming part of the new land mass.

The collision led to a rift valley developing along the Indus-Tsangpo suture at around 26–24 Ma. Soon after it was created, it was filled by a river, alluvial fan, and deep-water lake sediments (DeCelles et al. 2011). We know this because the lake sediment left behind by this process contains fish and turtle fossils, plant fragments, and pollen grains. These remnants also tell us that when it formed, its climate was warm and tropical, and, therefore, the rift was at a low elevation. Red soils in the young part of the sediment sequence show that uplift pushed the rift into a drier climate from about 20 Ma.

As the Himalaya's weight grew, a flexure (downward bend) in the crust to the south of the mountains produced the Gangetic trough. The trough contains up to 5 kilometers of sediment derived by erosion, mainly from the Himalaya (Wasson 2003; Ghosh et al. 2019). This Himalayan flexure has also produced a topographic bulge south of the Ganga about 450 meters high and 200–650 kilometers wide, with incised streams and large gullies. This trough and bulge formation changed the stress field in the Indian crust, causing many earthquakes in central India (Bilham and Szeliga 2008), and possibly triggering them in the Himalaya. The density differences caused by the Indian plate overrunning its own subduction slab have led to subsidence (gradual sinking of an area of land) in front of the Himalaya. This subsidence adds to the Gangetic trough's depth, creating as much as 1 kilometer uplift in the southern Himalaya and causing further subsidence of several kilometers in India south of the Gangetic trough (Husson et al. 2014).

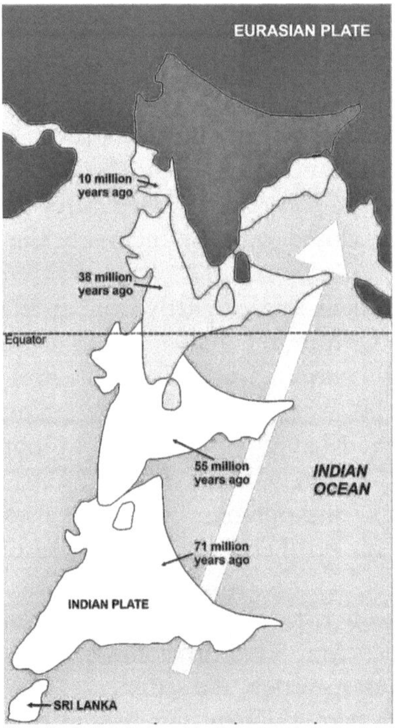

FIGURE 2.3 The northward drift of India from 71 Ma, about 50 Ma after leaving Gondwana. This simplified figure does not account for crustal shortening or that about 2,000 kilometers of northern India have been subducted below Eurasia. It is based on a figure from Chasker et al. (2023).

2.2 The Formation of the Himalaya

As the Indian plate subducted under the Eurasian plate, rocks from India's upper crust scraped off, folded, and faulted to form the Himalaya. The growing mountains incorporated the Tethys Ocean crust as ophiolites (Avouac 2015). Since the collision, there have been around 2,000 kilometers of convergence between India and Eurasia (Schellert et al. 2019).

Earth scientists took many decades to understand how these processes work. For a long while, they refused to accept that continental rocks could be subducted. It was not until 1915 that Alfred Wegener (1880–1930) proposed the continental drift hypothesis, and even then, other scientists did not embrace it enthusiastically. It took another decade for Swiss geologist Émile Argand (1879–1940) to produce the first tectonic model of the Himalaya using Wegener's idea and build on it to propose continental subduction. But it was not until the 1960s that the principle underpinning their work, the theory of plate tectonics and the idea that the solid outer part of Earth was composed of large tectonic plates that move around, became mainstream.

Since Argand's work, Earth scientists have proposed many models of the Asian Highlands' tectonic and geomorphic evolution (see Searle's 2017 review). The most likely variant of Argand's idea currently in circulation is

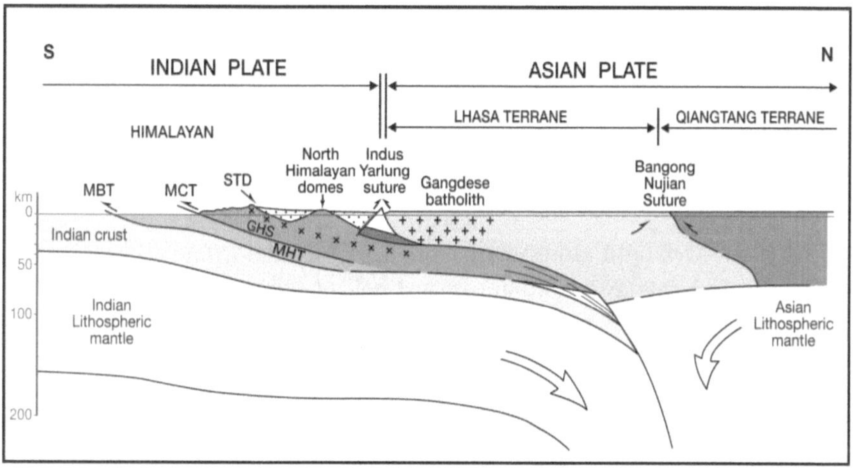

FIGURE 2.4 Geological cross-section through the central Nepal Himalaya and Tibetan Plateau as far north as the Qiangtang terrane (based on Searle 2017; and https://www.geologybites.com/mikesearle). MBT Main Boundary Thrust; MCT Main Central Thrust; STD South Tibetan Detachment; GHS Great Himalayan Sequence; MHT Main Himalayan Thrust. The Gangdese Batholith (or Khangtitse Mountains) is the remnant of the Gangdese Volcanic Arc.

that the Indian crust subducts and then thickens the crust below the Plateau. This model is useful because it not only explains the Himalaya's existence but also their formation.

The Asian Highlands are much more seismically active on their mountainous edges than they are at their center on the Tibetan Plateau. Their eastern margin, the Songpan-Ganzi terrane (Copley 2008), their northern margin, and their southern margin along the Himalaya are all seismically active. The Himalaya is particularly dynamically active, experiencing frequent earthquakes along the faults that fashion its topography. Some of these earthquakes are "great earthquakes" with magnitudes of eight or more and the mountains' tectonics have the potential to produce magnitude nine events (Wesnousky 2020). The mountains' crustal deformation is, therefore, episodic and often causes massive damage to people, property, and ecosystems.

The Himalaya's ongoing orogenesis (mountain creation) involves multiple interacting factors. As tectonics push the mountains up, climate and denudation processes (landslides, flooding, and glaciation) erode the rising crust. As the crust is eroded, a process known as isostatic adjustment moves the surface upward. Theoretically, isostatic adjustment can replenish about 80% of the eroded mass. In the Himalaya, however, it only produces about 20% of the highest peaks' height (Gilchrist et al. 1994).

The Himalaya's major geologic structural zones are:

1 The Indus-Tsangpo suture zone, which was caused by the closure of the Tethys Sea between the northern coast of India and the southern coast of Eurasia.
2 The Tethyan Himalaya sedimentary rocks (also known as the Trans-Himalaya).
3 The Great (or High) Himalayan Sequence of metamorphic rocks.
4 The Lesser Himalayan fold-and-thrust belt of sedimentary and metasedimentary rocks.
5 The folded and faulted sub-Himalayan Siwalik sedimentary basin.

The timing of major tectonic and metamorphic (alteration of rocks by heat and pressure) events from before collision to crustal thickening and uplift is approximately the same along the Himalaya's nearly 2,000 kilometers. At either end, however, syntaxes bend around the mountain range's corners—the Nanga Parbat Syntaxis in the northwest and the Namche Barwa Syntaxis in the southeast—disturbing the Himalaya's otherwise spatially regular geology and creating sites of geological speculation.

Earth scientists have presented several hypotheses for the processes that shaped the Himalaya in this way. Zeitler et al. (2001) posited that following the syntaxes' initial uplift, rivers and landslides eroded deep into them, and this led to crustal flow into the incisions, which produced more elevation and relief aided by isostasy. This isostasy induced even higher erosion rates and very young metamorphism. Others have called the process that created the syntaxes a "tectonic aneurysm," during which the co-creation of the gorge's incision, uplift, increased precipitation, and the exposure of young metamorphic rocks all worked together. Govin et al. (2020) and Searle (2017) have suggested a more straightforward process. In their view, tectonic forces produce uplift, which leads to the initiation of rivers and affects climate as air masses rise over the mountains. These climatic processes produce heavy rainfall that energizes the rivers below, producing more uplift isostatically. These processes were sufficient to shape the syntaxes. There was no need for an aneurysm.

Another important process in creating the Namche Barwa Syntaxis has been the high erosion rates of the rare megafloods that descended through the Yarlung Tsangpo Gorge (see Box 4a). On millennial time scales, these megafloods move 1,000 times more sediment than would otherwise be transported over millions of years (Borgohain et al. 2020). These events could, therefore, do more to shape the Syntaxis's topography in a short time than other processes that unfolded over much longer periods.

The interactions between mountain building, precipitation, and erosion are evident throughout the Himalaya and its syntaxes. Still, they are particularly spectacular in the Eastern Himalaya, where monsoonal precipitation intensified around 10 Ma and subsequently eroded 1–2 kilometers from the mountains' height. This denudation reduced the Eastern Himalaya's southward-directed gravitational potential energy (GPE) (see Box 3a), which enabled faster plate motion across the eastern margin of the Himalaya. This, in turn, increased plate motion, which is reflected in India's clockwise rotation (Iaffaldano et al. 2011) by at least 11 degrees (Gan et al. 2022).

2.3 The Formation of the Tibetan Plateau

While the Indo-Eurasian collision proved to be the Himalaya's defining orogenic event, the Tibetan Plateau's six litho-tectonic terranes—Qilian, Kunlun-Qaidam, Songpan-Ganzi, Qiangtang, Yidun, and Lhasa—began assembling earlier, nearly 300 Ma, and continued to interact with the Himalaya after the collision.

Earth scientists cannot agree on how high the Plateau was before the India-Eurasia collision. Ding et al. (2022) attribute this difference of opinion to a lack of paleo-height estimates for large parts of the Plateau and different methods with differing levels of robustness. Some scholars (England and Molnar 2022; Zhu et al. 2017; Spicer et al. 2021; Guillot and Replumaz 2013) argue that the Plateau was possibly as high as the present (~4.5 kilometers) before the collision. Guillot and Replumaz (2013) even suggest that the northern Plateau was at its current elevation by the

Eocene (55.8–34 Ma). By contrast, Wolf et al. (2022) contend that the entire Himalaya and the Tibetan Plateau reached their maximum height within 10–12 Ma of the India–Eurasia collision. Guillot et al. (2019) suggest the Plateau's southern margin had 2.5–3 kilometers elevations at ~50–60 Ma, with local heights of >4 kilometers. Most studies suggest that the Plateau probably attained its current form and elevation by the mid to Late Miocene (Miao et al. 2022). Although, Staisch et al. (2020) argue for its high elevation by the Mid Miocene.

The Highlands' paleotopography formed in stages. According to Ding et al. (2022), the Khangtise Mountains (also known as the Gangdese Batholith) were the first mountains to arise on the Plateau at ~55 Ma. Their high-elevation (~2 kilometers), east–west structure in the Plateau's south arose just before the India–Eurasia collision. Ding et al. argue that this collision occurred at 65–60 Ma, which is earlier than generally assumed. The Khangtise Mountains were then followed by the Changthang Mountains (Jang Tang Shan or Watershed Mountains) at ~45 Ma. A basin sitting at ≤2 kilometers separated the two mountain ranges. Both ranges were then further uplifted between 38 and 29 Ma.

The existence of fossil climbing perch (*Anabantidae*) and plant fossils dating to the Oligocene in the Plateau's north suggest it sat at an elevation of about 1,000 meters during this period (Wu et al. 2017). Then, millions of years later, during the Early Miocene, large fresh to brackish lakes stretched across the Plateau, almost as far north as the Kunlun Mountains (Zhenhan et al. 2008; Zhang et al. 2021). As the collision continued, tectonic deformation narrowed the lowland and uplifted the northern Plateau, including the Dangla (Tanggula) Range and the Hengduan Mountains to the east (Figure 2.2). The basin lowland rose to near-present elevations by the Mid Miocene when major tectonic deformation moved to the south of the Khangtise Mountains.

Ding et al. (2022) note that the Khangtise Mountains and the Himalaya may have been as much as ~1 kilometer higher than their present height in the Miocene (see also Iaffaldano et al. 2011). This assertion implies that there was post-Miocene crustal thinning. According to Molnar (2022), this thinning occurred by the east–west transport of the crust post-15 Ma. Such a process may help to explain why most of the Plateau has low relief today, along with mid to lower crustal flow and sediment deposition and glaciation (Xiao et al. 2014). Meng et al. (2008) suggest that uplift continued from the Pliocene to the Early Pleistocene in the upper catchment of the Langchen Tsangpo (Sutlej) in the Plateau's southwest. Global Positioning System (GPS) results, which we will return to later, suggest it is still rising.

In the Plateau's center, the crust beneath the Khangtise Mountains began to thicken from about 70 Ma before the India–Eurasia collision (Zhu et al. 2017).

Pre-collision crustal thickening, the breakoff of the Tethys Ocean subduction slab, and isostatic uplift all worked together to push the Khangtise Mountains upward. They may have reached as high as 4 kilometers before the collision, and this event pushed them even higher. They reached their maximum height between 20 and 10 Ma as India underthrust Eurasia. These mountains, with modern elevations up to 7.5 kilometers, are still rising and now form the southern margin of the internally drained part of the Plateau. They used to be the highest mountains in the Highlands with a wet southern face, but the Himalaya grew to overshadow them, casting a rain shadow that desiccated their environment, only leaving lakes behind.

After the Khangtise and Changthang ranges arose, the next major change to the Highlands' topography was the Plateau's contraction on its north–south axis between 50 and 19 Ma. Following this contraction, it deformed extensionally (stretching) by eastward and northern crustal flow. Earth scientists disagree about how this north–south contraction and north and eastern stretching occurred. In the pre-digital age, Molnar and Tapponnier (1975), then Tapponnier et al. (1982), developed a simple, analog, plasticine model that showed how the Plateau's crust could move up to 1,000 kilometers to the east and southeast, influencing the Chinese Lowlands and Southeast Asia. Their model stimulated much debate, and other scholars argued for and against their findings. Decades later, there is a developing consensus that extrusion occurred, but not as much as Molnar and Tapponnier first suggested (Searle 2017). Shellert et al. (2019) estimate the Plateau extruded eastward 260–360 kilometers along strike-slip faults (near-vertical faults that move blocks of rock on either side horizontally) (see Figure 2.2). GPS measurements show that flow (by plastic deformation where rocks deform without fracturing) over the last few decades is to the southeast around the Namche Barwa Syntaxis. However, it is essential to remember that while plastic deformation occurs, these flows are not solely responsible for the region's topographic formation. Crustal transport is mainly accommodated by giant strike-slip faults, often with large-magnitude earthquakes.

The deformation on the Plateau's eastern edge is one of many ways that this margin differs from the Himalayan margin. These two margins have different geomorphological histories and, therefore, different geomorphological formations. Data from the devastating Wenchuan magnitude 7.9 earthquake in 2008 showed that the Plateau's eastern crust is still thickening and rising (He et al. 2019). There is no equivalent of the Gangetic Plain at this margin. Instead, the abrupt fall from the Longmen Shan ends in the Sichuan Basin's dense rocks, which resist the Plateau crust's eastward flow. The Gangetic plain's soil bounty is only reproduced in a small part of the Sichuan Basin at the foot of the mountains. To the Sichuan Basin's south and north, there is no resistance to eastward crustal flow, and the land slopes more gently down from the Plateau.

The Plateau's extension east has produced stretching of the crust horizontally with normal faulting that defines grabens (troughs caused by faults on either side), which often contain small rivers. Over the past 19 Ma at most, this process has created many normal (nearly vertical) fault scarps in southern Tibet (Armijo et al. 1986; Mitsuishi et al. 2012). These are some of the region's most prominent landforms and are still evolving due to uplift and denudation. Chevalier et al. (2020) estimated the Plateau's eastern extension rate in southern Tibet to be about 9 millimeters per year during the Quaternary. There are also well-developed fault scarps in the Kunlun Mountains (Klinger et al. 2005) and central Tibet (Blisniuk et al. 2001), where many were formed in the post-glacial period and are therefore younger than ~15,000 years (15 Ka) (Tapponnier et al. 1982).

Lower or mid-crustal flow is another process that may have complimented the Plateau's eastern extrusion. As mountain and plateau accretions cause the crust to thicken, it can become hot and melt, causing parts of the crust to flow outward over hundreds of kilometers. The Himalaya and Tibetan Plateau began to melt like this around 30 Ma, and molten rocks from this process lie beneath the Plateau today (Jamieson et al. 2011). As we have already seen, Clark and Royden (2000) suggest that the Plateau's eastern margins result from similar flows. Crustal flow appears to be important on the southern side of the Plateau, where Hodges et al. (2001) posit the flow of middle to lower crust in a channel beneath the Plateau. This channel extrudes southward, producing high relief, uplift, and high erosion rates in the High Himalaya. However, crustal flow does not appear to have produced the Plateau's northeastern margin (Hu et al. 2022). There, indentation by India and upwelling of hot rocks from the asthenosphere (a weak zone of rock between 100 and 410 kilometers below the surface) produced uplift.

On either side of the Plateau, volcanism has added to the crust. Since 2.8 Ma at Ashikule, in the northwestern Kunlun Mountains, shear heating (a temperature rise as rocks are sheared by stress) has caused volcanism in a gap between the Indian and Tarim subduction slabs (Wei et al. 2017). In the Plateau's southeast, the Indian slab originates volcanism at Tengchong, where the youngest eruption occurred at only 3.6 Ka, meaning the site is still active (Li et al. 2020b).

2.4 The Asian Highlands in Global Mountain Formation and Possible Equilibrium States

A combination of tectonic plate velocity, crustal strength, and erosion efficiency determines the topographic evolution of collisional mountain belts. The first and last of these three factors (plate velocity and erosion) are the most important (Wolf et al. 2022). The Asian Highlands' topography is primarily determined by plate velocity, which induces crustal thickening and uplift that erosion cannot undo. Erosion rates are high at the Plateau's margins and low in its center.

The Highlands' tectonics have produced one of the thickest parts of the Earth's crust, extending to about 80 kilometers (Wang et al. 2021a). This thickening has created the world's highest peaks and plateaus. When Highland regions are not in a steady state, their uplift is faster than their erosion. Wolf et al. (2022) argue that the Plateau is in a steady state (changing little over time, a form of equilibrium; Renwick 1992), but the Himalaya (and probably other Highland mountain ranges) are not. Hodges et al. (2001) disagree, suggesting that the Tibetan middle crust's southward extension tends toward a steady state of extrusion and erosion, which has existed for about 20 Ma. And Lenard et al. (2020) argue that the Himalaya's uplift and sediment production by erosion have been in a steady state for at least the past 6 Ma that was only occasionally perturbed by short-term extreme erosion events.

Consistent change in a particular direction marks landform disequilibrium (Renwick 1992). In our description of the Plateau and Himalaya's construction and its river networks, we mention several examples of it: (1) the eastward and northern extrusion of the Plateau and the subsequent adjustments of river incision rates; (2) long time lags between uplift and the incision of river gorges; (3) drainage network changes by river capture and drainage divide migration that are still driving adjustments; (4) rivers entering transitional states as drainage networks change; (5) uplift rates that are higher than erosion rates; and (6) erosion rates that are higher than uplift rates. The disequilibrium of the Dri Chu-Yangzi and Yarlung Tsangpo-Brahmaputra river systems is expressed through their convex-up longitudinal river profiles (see Chapter 1). If these rivers were in equilibrium and not adjusting to uplift, erosion would have created concave-up profiles.

Studies that focus on different phenomena (and use other methods and periods) suggest that landform equilibrium occurs in some parts of the Highlands but not all. Over long periods, the Highlands and their rivers have been adjusting to the Indian plate's continuing indentation of Eurasia. In the shorter term, there is some evidence that human land use is adding to the disequilibrium between uplift and denudation rates through land use patterns that cause erosion.

2.5 River Network Development

The Highlands and its rivers co-emerged. Rising mountains led to increased precipitation, which led to the creation of rivers. Rivers formed within the mountains' topography and eroded paths through them down to the sea. The paths they took were many and changed over time. This section will give a sense of how these rivers formed and how they interacted with the mountains by considering the formation of three sets of rivers: (1) those that flowed across the region before the Plateau's uplift; (2) those that rose on the uplifting Plateau; and (3) those that transect the Himalaya.

The studies referred to in this section use plate tectonics theory as a starting point. Gregory (1925) summarized several studies performed before this theory's development, including his own, which begins with crustal buckling in the Cretaceous causing depressions in Asia, some of which he claims still exist on the Plateau. Rather than plate tectonics as a cause, he invokes polar flattening and argues that the uplift of the Himalaya after crustal buckling by an unspecified process produced folds and fractures (faults) that guided the large rivers from the Plateau to the southeast. In his view, the Yarlung Tsangpo first flowed into the Ayeyarwady, then the Dibang, and presumably its current track, through Arunachal Pradesh, where it is known as the Siang—although he does not mention this—as Assam became a trough. This sequence of changes is consistent in large part with more recent reconstructions by Zhang et al. (2019b), Liang et al. (2008), and Robinson et al. (2014) discussed below.

2.5.1 Pre-uplift Rivers

Rivers that existed before the collision of India with Eurasia and the uplift of the Tibetan Plateau were disrupted, then reformed, and their remnants continue to be rearranged. The earliest major river in the area, now largely occupied by the Tibetan Plateau, was a paleo-Da Chu (Mekong) channel. The collision between the Lhasa and Qiangtang terranes in the mid-Cretaceous disrupted it (Wang et al. 2022b). By the Late Cretaceous, the paleo-Da Chu derived most of its sediment from the Songpan-Ganzi terrane with subsidiary sediment from the Sichuan Basin and flowed to the ocean through the Khorat Basin in Thailand (Wang et al. 2021a). Scientists' observations of these sediments led them to infer that the Plateau's eastern margin uplifted early (although Zhao et al. 2021 infer late uplift). In the late Cretaceous-early Paleogene, another continental-scale river system drained southward to the Tethys Ocean. It formed at a low elevation and received sediment from the Songpan-Ganzi, Qiangtang, and Lhasa terranes to its west.

The existence of a southward-flowing river system with low relief for tens of millions of years implies tectonic stability, where an extensive low relief and largely erosional landscape could form. Some authors (Hetzel et al. 2011; Zhang et al. 2016; Clark et al. 2006; Haider et al. 2013) argue that this erosional surface extended over much of the Plateau. While Fox et al. (2020) largely agree, they also suggest—drawing on an idea previously developed by Yang et al. (2015)—that the loss of catchment area by river capture could preserve low-relief landscape fragments.

2.5.2 Post-uplift Rivers

The rivers that drain the eastern Tibetan Plateau, namely the Dri Chu, western Gyelmo Ngul Chu (Salween), and Da Chu, begin on different litho-tectonic terranes (Zhao et al. 2021):

1 The western Gyelmo Ngul Chu begins on the Qiangtang terrane.
2 The Da Chu, the Dri Chu, and its tributaries—the Nyak Chu (Yalong), eastern Gyelmo Ngul Chu (Dadu), and Min rivers—begin on the Songpan-Ganzi terrane.
3 The Ma Chu (Huang He or Yellow River) begins on the Songpan-Ganzi terrane.

Zheng et al. (2020) provide evidence that the Dri Chu came into existence before the Late Eocene after a major tectonic reorganization. Chen et al. (2021) used molecular data from cyprinid (carp) fishes to suggest that it formed at about 23 Ma as the Plateau rose and the monsoon strengthened. The river was then diverted to the northeast along a tectonic lineament to form the First Bend (Zheng et al. 2020). At the Plateau's eastern margin, the Dri Chu, eastern Gyelmo Ngul Chu, and Da Chu rivers are incised to depths of about 3 kilometers in gorges that began as the Plateau was uplifted 9–13 Ma ago. Most scientists agree that the three rivers cut headward into the Plateau (Cai et al. 2021; see Yuan et al. 2023 present a different idea) and that their gorges incised at an average rate of up to 0.2–0.3 millimeters per year (Godard et al. 2010).

At the Plateau's eastern margin at about 26°N latitude (Figure 2.5), three rivers, the Dri Chu, western Gyelmo Ngul Chu, and Da Chu, form a strange parallel pattern in an area known as The Three Rivers (Three Parallel Rivers, Ch. Sanjiang Bingliu). Hallet and Molnar (2001) showed that these rivers are ten times closer than rivers of comparable length worldwide. The three rivers' parallelism and pattern have incited interest since the late nineteenth century, and scientists have conducted long-standing debates on the region's formation. He and Chen (2006) used a molecular clock (the rate of mutation of biomolecules to estimate the time when life forms diverged) in fishes to provide times when rivers lost contact. They argued that the Yarlung Tsangpo and Ayeyarwady separated between 7.3 and 6.8 Ma, the Ayeyarwady, Da Chu, and western Gyelmo Ngul Chu between 7.1 and 6.6 Ma, and the Dri Chu, Da Chu, and eastern Gyelmo Ngul Chu between 6.8 and 6.2 Ma.

Another change in the region's hydrology occurred when the middle Yangzi reversed flow. The result was a reinvigorated Yangzi, which removed between 1 and 4 kilometers of rock from the Sichuan Basin. This erosion increased the relief and GPE on the Plateau's eastern margin, particularly near the Longmen Mountains, which, in turn, increased the rivers' incision into the Plateau post-Miocene. There were also profound changes to

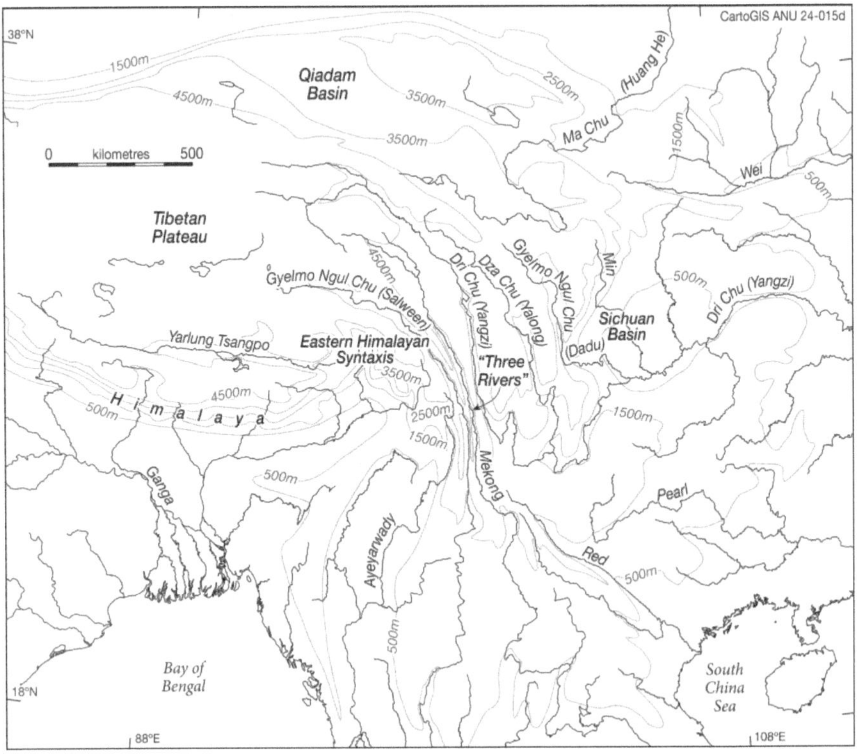

FIGURE 2.5 The major rivers leaving eastern Tibet with smoothed contours and arrows showing modern flow directions. Based on Clark et al. (2005).

the pathways of the rivers flowing east of the Dri Chu, the eastern Gyelmo Ngul Chu (Dadu), and Anning rivers. First, they were connected. Then, rapid uplift and incision from ~2.4 Ma allowed the paleo-Ngul Chu to "behead" the paleo-Ngul-Chu-Anning, capturing its upper course (Yang et al. 2020).

These rivers all have complex interactions with the region's tectonic faults. River-crossing faults offset channels and alluvial landforms such as river terraces and fans. Strike-slip faults in Southeast Asia and Yunnan have disrupted the Da Chu and Gyelmo Ngul Chu Rivers, creating strange hairpin loops in them. The Ganzi-Yushu-Xianshuihe Fault crosses the upper reaches of the Dri Chu Catchment. The Kunlun Fault sits below the Kunlun Mountains, where alluvial terraces are offset, and great earthquakes (magnitude >8) occur every 800–1,000 years (Woerd et al. 2002). The Fuyun Fault passes through the northeast Plateau, and the Altyn Tagh Fault sits on the Plateau's northern margin, both of which have disrupted rivers (Chen et al. 2021). In the Qilian Mountains, rivers that flow in valleys that traverse the ranges are capturing those that flow through fault-controlled longitudinal valleys as the Plateau extends to the north. Su et al. (2022) describe this process as transitional drainage reorganization. These formations indicate that the region's fault movements and accompanying earthquakes are young (e.g., Shi et al. 2022a).

2.5.3 The Yarlung Tsangpo and Its Gorge

The course and creation of the Yarlung Tsangpo have fascinated Earth scientists for over a century, partially because of its importance to regional geopolitics (see Chapter 8). Burrard and Hayden (1907), two of the earliest Earth scientists to work in the region, started speculation on the river's history by inferring that it used to flow from east to west across the southwestern Plateau and possibly descended through the gorge of the Langchen Tsangpo (Sutlej) in the Western Himalaya. They based their inference on their observation that the river's principal tributaries on the Plateau flowed in the opposite direction to it.

In the past two decades, geomorphologists have presented several new ideas about the Yarlung Tsangpo's formation. Liang et al. (2008) claimed that there had been a connection between the Yarlung Tsangpo and Ayeyarwady. Robinson et al. (2014) suggested that the Yarlung Tsangpo and Ayeyarwady connection probably occurred by the Mid to Late Eocene, and the combined paleoriver flowed into the Bengal Basin. Bracciali et al. (2015) found that the Brahmaputra had captured the Yarlung Tsangpo in the Early Miocene. Zhang et al. (2019b) have provided new evidence that supersedes many of these inferences. They show that the Yarlung Tsangpo first descended from the Plateau's southeast into a paleo-Hong He (Red River) before emptying into the South China Sea. Then, the Ayeyarwady captured it, and it flowed into the Andaman Sea. Finally, the Brahmaputra captured it, and it began flowing through Assam and out into the Bay of Bengal. The Brahmaputra captured the Yarlung Tsangpo through two stages. First, the Yarlung Tsangpo's waters flowed through the Dzayul Chu (Lohit), which is now one of its major tributaries, into the Brahmaputra in far-eastern Assam. Then, the Siang captured its flow, and the waters began to travel around the Yarlung Tsangpo's great bend and down to Assam.

The Yarlung Tsangpo has flowed along its current course since at least 10 Ma (Schmidt et al. 2015). Evidence for this timing includes the upstream migration of erosional waves in its bed, which began upstream of the Yarlung Tsangpo Gorge shortly before 10 Ma by a phase of uplift (Dai et al. 2021). These waves are also represented in sediment deposited 5 and 2 Ma and occurred either by uplift or incision through the Gyatsa (Jiacha) Gorge (the second deepest gorge on the river), which separates the provinces of Central Tibet and Kongpo. The river flowed through the Gorge after fault-induced abandonment of a more southerly course (Shen et al. 2021). The Gyatsa

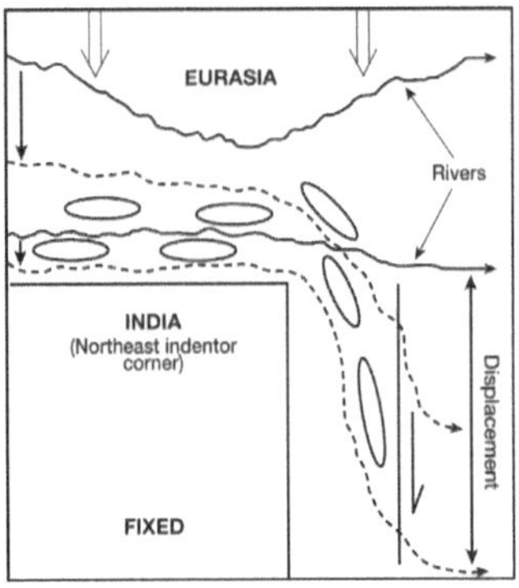

FIGURE 2.6 A cartoon based on Zhang et al. (2019b) and Hallet and Molnar (2001), showing the compressive strains radial to the syntaxis and to the east where there is roughly north–south right-lateral shear (where the right-hand block moves toward, and the left-hand block moves away from an observer on a strike-slip fault). The ellipses show deformation from a circle. The solid and dashed lines show the original and final positions of rivers.

Gorge then steepened between 2.5 and 2 Ma, which increased incision (Wang et al. 2014). This incision was, however, inhibited during subsequent ice ages by glacial damming (Korup and Montgomery 2008).

At a more local scale, Liu et al. (2020a) show that when the paleo-Yarlung Tsangpo flowed west of Gyatsa (Jiacha) Gorge, it emptied into paleolakes in western Tibet (Taylor et al. 2021a support this idea). Some of the river network patterns Burrard and Hayden (1907) identified as evidence of westerly flow may have originated at this time. Bhattacharya et al. (2020) provide evidence for the river's westerly flow by 27 Ma. Then, higher rainfall and stream flow east of Gyatsa Gorge enabled the Yarlung Tsangpo's headward erosion into the Gorge, which resulted in its capture of the river's western section. The river then flowed easterly from its headwaters in western Tibet, across the southern Plateau, down through the Siang to the Brahmaputra. This change in flow led to the demise of the western paleolakes.

Lakes also formed east of Gyatsa Gorge in Kongpo. Xu et al. (2020c) show that between around 51 and 13 Ka, there was a closed saline lake that stretched from Gyela (Jiali) village upstream of the Yarlung Tsangpo Gorge to near Menling (Mainling) village, around 100 kilometers upstream. It was probably dammed by a fault and drained by flow cutting through the dam. There is also some evidence for another smaller lake that formed in Kongpo around 1500 years ago (Xu et al. 2020c). This lake would have existed slightly before the Tibetan Empire and the invention of writing, but as we will explore in Chapter 6, memories of it may be woven into many of the Tibetan stories of their early kings.

2.5.4 Himalayan Rivers

The rivers draining the Himalaya have a different origin and history from those that rise on the Plateau. About 11 out of 16 major Himalayan rivers originate north of the Himalaya and cross the range (Seeber and Gornitz 1983). Four of these are in the upper Brahmaputra catchment: the Drangme Chu, Torsa Chu, Subansiri, and Yarlung Tsangpo. Four other rivers, including the Teesta, exclusively drain the catchment in front of the Himalaya. Burrard and Hayden (1907) made an early attempt to understand the trans-Himalaya rivers' flow, suggesting that they were all originally the Yarlung Tsangpo's north-flowing tributaries that were then captured by headward erosion and began draining the Himalaya's southern flank. Later modeling based on more sophisticated technologies than that to which Burrard and Hayden had access suggests that if this were true, the rivers would have produced irregular watersheds, which they do not. It would also lead to the Yarlung Tsangpo being captured in a similar way, which has not happened. Therefore, instead of accepting Hayden's assertion, later geomorphologists began suggesting that these trans-Himalayan rivers are antecedents; they pre-date the Himalaya's uplift.

Along with similar long histories, these trans-Himalaya rivers also share similar characteristics. They have low gradients on the Plateau, descend through steep Himalayan gradients, and then re-enter low-gradient regions downstream in the Lesser Himalaya, Siwalik Hills, and on the Indo-Gangetic Plain. This pattern of different gradients is unusual in river profiles as rivers tend toward equilibrium. They still exist on these trans-Himalayan rivers because the Himalaya's rapid uplift has maintained steep gradients in the rivers' middle sections (Eizenhöfer et al. 2019).

Tremblay et al. (2015), Laskowski et al. (2017), and Taylor et al. (2021) present various parts of an argument for the existence of several other paleo-trans–Himalayan rivers that once rose on the Lhasa terrane, including a paleo Kyi Chu. Given that they flowed down the front of the paleo-Himalaya, they would have been steep rivers, and their valleys may have funneled precipitation to their upper catchments, thereby providing strong erosive flows. As the Himalaya rose, these rivers stopped crossing the Himalaya at ~10 Ma and became northward-flowing tributaries of the Yarlung Tsangpo, with low gradients and low erosive power.

The current scientific view is that the rivers now flowing down the Himalaya's southern front formed as the front lengthened, and their catchments enlarged as the mountains grew (Castelltort and Simpson 2006 disagree). This enlargement may have occurred by river capture and divide migration. Both events have biological implications, which should be examined.

2.5.5 River Network Reorganization and Aquatic Biota

As the timescales of aquatic species' evolution are comparable with long-term landform change (Stokes and Perron 2020; Craw et al. 2016), it can be possible to use data from one to explain the other. One example of how this works comes from Aotearoa (New Zealand), where fish genomes (the complete genetic material in an organism) have retained evidence of tectonic landscape development. A correlation between geologic age and DNA sequence divergence (the rate of molecular evolution) shows that "landscape evolution has controlled ongoing biological diversification over the past 25 million years" (Craw et al. 2016). Lyons et al. (2020) show that drainage reorganization can increase aquatic species' diversity by nearly five times.

The Asian Highlands' geomorphic processes changed much of its habitat connectivity. This change may have produced new species, new dispersal corridors, and barriers that prevented aquatic organisms from moving. Stokes and Perron (2020) have shown that a catchment area primarily controls species richness, and river capture temporarily increases species richness. Elevated speciation and extinction rates can provide a persistent biological record of river network reorganization. Wang et al. (2013a) argue from molecular data that a major reorganization of drainage networks explains the speciation and distribution of fishes in the Tibetan Plateau's southeastern rivers. Further investigation of these connections would give us greater insights into the Highlands' formation.

2.5.6 Glaciation and the River Network

Another important part of Himalayan river history has been their periods of glaciation. The extent and influence of these glaciation events have been a cause of arguments between Earth scientists. For decades, scientists have argued over the extent of the Tibetan Plateau's glaciation during the Last Glacial Maximum. Kuhle (1995) suggested that the entire Plateau, including the Yarlung Tsangpo valley, was glaciated with ice up to 3 kilometers thick (Figure 2.7). Kuhle estimates that the glacial-isostatic effect would have depressed central Tibet by up to 700 meters and then uplifted it by 600 meters after deglaciation. He also suggested that the ice sheet would have produced forebulges in the Himalayan foreland, the Tarim Basin, and the Qaidam Basin with subsequent isostatic depression after deglaciation.

This account of the extent of glaciation is no longer accepted. If all of the Tibetan Plateau had been glaciated, this glaciation would have laid down extensive sheets of glacial sediment that would have obliterated the smallest rivers, smoothed fault scarps, and produced significant post-glacial changes such as meltwater channels, glacial lake outburst floods (GLOFs) and when rivers mobilized the debris left behind by glacial retreat, the deposition of paraglacial sedimentation downstream. There is no evidence for this. Instead, Owen et al. (2008) concluded that Li et al. (1991) provided the best estimate of the extent of glaciation in the Plateau during the Last Glacial. Li et al. showed that only ~10% of the Plateau had been glaciated, and this glaciation would have disrupted the Yarlung Tsangpo's tributaries but not its mainstream.

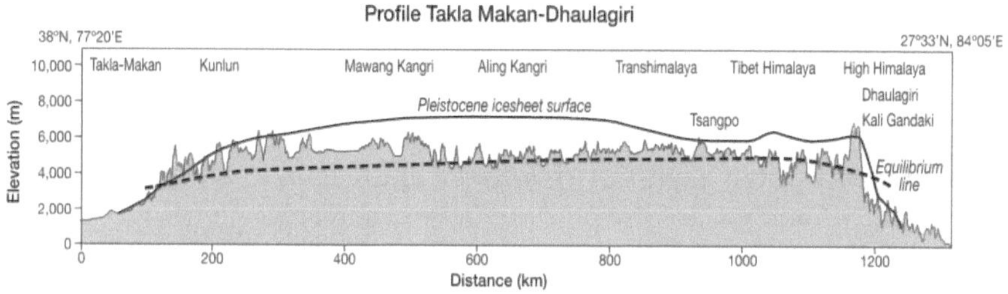

FIGURE 2.7 A cross-profile of Tibet's proposed extreme Pleistocene glaciation from the Taklamakan Desert (in the Tarim Basin) to the Himalaya, based on Kuhle (1995). There is no evidence for this ice sheet in the most reliable mapping by Li et al. (1991).

2.6 The Contemporary Tibetan Plateau and Its Future

Frank Kingdon-Ward (1885–1958), an English botanist, plant hunter, spy, and explorer, said of the Plateau:

> There are, as it were, two Tibets. There is the Plateau country including the lake region, called Chang Tang [Chang Thang], and the upper courses of the great Tibetan rivers, where they flow eastwards or south-eastwards in comparatively wide shallow valleys; and there is little known and far more formidable country comprising the middle courses of these rivers, where, having dug themselves in, they change direction to the south and force the barrier ranges to flow down to the plains of India and China (1926).

Although he relied on his eyes and memories rather than GPS to make this assessment, Kingdon-Ward's appraisal is reflected in many recent high-tech maps of the Asian Highlands. As the Geomorphological Map of China (Xiao et al. 2014) shows, much of the Plateau is internally drained and has shallow relief, with average slopes of ~5° and relief of ≤1 kilometers. The only area with a higher relief is the Plateau's eastern margins, reaching 6 kilometers (Xiao et al. 2014). The elevation also changes somewhat across the Plateau; the southwest is its highest region, and the northeast is its lowest.

A combination of factors worked together to flatten the Plateau. Rivers descending from hillslopes and glacial processes removed some uplands and deposited sediments in the lowlands. Crustal flow (Fielding et al. 1994) moved high ground to lower areas, and wind erosion (see Box 3b) further flattened some areas. According to Liu-Zeng et al. (2008), the major causes of flattening were sediment infilling of depressions in the Plateau's internally drained area and uplands erosion. This smoothed the Plateau's relief after its major uplift had ceased.

Despite this overall low relief, several major mountain ranges cross the Plateau. Most are orientated east–west or northwest–southeast and separated by large basins. These ranges with their highest elevations are Khangtise (Gangdese) (6,714 meters), Dangla (Tanggula) (6,621 meters), Kunlun (7,120 meters), Altyn Tagh (7,167 meters), and Qilian (5,547 meters). The Hengduan Mountains are oriented southwest–northeast and have a maximum elevation of 7,556 meters.

The modern Plateau is a result of a long history and a relatively recent—15–10 Ma—major shift in its tectonic regime, which changed the entire Plateau's deformation style (Gan et al. 2021). Recently, studies using GPS have plotted this deformation as it happens. Active deformation is now dominated by eastward transport (up to 14–20 millimeters per year) and the crust's clockwise rotation around the Namche Barwa Syntaxis, extending as far north as the Jinsha suture (Figures 2.2 and 2.5) and as far south as Yunnan (Tao et al. 2020). Westward transport dominates the Plateau north of the Kunlun fault, and there is little transport in the Songpan-Ganzi terrane. This deformation pattern coincides with crustal thinning and east–west extension by normal faulting (near vertical faults on either side of which blocks move approximately vertically) in the Plateau interior since ~10 Ma (Gan et al. 2021; Conrad 2000).

In the Three Rivers (or Three Parallel Rivers) Region, a decadal uplift occurs at 2–3 millimeters per year (with some areas of subsidence) (Gao et al. 2023). There has been between 0.5 and 4 millimeters of uplift per year over the past 100 Ka (Shen et al. 2021) and between >6 and –0.5 millimeters per year over more extended periods (Kirby et al. 2003). The spatial variation of river incision rates is related to thrust faults that have disrupted rivers (Yang et al. 2015). More broadly, the decadal rates of uplift in the Himalaya are 2 millimeters per year and between 1 and 2 millimeters per year over much of the Plateau (or 0.06–1.97 millimeters per year when corrected for the effect of the weight of water in lakes) (Shi et al. 2022b). There is subsidence of up to 3 millimeters per year in parts of the internally drained Plateau (Liang et al. 2013). Also, based on recent GPS measurements, the Khangtise Mountains are rising surprisingly fast, at about 3 millimeters per year (Taylor et al. 2021).

On the central Plateau, GPS analyses indicate that uplift rates are higher than denudation rates, which are 0.01–0.02 millimeters per year (Li et al. 2014). This shows that the Plateau is still slowly rising. In the Kunlun, Amnye Machen (Anyemaqen), Qilian, and Mustagh-Ata-Khangar Mountains, denudation rates range from 0.07 to 0.4 millimeters per year (Li et al. 2014). The Kunlun Mountains, for example, are rising at about 30 times the denudation rate and isostasy following deglaciation accounts for 40% of this uplift (Liu et al. 2022a). Mount Yumu (in the Qilian Mountains) is also outpacing denudation. Its uplift is between 0.5 and 0.8 millimeters per year, and its denudation is between 0.18 and 0.28 millimeters per year (Palumbo et al. 2009).

We can also use these long-term denudation rates as a baseline for comparison with contemporary rates to infer how much impact contemporary land use practices are having. But this usage has its problems, such as its comparison of rates on very different timescales. In the Yarlung Tsangpo Catchment upstream of the Gorge, the average denudation rate over the past 12–18 thousand years is 0.04–0.05 millimeters per year (Lupker et al. 2017), and the contemporary rate is 0.06 millimeters per year (Liang et al. 2022), the slight difference possibly a result of land use. The river's tributaries have such high rates of denudation, which are most likely due to differences in their relief. The increase from the long term to contemporary denudation rate is about 1.75 millimeters per year, probably caused by land use changes. Modeling studies in this region have estimated the area's rate of sheet erosion by water on hillslopes—when a thin sheet of soil is removed—at ~4 millimeters per year. This rate will likely increase as the climate changes (Wang et al. 2020b). Wind erosion occurs between 1.8 and 4.3 millimeters per year, a response to the Plateau's arid climate and overgrazing (Ping et al. 2001). Across the Plateau, the average is only around 1.1 millimeters per year. Where land use-induced erosion is more significant than uplift, tectonic forces are negated by human action.

While most Earth science models examine the past, they can sometimes offer glimpses into the future. By extrapolating current tectonic plate motions and climate data, and including the precipitation barrier caused by the Himalaya, Rey et al. (2022) suggest that over the next 10 Ma, India and the Tarim Basin will collide. Once Tibet has docked with the Tarim Basin, India will no longer underthrust Eurasia, and when this occurs, the Plateau will contract south to north. As an eastward shear moves across the crust, the Ayeyarwady, Gyelmo Ngul Chu, Da Chu, and Dri Chu rivers may merge. This merging pattern had already occurred in the Western Himalaya during the past 5 Ma when the rivers of the Punjab were rerouted from the Ganga to the Indus (Clift and Blusztajn 2005). The Yangzi's middle reaches may also be rerouted either through or to the north of the Sichuan Basin, and the Brahmaputra will continue to cut through the Himalaya. These tectonic and landscape changes will increase erosion rates in the Brahmaputra catchment, further augmenting sediment transfer to the oceans.

2.7 Conclusions

The construction of the Tibetan Plateau and its adjacent mountain ranges—the Hindu Kush, Karakoram, Pamir, Kunlun, Hengduan, Dangle, Khangtise, and Tian Shan—occurred over almost 300 million years. Before the India–Eurasia collision, six litho-tectonic terranes joined on Eurasia's southern margin. The subduction caused by their coming together closed intervening oceans caused volcanic island arc formation, subduction slab breakoff, and intense faulting and deformation along the sutures between the terranes.

The Himalaya-Tibetan Plateau-Tian Shan Highlands result from India's high-velocity indentation and its underthrusting of Eurasia from ~50 Ma. This produced extreme crustal thickening, uplift, and elevations. Over much of the Highlands' evolution, erosion has not exceeded uplift. The collision continues to move the Plateau's crust to the north, east, and southeast around the Namche Barwa Syntaxis. Subsurface mid-crustal flows may have also aided eastward crustal movement, raising mountains where they were obstructed by strong crust.

The rise of the Highlands, combined with a strengthened monsoon around their margins, formed the region's largest rivers. The rivers leaving the Plateau took their current form through uplift-induced erosion of deep gorges and eastward-directed crustal movement that carried some rivers closer together. River capture and catchment divide migration reorganized the river network, creating the Highlands' contemporary watercourses. These changes probably impacted the spatial patterns of fish genomes and their diversity.

The Dri Chu (Yangzi) began to form at about 23 Ma as the Plateau was uplifted. Its current gorge formed after renewed uplift at 13–9 Ma. Upstream of its largest gorge through the Namche Barwa Syntaxis, the Yarlung Tsangpo has flowed through approximately the same place for about 10 Ma. But it only connected with the Siang and began emptying into the Brahmaputra relatively recently, probably in the Late Miocene or early Quaternary.

At present, uplift rates are outpacing the Plateau's low erosion rates, and the Plateau is still rising. In the Himalaya, uplift and erosion rates are about the same. Although downstream-directed incision may have dominated in the past, headward-extending rivers are eating away the eastern edge of the Plateau and part of its center. In the Yarlung Tsangpo Catchment, contemporary land use has increased erosion rates to values that are sometimes higher than uplift rates. This is an example of human action overcoming tectonics. These and other observations show that parts of the Asian Highlands are in equilibrium, where minor variations occur around a constant value, while others are in disequilibrium.

Fast-acting processes such as enormous floods, landslides, and earthquakes are superimposed on all these long-term geomorphic changes to topography and rivers, whether steady or not. These events, which also impact ecosystems and people dramatically, deplete the enormous GPE the mountains and Plateau have stored in their thick crust (see Box 3a). Over the next 20 Ma, the rate of GPE expenditure will increase as the Plateau docks with the Tarim Basin, increasing elevation and erosion rates. It will be a long wait to discover how the region's other geomorphic components, climate, rivers, ecologies, and human inhabitants respond to these ongoing processes.

EON	ERA	PERIOD		EPOCH		Ma
Phanerozoic	Cenozoic	Quaternary		Holocene		0.011
				Pleistocene	Late	0.8
					Early	2.4
		Tertiary	Neogene	Pliocene	Late	3.6
					Early	5.3
				Miocene	Late	11.2
					Middle	16.4
					Early	23.0
			Paleogene	Oligocene	Late	28.5
					Early	34.0
				Eocene	Late	41.3
					Middle	49.0
					Early	55.8
				Paleocene	Late	61.0
					Early	65.5
	Mesozoic	Cretaceous		Late		99.6
				Early		145
		Jurassic		Late		161
				Middle		176
				Early		200
		Triassic		Late		228
				Middle		245
				Early		251
	Paleozoic	Permian		Late		260
				Middle		271
				Early		299

FIGURE 2.8 Geologic time scale based on the simplified time scale from the International Commission on Stratigraphy.

BOX 2a HIGHLAND ORIGIN STORIES

The communities of the Dri Chu (upper Yangzi) and upper Brahmaputra catchments maintain diverse origin stories. Unlike the stories told by scientists, these are handed down from generation to generation, either orally or written on manuscripts. Although these stories change over time, the point of their retelling is not to improve on earlier storytellers' work, but to retain wisdom from the past. Most of these origin stories have three components: they tell how matter arose from the void, how the Highlands emerged from an ocean (connecting them to the Earth scientist's story), and how this new land was

(*Continued*)

peopled. The stories reflect specific hydrologies, geographies, ecologies, and genealogies. The most broadly influential stories are:

1 Bon cosmological origin stories that feature *lu*;
2 local origin stories from across the Tibetic-speaking Highlands, including the *Epic of King Gesar of Ling*;
3 local origin stories from non-Tibetic communities, such as the Rong (Lepcha) in Sikkim;
4 Buddhist cosmological origin stories; and
5 origin stories from the *Mentsi* (medicine-astrology-geomancy) tradition.

Most communities accumulate rather than replace origin stories. They maintain stories from Indigenous and local lineages, along with Buddhist and *Mentsi* accounts and even contemporary-science origin stories. The newer stories have intersected with and modified the older tales, and over the centuries, individual stories' popularity has waxed and waned. What follows is a brief overview of some of these origin tales, focusing on their descriptions of water and rivers.

2a.1 Origin Stories Preserved in Bon

The oldest Tibetan stories about the origin of the universe, the Tibetan landscape, and the Plateau's peopling are preserved within the storytelling and texts of Bon, a religion that developed in Tibet. Bon practitioners reformulated their religion in the tenth and eleventh centuries in response to Buddhism's growing influence (Powers 2007, 497), but this new form of the religion retained elements of pre-Buddhist, pre-sixth century Tibetan stories and rituals (Karmay 1998; Dotson 2013), including origin stories.

The Bon canon's extant collection of origin stories are found in the *Lu Compendium*, a multi-volume work associated with a class of being called *lu* (see Chapter 1 and Box 5b). It contains numerous origin stories probably collected from various oral storytelling lineages. Most of the stories start with a primordial prehistoric void and connect the world's origin with the appearance of light, followed closely by the appearance of *lu*, who often manifest as turtles.

One such story tells of a *lu* that appeared from a blue light in the middle of a primordial lake as a golden turtle (Zeren 2021, 57). The world's components—the five elements (earth, water, fire, wind, and space), the sun, the moon, the 28 constellations, the *no* (container or environment), and all its *cu* (contents or being)—all evolved out of the turtle's body. When the elements had arisen, four seasonal *lu* queens developed from its softer, segmented belly, along with the primordial form of Bon, Yungdung Bon.

(*Continued*)

Another story tells of a *lu* queen who comes into being from light caused by the wind of emptiness. She created white light from her lungs and black light from her liver, which, in turn, produced the white and black *lu*. Elsewhere in the *Lu Compendium*, a *lu* that called itself Sipa Chaken (Diviner of Existence) appears spontaneously in the void within a mustard-seed apparition. (Mustard seeds are power substances in both Bon and Buddhism and are often used in rituals). After its appearance, Sipa Chaken's *rigpa* (pure presence-awareness) gave rise to 37 syllables, out of which the five elements, the stars, the five poisons, and a healer named Tsoche Zhonnu arose (Zeren 2021, 44, 57–59).

These stories of the Highlands' creation and evolution from animated beings often insist, like geomorphologists, that the mountains rose out of a primeval sea and that this rising gave the Highland valleys their specific geographies. The sixteenth-century Tibetan historian Tsuklak Trengwa (1504–1566) repeated the story of this rising from older Bon narratives.

> It is said that eventually, the ocean stopped boiling, completely cooled, and its waves subsided. The water that had gathered in the four horns [provinces of Central Tibet] evaporated, as did the waters in other places [to its west and east]. In [the waters'] place, Tibet and Kham manifested clearly.
>
> (Gtsug lag 'phreng ba 1986, 150)

The *lu* and other *lha-lu* (supernatural beings) began inhabiting the Highlands as they grew out of the sea. Then came the animals, who evolved with the land.

> In the west, the three divisions of Ngari arose as the land of deer, ass, and antelope. In the center, the four provinces arose as the land of wild animals, tigers, and leopards. And in the lowlands, the six ridges (Kham) arose as the land of birds.
>
> (Gtsug lag 'phreng ba 1986, 150)

Humans were the last to arrive in the Highlands. They evolved from human, human-deity, human-demon, or animal-demon ancestors.

2a.2 Local Tibetic Origin Stories

Local origin stories describe the creation and peopling of individual valleys. They often describe a valley's physical features and create a kinship association between local human rulers and the valley's long-lived supernatural residents, the *lha-lu*. Many local chiefs claim descent from the deities of the mountain at the top of their valley. The Tibetan kings and emperors were no exception, claiming descent from the mountain deity Yarlha Shampo, who lives at the top

(*Continued*)

of the Yarlung Valley (see Chapter 1 and Chapter 6). Local rulers on the plains north of Lhasa claimed a link with Nyenchen Thanglha, the god for whom the mountain range from which the Kyi Chu (Lhasa River) flows is named. Nyenchen Thanglha is depicted in stories and images as a warrior chief, armed with a bow and arrow, wearing a feather, and guided by a golden *khyung* (a *garuda*-like creature) (Rikey 2011). *Khyung* play a significant role in gods' stories as they traverse the highest realms where deities live and oppose—or are possibly the mirror image of—the *lu*, who inhabit the underworld (Ramble 2013; Box 5b). Nyenchen Tanglha is the god of hail, and his celestial partner is the sacred lake Nam Tso, whose waters are said to mingle with the ocean. Local stories warn humans and animals of her ability to transform them into aquatic monsters if they fall into her waters (Bellezza 1997, 102).

Eastern Tibet or Kham is more closely associated with another compelling figure, King Gesar of Ling (see Figure 2a.1 *bottom left*). The open collection of stories about him, the *Epic of King Gesar of Ling*, still has a significant living oral tradition and has been listed by UNESCO. It recounts Gesar's life story and heroic adventures, from his magical birth to the horse race he won as a teenager, his enthronement, and the rescue of his wife Drugmo (Dragon Woman) from a jealous suitor. Gesar is a divinity in human form. His father descended from *lha* (gods) and *nyen* (mountain deities), and his mother is a *lu* princess. This semi-divinity gives him a special relationship with mountain deities like Magyel Pomra, who grants him magical weapons and armor (Fitzherbert 2017).

Khampas (inhabitants of Kham) consider Gesar of Ling, a divine ancestor, and his story plays out within identifiable settings. He spent three weeks, for example, in a hill cave called Drakri Yangkhar (Cliff Stronghold in the Stony Mountain), about 50 kilometers north of Karze, a town in eastern Tibet, where a tributary of the Nyak Chu (Yalong) runs through the hill. A self-arising imprint of his right hand reportedly appeared near the cave's entrance centuries ago, and pilgrims circumambulate the entire hill to create the right merit to pass through the *bardo* (the space between death and rebirth) without issue.

2a.3 A Rong (Lepcha) River Story

Many Tibeto-Burman–speaking communities living on the southern side of the Himalaya, among the Brahmaputra's tributary catchments, have their own river-orientated origin stories. These groups speak many languages, and each language group has multiple stories of the rivers that descend through their homelands. Two Rong (Lepcha) scholars, Rongnyoo Lepcha and Mongfing Lepcha (2021), have combined words and artwork to articulate their community's relationship with the Teesta River. Within this work, they tell the following story of how the Teesta River was formed. We hope retelling their story will give some sense of the diversity of living river tales across the Eastern Himalaya.

(*Continued*)

They tell of two water spirits, Rongnyoo and Rangnyit, who lived in lakes at the base of Mount Kangchenjunga and were in love. When an earthquake shook them out of their lake homes, they made a compact to flow down through the mountains until they found a place to merge. To make the descent more interesting, they made it a race, and each chose an animal spirit to guide them. Rongnyoo chose Paril Bu, a python, as her guide, and Rangnyit chose Tut Fo, a pheasant. Tut Fo was a fast guide but became easily distracted by flowers and insects, zigzagging here and there, creating a dynamic, twisting river. Paril Bu was slower but steadier, creating a meandering river and ensuring Rongnyoo arrived at the meeting place first. As he came toward their agreed meeting place, Rangnyit grew agitated, realizing that Rongyoo had beaten him. He yelled out, "*Thee sutha*?" ("When did you arrive?"), and by doing so, granted the river its name, Teesta, is a shorter form of *thee sutha* (see Figure 2a.1 *top*).

2a.4 Buddhist Origin Stories

As the most influential knowledge lineage in the Highlands, Buddhist cosmologies have influenced Highlanders' understanding of their origins. Buddhists assert that all existence arises through *tendrel* (dependent origination, see Chapter 1), which is to say, each phenomenon arises in dependence on other causes and conditions and continues to exist while those causes and conditions persist. The principles of *tendrel* mean all phenomena, including beings and their worlds, pass through cycles of birth, existence, disintegration, and rebirth. This is a spacious cosmology, including many multiple worlds, and not all exist simultaneously. One universe may be evolving while another is dissolving.

The different forms of Buddhism—including early Buddhism, Mahayana, and Vajrayana—outline a variety of origin stories, and many of the individual *tantras* (textual cycles associated with specific Vajrayana Buddhist deities) provide their own detailed version of why and how worlds come into being. The *Wheel of Time* (Skt. *Kalacakra*) cycle of tantric texts, for example, explains that a new physical world arises because beings have the collective *karma* to be born in it. This force leads to the ever-grosser congregation of elemental atoms. The subtlest particles of space give rise to the more subtle element of wind energy, which provides the conditions for fire, then water and, finally, earth (Berzin 2011). *Karma* is the underlying impetus that moves these processes forward, determining the main features of beings' lives and their myriad worlds. As both these environments and our mental habits arise from past *karma*, they interactively create our experiences, shape our worlds and worldviews, and, therefore, the *karma* we go on to create. Because beings tend to become habituated to the deliberate actions they perform, there is a general tendency toward consistency, but this can be changed if someone becomes aware of the dynamics of this process and decides to make conscious changes in their behavior.

Along with these more generalized origin stories within Buddhism, Vajrayana Buddhists also maintain specific lore about the Asian Highlands' origins. This

(*Continued*)

lore holds the Highlands to be the abode of the Bodhisattva Chenrezik (Skt. Avalokiteshvara). His mantra *"Om mani padme hung"* ("Hail to the Jewel-Lotus One") is ubiquitous across the Buddhist Highlands, and the Dalai Lamas are among his many manifestations. Chenrezik first visited the Highlands millennia ago, just after they arose from the primordial ocean, and committed himself to protect the region and its inhabitants. He was not impressed by what he saw on the Plateau and worked to make it more inhabitable. One of the ways he did this, the stories say, was by manifesting as a monkey and mating with a *drag sinmo* (rock demoness) who gave birth to six children, the progenitors of the first six Tibetan clans (Ahmad 1995).

FIGURE 2A.1 *Top Left*, Rongnyoo, the female water spirit, being guided by Paril Bu. *Top Right*, Rungnyit, the male water spirit, being guided by Tut Fo. Drawings by Mongfing Lepcha. *Bottom Left*, Statue of King Gesar of Ling. Photograph by John Powers, 2012. *Bottom Right*, A traditional Tibetan image of the *drag sinmo* (rock demoness) and bodhisattva ape, progenitors of the first Tibetan clans. Photograph by John Powers, 2012.

(Continued)

2a.5 Turtles of the *Mentsi* Lineage

The Tibetan *mentsi* tradition also recounts numerous stories about the destruction and re-creation of the world, many of which involve turtles. We have encountered world-making turtles in one of the *Lu Compendium*'s origin stories. The motif of turtles re-creating the world can also be found in origin stories from Buddhist, Indian, and Chinese stories, which influenced the *mentsi* lineage and introduced more turtle origin stories to the Highlands. In the Indian text, the *Shatapatha Brahmana*, the turtle, is an avatar (physical manifestation) of the god Vishnu. Its plastron (underbelly) symbolizes the earth, its shell the sky, and its body intermediate space. In the *Leizi*, a Daoist text from China, five turtles with cosmic origins are connected to China's five sacred peaks and divination. In Chinese and Tibetan divinatory traditions, the trigrams on turtle shells are considered divinatory.

In Buddhist Tibet and Chinese divinatory lineages, turtles work with the Bodhisattva of Wisdom, Jampelyang (Skt. Manjushri, Chin. Wenshu Pusa), to re-create the world at the beginning of an eon. These texts were translated into Tibetan from Chinese in Dunhuang (van Schaik 2008), sometime between the seventh and ninth centuries. Desi Sanggye Gyatso (1653–1705), who was regent to the fifth Dalai Lama, Ngawang Lozang Gyatso (1617–1682) and a leading figure in the *mentsi* tradition, retold many of these stories in his text, the *White Beryl*. In one of these re-creation stories, he explains that after existence has dissolved into a watery void, Yeyo Nepe Rubal (Turtle Abiding in the Void) emanates from Jampelyang's mind and floats around. When it moves high into the void's upper realms, Jampelyang shoots it with a golden arrow that pierces its right side. Yeyo Nepe Rubal turns onto its back, sinks to the ocean's bottom, and forms a base for the universe. This turtle base, which is divided into the segments of the turtle's plastron, provides existence with divinatory structures and paradigms.

Sanggye Gyatso presents a more complicated version of this story in another chapter of the *White Beryl*. There he describes a series of five turtles that represent the evolution of existence. This process begins when Jampelyang appears, holding his flaming sword (symbolizing the wisdom that cuts through ignorance). This encourages Nepe Rubal to emerge from the great primordial ocean. After this emergence, Nepe Rubal remains outside space and time for eons, containing the potential substance of all creatures, including buddhas and ordinary beings. This potential becomes reality when Chakpe Rubal (Appearing Turtle) arises out of Nepe Rubal.

Along with his activation of Nepe Rubal's potential for beings' creation, Chakpe Rubal manifests air in space, mountains on earth, and waters for life. Eventually, these two turtles meet a third turtle, who has two names, Sipe

(*Continued*)

Rubal (Life Turtle) and Maha Sergi Rubal (Great Golden Turtle). Sanggye Gyatso insists Maha Sergi Rubal originated from Jampelyang's saliva and represents the bodhisattva's essential nature and the universe's divinatory dimension. The fifth and last turtle is Ringsel Rubal (Turtle of Relics), which emanates from Jampelyang's compassion and allows the bodhisattva to explain divinatory calculations on its plastron (Sangs rgyas rgya mtsho 1997, *stod cha*).

FIGURE 2A.2 Splayed turtle below a nineteenth-century divinatory chart, Tibet nineteenth century. Wellcome Library, London, Oriental Tibetan 114, L0035124, used with permission.

3

CLIMATIC RIVERS

The evolution of the Asian Highlands not only changed Asia's geology and topography but also influenced how air and water cycle around Asia and Earth's atmosphere. At this moment in their deep-time geological cycle, the Highlands sit at a mean elevation of 4,000 meters above sea level. Despite their sub-tropical latitudes, they are home to extensive ice fields that cover 5 million square kilometers, including about 46,000 glaciers. Their height, mass, position, and ice generate complex thermal and mechanical processes. The interactions of these processes sustain climate and hydrological systems that shape billions of human lives and some of Earth's most biodiverse regions (see Chapters 4 and 5). Highland rivers are produced by these thermal and mechanical processes and sit within these climate and hydrological systems.

The Asian Highlands' presence is not, however, the only force causing the rivers to flow. The Highlands' atmospheric influence is wide-ranging, and the climatic drivers that determine flows in Asia's great rivers include the global modes of climate variability that control seasonal and inter-annual, decadal, centennial, and even millennial climatic patterns as well as episodic extreme events. Among these global modes is the well-known El Niño Southern Oscillation (ENSO) and other less well known but equally influential systems such as the Indian Ocean Dipole (IOD), the North Atlantic Oscillation (NAO), and the Arctic Oscillation (AO). Each of these drivers impacts key

DOI: 10.4324/9781003392033-3

weather systems that deliver or moderate precipitation to the region: the monsoons from the south and east, and mid-latitude Westerlies from across Eurasia. The impacts of these drivers on precipitation and temperature can be stronger or weaker when one or more of them are in a positive or negative phase. Critically, when phased coupling or synchronizing of these systems occurs, their compounding effects cause extreme weather conditions, which are associated with floods, droughts, or tipping points at which abrupt changes in climate occur. All these systems operate together to create the Highlands' precipitation, snow accumulation, and snowmelt. These, in turn, determine the flow regimes of the rivers that drain the Highlands.

This chapter reviews these dominant weather systems and the modes of variability that influence them. It then describes how they define the Highlands' hydro-climate and places it in a historical context, exploring the evidence for variability and extreme events in the climate record. The datasets with which climatologists work connect our deep-time knowledge to historical events. While geologists and geomorphologists look to the Asian Highlands' rocks for information about the great geological dramas of millions of years ago, climatologists use other geoarchives such as ice cores, pollen deposits and tree rings to explore climate patterns over the past few thousand years. The historical climate baselines they find in these datasets provide a comparator for recent anthropogenic changes to the climate. Understanding the interactions between climate, water, and human lives in this region is particularly important as they have grave consequences. This is, after all, "one of the most active centers in the world water cycle" (Curio et al. 2015), within the Earth's most population-dense regions, so the way climate change unfolds in the Highlands will have global repercussions.

3.1 Weather Systems and Precipitation Pathways

On a broad scale, the Asian Highlands have three main climatic zones:

1. Mid-latitude Westerlies dominate in the north and west;
2. Monsoons dominate in the south and east; and
3. A transition zone at the Tibetan Plateau's center, where the climate is defined by both the monsoons and westerlies (Thompson et al. 2018; Yao et al. 2013) (Figure 3.1).

The boundary between these climatic zones and their variant precipitation patterns lies at latitude 32–33°N (Wang et al. 2007). What is more, evidence from ice cores extracted from glaciers situated north and south of this latitude suggests that this division has existed for at least 500 years. Ice accumulation rates in the Khariya (Guliya) and Dunde glaciers on the Plateau's northern edge in the Kunlun Mountains and the Dachuphu (Dasuopu) ice core from near Mount Shishapangma in the central Himalaya show that these glaciers have existed in these different climatic zones (Thompson et al. 2000; Wang et al. 2007; Thompson et al. 2018).

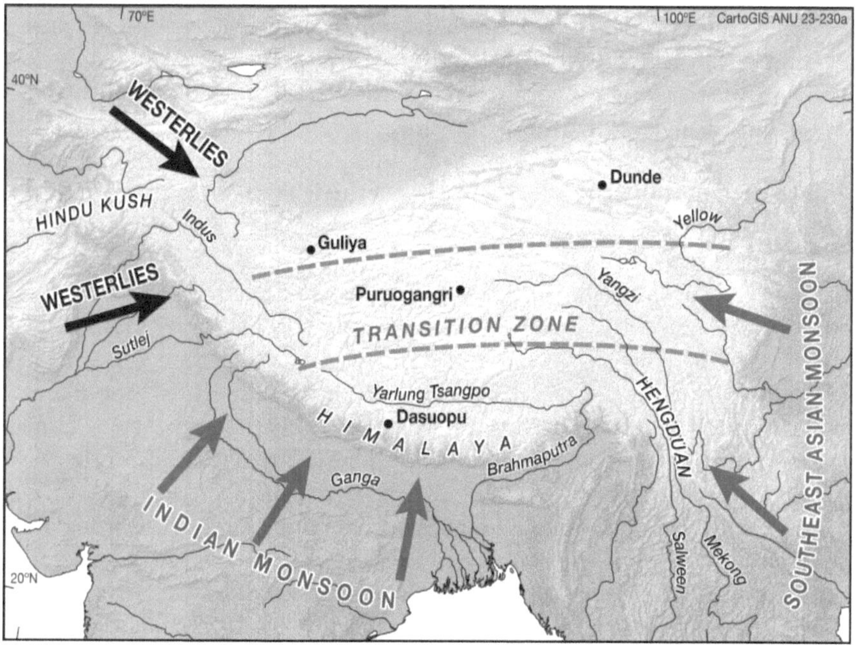

FIGURE 3.1 Major climate systems that influence precipitation in the Third Pole region. Boundaries are shown as dashed lines. Published ice-core sites are marked by squares, and the major river systems are marked by blue lines. Schematic is based on Yao et al. (2013). Image from Thompson et al. (2018).

The monsoons bring the most precipitation to the Highlands. The word "monsoon" entered English via Portuguese and perhaps derives from the Arabic word *mawsim*, which means season. In the Highlands, it is known as *chartu* in the Tibetic languages and *varsha ritu* in Nepali and Hindi. Both terms mean "rainy season." Colloquially, the season is called *charpa* or *varsha*, "the rains."

Monsoons develop when a seasonal reversal of winds occurs over the tropics, creating low atmospheric pressure that draws moist air from the land, bringing rain. In the Highlands, the monsoons dominate precipitation variability during the summer months (June–August) (Thompson et al. 2000), when 60–90% of total annual precipitation occurs (Dong 2016). The Indian Monsoon brings moisture from the Bay of Bengal and the Arabian Sea. It has the strongest influence on precipitation in the Yarlung Tsangpo and upper Brahmaputra catchment. The Southeast Asian Monsoon brings moisture from the South China Sea (Simmonds et al. 1999) and affects precipitation in the eastern Highlands, including the Dri Chu's catchment (Tian et al. 2007; Yao et al. 2013). The Indian Monsoon uplifts moisture on the windward side of the Himalaya, where it is cooled and condensed, generating orographic rainfall (rainfall produced by air moving up mountain ranges) in the eastern and central Himalaya (Barros et al. 2000). The Southeast Asian Monsoon operates similarly on the windward side of the Hengduan and Longmen Shan.

Not surprisingly, this region experiences one of the highest rates of orographic rainfall on the planet (Liu et al. 2009; Dong 2016), and much of this precipitation typically occurs in the high Himalaya, in their center and east (Bookhagen and Burbank 2006). As most precipitation cannot cross the Himalaya, it creates an effective rainshadow. Still, some precipitation does find its way through the mountains along river valleys, and some precipitation is carried across the boundary by the "up-and-over" transport of moisture, where convective systems lift the moisture over the mountains. At least 50% of the total summer rain falls on the southwest Tibetan Plateau (Dong et al. 2016). Precipitation travels over the central and eastern Himalaya before being swept by the mid-tropospheric circulation into otherwise arid southwestern Tibet (Dong et al. 2016).

There is less precipitation in winter, but that which does fall is brought by the Westerlies (December–February). The Westerlies are the prevailing winds that blow west to east in the mid-latitude regions from the "horse latitudes" (~35° latitude) to the poles (~65° latitudes) (Harman 1987). They typically bring extra-tropical cyclones that affect moisture transport in the northern and western parts of the Tibetan Plateau (Figure 3.1) (Simmonds et al. 1999; Thompson et al. 2000; Yao et al. 2013). Their primary sources of moisture are much further afield: the northern Atlantic Ocean and the Mediterranean Sea (Thompson et al. 2000). The Westerlies do not have the same influence on the Highlands' precipitation as the monsoons. Not only do they bring less precipitation, but their influence is only strong on the northern and western Highlands (Simmonds et al. 1999). However, this lack of strength and a smaller sphere of influence does not mean they are unimportant. As the Westerlies influence the climate in the Highlands' most arid and coldest areas, any disruption to them can have profound impacts on these climatic zones. Disruptions to the Westerlies can, for example, increase the chance of drought (Zhang et al. 2015; Zhu et al. 2021a). Perhaps even more critically, any disruption to the Westerlies changes the mass balance of glaciers in the western, southern, and central Highlands. The glacier mass balance is the difference between the mass *gain* from the accumulation of solid precipitation, surface deposition, and internal refreezing and the mass *loss* from surface sublimation and surface and subsurface melt (Mölg et al. 2014). The contribution to glaciers of the various climate systems varies across the Highlands, but as most of the precipitation the Westerlies bring falls in cold places as winter snow, they can make an outsized contribution to glacier mass. As Mölg et al. (2014) found from their work on the Jathang (Zhadang) Glacier in Central Tibet in the Nyenchen Thanglha Range (at the end of the Khangtise Mountains), for example, the Westerlies were responsible for 73% of the annual glacier mass balance compared to only 16% from the Indian summer monsoon.

Variations in the influence of the monsoons and Westerlies across the Highlands determine river flow regimes. There is more streamflow, for example, in the monsoon-dominated zone than in the Westerlies-dominated or Transition zones. The highest river discharges are found in the Ganga (Ganges) River (26.9%) and the Yarlung Tsangpo-Brahmaputra System (25%) (Wang et al. 2021b). By contrast, the most significant contributor of total river runoff from the westerly domain is the Indus River which only accounts for 11.6%, and the lowest contributor is from the transition region in the eastern Tibetan Plateau, the Yangzi Catchment (3%) (Wang et al. 2021b).

In addition to these dominant weather systems, several transitory "atmospheric rivers" or "sky rivers" also shape the Asian Highlands' climates and feed its terrestrial rivers. These "atmospheric rivers" are long (>200 kilometer), narrow (<300 kilometer), and horizontal fluxes of water vapor accompanied by strong winds that occur in the lower atmosphere within the warm conveyor belt (WCB) of extra-tropical cyclones, generally in winter. We call them "rivers" because they have the same volumetric flow rates as the world's largest terrestrial rivers (Gimeno et al. 2014). Globally, these atmospheric rivers transport >90% of atmospheric water vapor in the mid-latitudes (Ralph and Dettinger 2012), bringing moisture from the oceans toward the poles. They produce heavy rain, particularly in mountainous regions where orographic uplift occurs, and the storms associated with them have extreme volume and intensity. These rains can act as a drought breaker but are more often associated with large-magnitude historical floods, debris flows, and landslides (Dettinger et al. 2018).

Atmospheric rivers and the extreme rainfall and flooding events associated with them have been documented in the US, Europe, Japan, India, and China (Ralph and Dettinger 2012; Gimeno et al. 2014; Kamae et al. 2017; Lavers and Villarini 2013; Lakshmi and Satyanarayana 2019). They are now recognized as a salient factor influencing regional hydrology, but we do not know enough about how they operate in the Asian Highlands. There is some evidence that atmospheric rivers have played a role in extreme precipitation events in western Nepal (Thapa et al. 2018) and parts of the Indian subcontinent (Lakshmi and Satanarayana 2019). Recent and emerging studies suggest that atmospheric rivers—distinct from the monsoons—transport moisture from the Bay of Bengal to the Brahmaputra and eastern Ganga catchments, generating catastrophic flooding and short-duration, high-intensity rain and snowfall. They may also operate from the Arabian Sea to the Indus

Basin, with heterogeneous timing across the Himalaya from west to east (Nayak et al. 2021). In Nepal, at least ten atmospheric rivers generate heavy orographic precipitation every year. As they meet the barrier of the mountains, these are uplifted and then penetrate the Himalaya onto the Tibetan Plateau as high-energy, "filamentous" streams (Thapa et al. 2018). These systems still need to be better understood, including how climate change will influence them.

3.2 Key Climate Drivers: Ocean–Atmosphere Interactions

While the monsoons and the Westerlies tap out the Highlands' seasonal rhythms, their strength, duration, and timing are controlled by much larger-scale climatic systems in distant parts of the earth. These "modes of climate variability" are generated through interactions between the atmosphere and oceans, and they generate profound impacts on climate variability at large spatial and temporal scales. Driven by these interactions, four key modes of interannual climate variability shape Highland climates:

1. ENSO;
2. the IOD;
3. the NAO; and
4. the AO.

Each of these works individually and, when they coincide, collectively on Highland climate systems.

3.2.1 ENSO and IOD

ENSO is the most well-known international climate system. It occurs through the coupling of atmosphere-ocean processes and manifests in three phases:

1. El Niño, when there is ocean surface warming in the east and central Pacific Ocean;
2. La Niña, when there is sea surface cooling in the east and central Pacific Ocean; and
3. Neutral when sea surface temperatures are close to average.

Although ENSO is considered the dominant climate mode in the Pacific Ocean (Rasmusson and Wallace 1983), it also interacts with other ocean basins, such as the Indian and Atlantic Oceans (Cai et al. 2019; Wang 2019). Consequently, the sea surface anomalies associated with ENSO have far-reaching impacts on climate, including across the Asian Highlands. The Indian Monsoon tends to be weaker during El Niño events and stronger during La Niña events (Wang et al. 2003). This contributes to variability in rainfall across South Asia, including the Himalaya, the southeastern Tibetan Plateau, and the Hengduan Mountains. Evidence for the connection between ENSO and the Southeast Asian Monsoon was found in Mile, southern Yunnan, southwest China, in speleothem data. Speleothems are mineral deposits precipitated from groundwater and found in underground caverns. They are annually banded and can be dated and used to determine past climate. The speleothem found near Mile showed that in this region, just below the southeastern Plateau protuberance, tropical ocean sea surface temperatures and ENSO both affected monsoon precipitation (Tan et al. 2017; Hu et al. 2021). Historical and recent evidence from the Khariya (Guliya) ice core (1690–1990) in the northern Tibetan Plateau also suggests that negative precipitation anomalies (based on net ice accumulation) tend to be associated with El Niño years. However, there is no clear correlation throughout all ENSO periods (Yang et al. 2000). The geochemical composition of summer precipitation derived from satellite data in Lhasa, Tibet, also shows strong correlations with ENSO (Gao et al. 2018). These multiple pools of evidence suggest that ENSO is a dominant driver of climate variability through multi-year timescales across the monsoon-affected Highlands.

The IOD is a less well-known but profoundly important climatic driver associated with ocean-atmosphere interactions over the Indian Ocean (Saji et al. 1999; Webster et al. 1999). It also manifests as extreme events in regions remote from the Indian Ocean, especially when it is in phase, or teleconnection, with ENSO. Similarly to ENSO, the IOD has three phases:

1 The neutral phase occurs when warm water in the western Pacific flows into the Indian Ocean between the islands of the Indonesian archipelago, generating higher surface sea temperature in the Indian Ocean;
2 The negative phase occurs when westerly winds strengthen with surface sea temperature increasing in the east; and
3 The positive phase surface sea temperature cools in the eastern Indian Ocean as the westerly winds weaken along the equatorial Indian Ocean (Bureau of Meteorology 2020).

As the IOD's positive phase causes greater surface sea temperature anomalies, it has greater and more widespread impacts than its negative phase (Abram et al. 2020).

Synchronizing ENSO and IOD compound each other's effects. For example, El Niño and positive IOD phases combine to increase snowfall on the Tibetan Plateau. This increase is partly because the positive IOD produces westward-moving, equatorial, oceanic Rossby waves (Jiang et al. 2019; Shi et al. 2021; Zhou et al. 2021). These propagating waves can induce anomalous cyclonic activity that, in turn, provide ideal conditions for transporting more moisture from the Bay of Bengal and the Arabian Sea up into the mountains (Jiang et al. 2019; Shi et al. 2021). The transported moisture causes heavier snowfall during early winters, with consequent implications for soil moisture storage, snowmelt inputs to river flows, and agricultural productivity (Shen et al. 2021). The connection between the IOD and ENSO can also transform climate patterns on the eastern side of the Plateau. Anomalous heavy rainfall and severe flooding events in 2020 in the Sichuan and Chongqing regions of the Yangzi River, for example, were linked to the strong positive IOD and weak El Niño events during 2019 and 2020. (Zhou et al. 2021).

3.2.2 NAO and AO

Since the primary source of moisture for the western and northern Tibetan Plateau regions is the northern Atlantic Ocean (Thompson et al. 2000), the NAO, which is the dominant system in the Atlantic Ocean, may also affect the Asian Highlands' precipitation. The NAO can be defined as a north-south fluctuation in atmospheric pressure between the sub-tropical high over the Azores Islands and the polar low over Iceland (Walker 1924; Walker and Bliss 1932; Wallace 2000). The NAO is associated with changes to the Westerlies (Wallace and Gutzler 1981), which, as we have already learned, is critical to the Highlands' winter precipitation. The NAO not only drives the Westerlies but also can be used to measure and interpret their strength (Wanner et al. 2001).

Like the other oscillations, the NAO also consists of positive and negative phases (Wanner et al. 2001). The positive phase represents a stronger-than-usual pressure difference between the Azores and Iceland, resulting in more forceful westerly winds and increased storminess. These, in turn, cause warmer and wetter conditions across northern Europe and eastern North America, as well as drier and colder conditions over southern Europe (Visbeck et al. 2001). Conversely, NAO's negative phase consists of lower atmospheric pressure differences between the Azores and Iceland that can generate weaker Westerly winds. These create drier and colder conditions in northern Europe and eastern North America, and warmer and wetter conditions in southern Europe (Visbeck et al. 2001). When the NAO is *powerful*, its impact can extend far into Asia, including the Asian Highlands, where it will cause less precipitation during a positive phase and the converse during a negative phase (Wang et al. 2002b, 2003).

The ice-core datasets extracted from the Khariya (Guliya) and Dunde glaciers, both situated on the Tibetan Plateau's northern rim, and the Dachuphu (Dasuopu) Glacier, located in the southern region of the Himalaya, consistently exhibit NAO's influence. They all have strong periodicities for NAO (~7.6 years) and ENSO (3–7 years) variability (Wang et al. 2007). At a seasonal timescale, the chlorine (Cl^-) and sodium (Na^+) concentrations in the Dachuphu ice cores show positive correlations with the NAO index during monsoon (June–August) and non-monsoon (September–May) periods, indicating a higher transport of Cl^- and Na^+ from drier regions when the Westerlies are strong (Wang et al. 2002b). A further correlation exists between the NAO and the annual $\delta^{18}O$ (a stable isotopic signature) of the Malan ice core, which was taken from a glacier on Mount Phukula in the Kunlun Mountains. The Malan ice-core data indicate a very weak to weak positive relationship to summer air temperature in the northern Tibetan Plateau for the century between 1887 and 1998 (Wang et al. 2003). Xin et al. (2010) also suggest a correlation between winter NAO and snow depth cover in the central and eastern Tibetan Plateau based on observation datasets from 1958 to 2001.

There is more explicit evidence of NAO's influence on the east of the Tibetan Plateau. Linderholm et al. (2013) found a stronger negative correlation between the summer NAO and tree-ring width over the last four centuries, with the correlation being stronger from interannual to decadal and longer timescales (Linderholm et al. 2013). Liu et al. (2015a) showed that NAO's positive phase brings warm, moist air from the North Atlantic to the northeast Tibetan Plateau via a cyclonic anomaly in East Asia. Once this warm air meets the cold air at high elevations, more convection and consequent precipitation occur. This cyclonic anomaly will subsequently occur in northwestern India and Pakistan, preventing moisture transport from the Arabian Sea to the southeast Tibetan Plateau and resulting in *less* precipitation (Liu and Yin 2001; Liu et al. 2015a). These interactions generate a seesaw pattern of summer precipitation between the northern and southern Tibetan Plateau that can be attributed mainly to the NAO.

Like the Pacific ENSO and IOD, the NAO works with other associated oscillations—for example, the AO. This similar meridional (or north-south) oscillation modulates changes in pressure between the Arctic and mid-latitudes of the North Atlantic and North Pacific (Thompson and Wallace 1998). During the positive phase of AO, there is lower pressure over the Arctic but higher pressure over mid-latitude regions. This results in a northward shift of the mid-latitude jet stream and associated storm tracks, causing fewer cold air outbreaks over the mid-latitude regions of North America, Siberia, Europe, and East Asia (including the Tibetan Plateau) (Elias 2021; Talley et al. 2011). The opposite happens during AO's negative phase, which typically produces more polar air outbreaks over mid-latitude regions (Elias 2021; Talley et al. 2011, Chapter S15, 1–36). Although there is debate over the relative dominance of the NAO and AO and the different mechanisms by which weather is manifested, the AO appears to play a salient role in autumn/winter snow cover at interannual and interdecadal timescales on the Tibetan Plateau (Lu et al. 2008; You et al. 2011; Liu et al. 2021b). During a positive AO phase, snow depth decreases during autumn but increases in the following winter, with the converse occurring during a negative phase. When positive NAO and AO synchronize, there is stronger cyclonic activity near Lake Baikal in Siberia, resulting in more moisture transport and, therefore, more winter snowfall on the eastern and central Tibetan Plateau (You et al. 2011; Xin et al. 2010).

Understanding the influences that these climate modes impose on precipitation (as rain and snowfall) provides a foundation for examining the climate record over time and across space. Notably, such knowledge tells us that the weather influencing the flows of the great rivers that drain the Tibetan Plateau is not just a local manifestation but is drawn from complex planetary processes operating at a vast scale and originating from distant regions.

3.3 Climate Variability and River Flows, Seventeenth to Twentieth Centuries

All these climate factors influence the Highland rivers' flow regimes. We can see this by comparing the historical evidence for key climate indices with streamflow reconstructions for the Brahmaputra River and precipitation in the Yangzi Catchment (Figure 3.2).

The Brahmaputra's July–September streamflow for the period 1309 to 2004 can be reconstructed from tree-ring networks in its catchment (Rao et al. 2020). The Yangzi Catchment's February–April precipitation from 1856 to 2013 can be reconstructed from tree-ring networks in the Xianxia Mountains, southeastern China (Shi et al. 2015). In each case, tree-ring data for the last century were calibrated against the existing hydrological data, and this relationship was used to reconstruct past streamflow and precipitation records.

The reconstruction of the Brahmaputra streamflow indicates periods of *low* streamflow across four centuries, indicative of drought conditions. The Brahmaputra streamflow, Dachuphu ice core, and scPDSI (Palmer Drought Index) all provide evidence of a severe, prolonged drought period at the end of the seventeenth century, which was, perhaps not coincidentally, a tumultuous political time in the Asian Highlands (MacCormack 2018). This drought period coincides with the most extreme positive IOD in the last millennium, which occurred in 1675 (Abram et al. 2020) and was accompanied by a strong positive NAO. Speleothem data from Northeast India support this connection. It shows more enriched $\delta^{18}O$ during this period, which indicates a decrease in monsoon precipitation (Sinha et al. 2011). Other tree-ring records of this period also characterize the broader seventeenth century as drought prone. Earlier in the century, the Ming Dynasty Megadrought (1637–1643) contributed significantly to the Dynasty's collapse and the rise of the Qing Dynasty (Fan 2023). Climatic deterioration during this period generated food shortages and associated social instability in the Yangzi Catchment (Zheng et al. 2014).

Tree-ring records from the Pobor Gang (Shaluli Shan) Mountains in Kham, eastern Tibet, and speleothems from southeastern Yunnan suggest there was another dry period in 1790–1796 (Tan et al. 2017), which registers as a very strong drought in the scPDSI (Palmer Drought Index) annual data-set (Figure 3.2g). This dry period on the Tibetan Plateau's southeastern edge occurred during a time of intense drought in South Asia, generally known as the East India Drought (1790–1796) (Cook et al. 2010). These droughts, which stretched from East India into Southeast Asia, occurred in response to a very strong El Niño in 1790–1793 (Figure 3.2c) and a moderate El Niño in 1794–1797 (Ortlieb 2000). The Asian Monsoon also failed several times in the 1790s (Thompson et al. 2000; Cook et al. 2010). What is more, they were particularly damaging as they came on the back of an earlier series of droughts that historian Victor Lieberman coined the "Strange Parallels Drought" (Lieberman 2003; Buckley et al. 2010; Lieberman and Buckley 2012). This series of severe droughts occurred across South Asia, Southeast Asia, and Southern China from the 1740s to the late 1760s.

The late eighteenth-century dry periods and droughts were harbingers of another abrupt climatic shift that occurred just a few years later, in 1810. In that year, snow accumulation at Dachuphu (Dasuopu) Glacier began to increase markedly, and it continued to increase for the next 70 years, and this is supported by Dachuphu Glacier's ice-core isotopic evidence during this same period which also indicates change (Gabrielli et al. 2020). The moisture that caused this precipitation originated from the tropical Indian Ocean, probably in response to negative NAO conditions that displaced winter Westerlies. The thesis that the Westerlies were displaced is supported by (1) ice-core evidence from the Purok Khangri (Puruogangri) Glacier of a contemporaneous drought in Central Tibetan and (2) ice-core evidence from the Dachuphu Glacier, which shows no correlation between snow accumulation and monsoon precipitation (Gabrielli et al. 2020). After this anomalous period, the correlation between the NAO and snow accumulation and between the summer monsoon and snow accumulation become significant again.

The scale of this anomaly is evident in data showing the departure from long-term mean accumulations (Figure 3.3). Here it is evident that this hydro-climate anomaly generated a large store of ice and snow that may have buffered the impact of global warming in the Highlands for a while. It will also have generated higher snowmelt-associated streamflow that will continue until the store is depleted.

The stores accumulated during the anomaly were somewhat depleted by another dry period from the late 1950s to the 1970s. The lowest flows in the Brahmaputra's reconstructed streamflow data during this time were in 1959. Streamflows remained low for a prolonged period after this, coinciding with low precipitation in the Yangzi Catchment (Figure 3.2b). Indicators

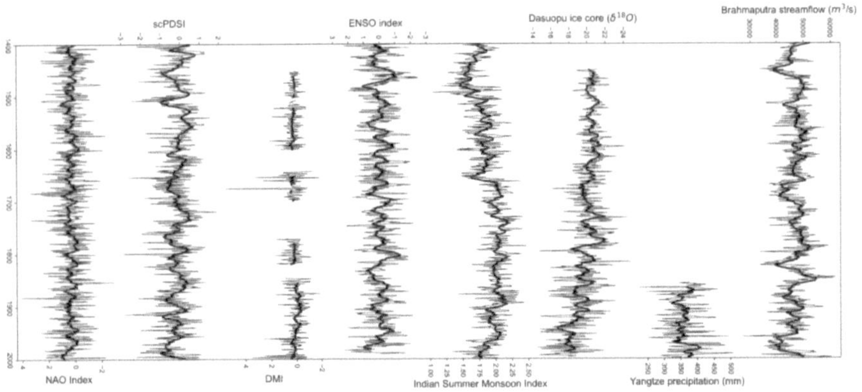

FIGURE 3.2 Comparison of annual reconstructions from (a) Brahmaputra streamflow (Rao et al. 2020); (b) Tree-ring-based February-April precipitation reconstruction for the lower reaches of the Yangzi River (Shi et al. 2015); (c) Dachuphu (Dasuopu) ice core (Thompson et al. 2000); (d) Indian Summer Monsoon Index (Shi et al. 2017); (e) ENSO index (Dätwyler et al. 2020) and (f) DMI (Abram et al. 2020); (g) Self-calibrating Palmer Drought Severity Index (scPDSI) from southeast Tibetan Plateau (Wang et al. 2020a); and (h) NAO index (Cook et al. 2019).

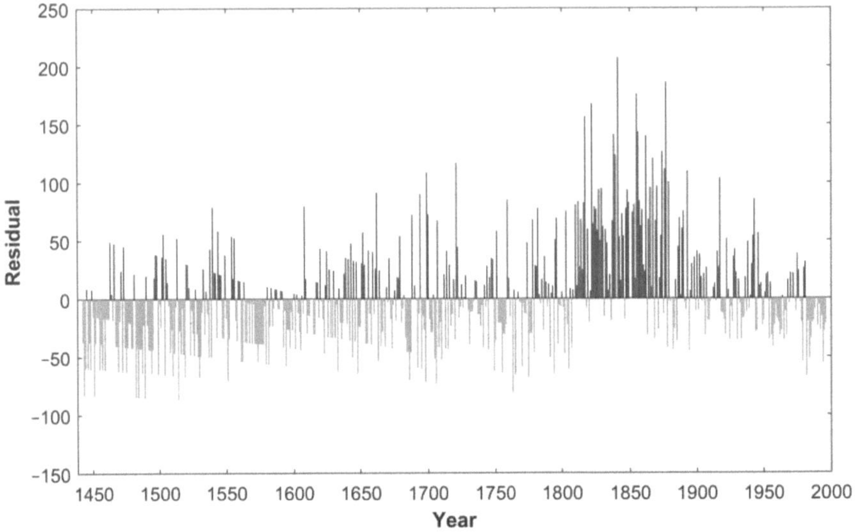

FIGURE 3.3 Departure of ice accumulation from the long-term mean, Dachuphu (Dasuopu) ice core, 1442–1996.

of dry periods in both catchments suggest that the dry conditions extended across the southern and eastern Tibetan Plateau. Tree-ring data from 14 trees around the southeastern Tibetan Plateau reflect the same event. Together, this data records the longest drought in the last 900 years, which occurred relatively recently, from 1958 to 1976 (Wang et al. 2020a). Dachuphu ice core's isotopic data ($\delta^{18}O$) confirm the drier conditions (Figure 3.2c). The most likely causes of this drought were the strong El Niño and positive IOD events during the same period (Figure 3.2c and 3.2d). The NAO index may have also been a driver of lower precipitation in the southern and eastern Tibetan Plateau during the same period (Figure 3.2h). As we will explore more in Chapter 8, this period was coincidently a time of political and social upheaval in the Highlands, as the Chinese state asserted control over Tibet and brought in much social and cultural change.

3.4 Anthropogenic Climate Change

All these historically significant developments have been overshadowed, however, by the anomalous shifts that have occurred in the Asian Highlands' climate over the last few decades of anthropogenic climate change. Anthropogenic climate change is recognized as a global phenomenon with broad-scale and catastrophic consequences. This global warming began with the Industrial Revolution in the first half of the nineteenth century but has risen exponentially since the middle of the twentieth century. High-elevation regions such as the Asian Highlands are especially vulnerable to climate change impacts. The type of warming they are experiencing is called "significant and elevation-dependent" (You et al. 2016). It causes temperature extremes and dramatic changes in precipitation (Li et al. 2017a). The impacts from these changes include—but are not limited to—decreases in snow cover and permafrost, glacier retreat, and cascading implications for high-elevation ecosystems.

Greenhouse gas emissions that cause atmospheric warming are the most well known but not the only cause of the Asian Highlands' changing climate. Salient drivers also include increased incoming solar radiation, cloud cover changes, and dust and black carbon deposition. The increase in dust and black carbon deposits is particularly noteworthy. The Highlands represent the Earth's greatest snow and ice concentrations within a relatively short distance from large, growing populations in countries with high economic growth. Dust and black carbon (or soot) are transported by wind from these population centers and deposited on snow. This causes a reduction in surface albedo, which is a measure of the proportion of light or radiation reflected from a surface. Snow and ice both have a high albedo. When they are darkened by dust and black carbon, this reduces the reflection of light from them and increases the surface and near-surface temperatures (Yao et al. 2012a).

Throughout the period for which there is climatic data, there is convincing evidence of anthropogenically forced climate change across the Asian Highlands. Early signatures of change emerge in the mid-twentieth century, but they commence at slightly different times across the region. A warming trend is evident as early as the 1960s on the eastern side of the Plateau, with a rate of 0.36°C per decade or 1.8°C over the 50 years between 1961 and 2007 (Liu and Chen 2000; Wang et al. 2008). The northwestern Tibetan Plateau, by contrast, only began warming in the 1970s at a rate of 0.74 ± 0.12°C per decade (An et al. 2016). Regardless of this heterogeneity, the region's warming rate is considerably higher than the global average (Qiu 2008). Long-term data from the Dachuphu (Dasuopu) ice core provides additional supporting evidence of a warming trend beginning in the mid-twentieth century (Thompson et al. 2011).

Warming changes the Highlands' climate and their rivers' stream flows in multiple ways. In the southeast, in the Bobor Gang Mountains, tree-ring evidence suggests it has brought wetter conditions (Li et al. 2017a). Precipitation data from the broader Yangzi Catchment (Figure 3.3b) has also increased in the twentieth century (Shi et al. 2015), as it has in the Plateau's north, where the Khariya (Guliya) ice-core data suggest a warming trend since the mid-nineteenth century and increased precipitation (Thompson et al. 2018). This increase may result from more convection caused by higher temperatures (Li et al. 2017a). This greater precipitation will augment river flows.

The global phenomenon of climate change is not operating only in the Highlands. At a broader scale, around the Highlands, changing monsoon and Westerlies' patterns are impacting the region's climate. The mechanisms and directions of these changes and their spatial variability are open to debate. Although significant uncertainties in future projections exist, warming over the Indian Ocean and stronger temperature gradients between land and sea are generally expected to increase monsoon rainfall (Turner et al. 2012). By contrast, the Westerlies operate differently, bringing more complex climatic changes to the Highlands' sub-regions (Cannon et al. 2015). There have been increases, for example, in the intensity and frequency of winter Westerlies since 1979, and these have brought increases in heavy winter precipitation to the western Himalaya. But in the eastern Himalaya, weakening winter Westerlies have caused a decrease in local heavy precipitation in winter. Similar trends in glacial mass balance have accompanied these changes in precipitation. They have increased in the western Himalaya and decreased in the eastern Himalaya (Bookhagen and Burbank 2010; Bolch et al. 2012).

Despite the western Himalaya's against-the-grain higher winter glacial mass balance, the overall trend in the Highlands is glacial loss and reduced snow coverage. The IPCC's *Special Report on the Ocean and Cryosphere in a Changing Climate* (2019) noted that observations indicate decreased low-elevation snow cover (high confidence) and glacier extent (high confidence).

This has caused changes in the amount and seasonality of river runoff (very high confidence) in the Asian Highlands region (IPCC 2019). Future projections indicate that the impacts of increased glacier melt are not uniform across the major rivers draining the Tibetan Plateau (Immerzeel et al. 2010). Models suggest that climate change will cause more damaging consequences in the Brahmaputra catchment than the Ganga (Ganges), Yangzi, and Huang He (Yellow) catchments as there is a greater risk of damaging floods and its population's food security is threatened. There are, however, caveats to this modeling. The relationship between moisture and temperature on the Tibetan Plateau is underestimated in general circulation modeling, resulting in uncertainty in this region's future moisture projections and precipitation (Wang et al. 2020a). At present, for example, the Brahmaputra streamflow reconstruction (Figure 3.2a) does not show a clear trend in the direction of increased flow from glacial melt, but it may in the future.

The other element of regional climate about which we know too little is atmospheric rivers. Climate models predict that increased atmospheric moisture from global warming will amplify the intensity of atmospheric rivers. This will, in turn, cause increased precipitation and other extreme events associated with atmospheric rivers: catastrophic floods, debris flows, and landslides (Payne et al. 2020). We have yet to determine, however, how these global trends will play out in the Highlands. No trends toward the proliferation or strengthening of atmospheric rivers have been detected in existing data over the past few decades. This may happen in the future, or other yet unknown factors may mitigate the tendency toward their development. Any increased flooding risks in the Asian Highlands include Glacial Lake Outburst Floods (GLOFs) and Landslide Lake Outburst Floods (LLOFs) (see Box 4a).

3.5 Conclusion

This chapter has provided an overview of the three main weather systems and precipitation pathways on the Tibetan Plateau: (1) the mid-latitude Westerlies that are dominant on the Plateau's north and west; (2) the monsoons that are dominant on the south and east, and (3) the transition zone at the Plateau's center. It also explained the interactions between these systems and key climate drivers such as ENSO, the IOD, the NAO, and the AO before outlining how these intersecting systems have shaped the Highlands' climate history.

These multiple sources provide a basis for understanding what drives weather and the patterns, trends, and variability of the Asian Highlands' hydro-climate. The scientific evidence is, furthermore, reflected in documented historical and literary sources, as well as local knowledge and belief systems. Our understanding of the Highlands' climate continues to build and contributes to an evolving narrative about their past and future. This narrative is developing, however, amid concerns about these high-elevation regions' vulnerability to climate change, with the raft of implications it brings for the environment and human populations. The enormity of those impacts' spatial and temporal scales amplifies our concerns.

BOX 3a GRAVITATIONAL POTENTIAL ENERGY AND STREAM POWER

Rapid and damaging erosion and flood events in the Highlands are commonly linked to high relief, steep slopes, and—in some areas—intense precipitation, but they have a more fundamental cause. Because the Indian Plate's collision with the Eurasian Plate created crustal thickening and uplift, the Tibetan Plateau stores Earth's largest concentration of GPE. On the Plateau's margins, this GPE is dissipated to areas of low energy concentration at lower elevations. The presence of this high GPE has positive and negative impacts on humans. It multiplies hazards, but it also explains the region's huge hydropower potential.

GPE per unit area of a rock column in the lithosphere (the crust and uppermost solid mantle) is the integral (the continuous analog of a sum) of weight multiplied by height above the level of isostatic compensation (the depth at which the pressure of overlying rock is constant globally) and is given by:

$$\text{GPE} = \int_0^{h+L} \rho(z') g z' \, dz' \quad (1)$$

where z' is the height above the compensation depth L, h is the topographic elevation, g is the gravitational acceleration, and $\rho(z')$ is the rock density at height z' above L (Jones et al. 1996).

The GPE stored in the relief of that part of the crust above the average surface of the land surface is given by Artyushkov (1973):

$$\text{GPE} \sim g(h)^2 / 2 \quad (2)$$

where ρ_c is the average crustal density and g and h are as in (1). Because h is squared, GPE will increase mostly as h increases, although there will be some effect from changes to rock density as uplift occurs.

The Tibetan Plateau has the highest GPE on the planet, up to 1.6×10^{14} N m^{-1} (Ghosh et al. 2009). This value varies across the Plateau and causes deformation of the central and southern Plateau as the crust extrudes to the east with some smoothing of topography (Zhang and Shi 2002). Extrusion of the crust around the Plateau's southern, northern, and eastern margins is one of the processes of

(Continued)

GPE dissipation (Hodges et al. 2001), but if we want to understand the potential impacts on humans, other processes are more important.

In a landscape of high-altitude mountains, steep gradients, intense rainfall, and great earthquakes, GPE is converted to kinetic energy by downhill flows: rivers, landslides, avalanches, glacial lake outburst floods (GLOFs), and landslide lake outburst floods (LLOFs) that produce megafloods. GPE increases as the Tibetan Plateau and Himalaya rise. A simple calculation (from the product of mass, *g* and *h*) shows an increase from 9.8 to 39.2 million joules (the metric unit of measurement for work) when a tonne of rock or water is lifted from 1 to 4 kilometers elevation; a 300% increase of both elevation and GPE.

As a body of water moves downhill in a river, its GPE is converted to kinetic energy, and this energy dissipation can be described by stream power. Unit steam power (SP) is expressed as:

$$\mathrm{SP} = \frac{\gamma g Q S}{w} \tag{3}$$

where γ is the density of water, g is the acceleration due to gravity, Q is the river discharge, S is the channel slope, and w is the channel width. In words, SP is the product of water density, gravitational acceleration, discharge, and channel slope divided by channel width to allow comparison between rivers.

SP indicates the amount of energy exerted on the banks and bed of a river, a small fraction of which is used to transport sediment. It can be used to estimate incision into bedrock, channel erosion rate, sediment transport and deposition (Bizzi and Lerner 2015), riverbank erosion rates, and channel width (Akhtar et al. 2011). It helps to explain differences in river and floodplain types (Nanson and Croke 1992) and correlates with key features of the river continuum to which aquatic biota (animal and plant life) are adapted (Zhou et al. 2017). A Tibetan example of the ecological effects of SP units comes from the modern Yarlung Tsangpo. As SP increases along the river, the density of macroinvertebrate (animals lacking backbones large enough to be seen without a microscope) declines. The riverbed substrate changes to uniform cobbles, and floods wipe out organisms (Zhou et al. 2017). This set of processes also affects the richness of fish diversity (Li et al. 2022).

In the modern Yarlung Tsangpo watershed, the average annual SP on the Plateau is <500 W m^{-2}, but it rises to a peak of 3,000–4,000 W m^{-2} in the Tsangpo Gorge, then falls to <500 W m^{-2} as the Brahmaputra enters the Assamese plains (Finnegan et al. 2005; Larsen and Montgomery 2012). The differences in SP are largely a result of stream gradient, which in the Gorge is about 10–12 times higher than on the Plateau or the Assamese plains. Higher values of SP are in the section of this waterway between the Gorge and the confluence with the Brahmaputra, which the Adi people call the Siang River. Similar patterns are likely

(*Continued*)

in the Dri Chu and other rivers, incising the Plateau's eastern margin. However, in this region, river incision and SP are affected by large landslides and their deposits in the river (Ouimet et al. 2007). The average annual SP values in the Yarlung Tsangpo Gorge are unusually high, indicating the extreme dissipation of GPE. They are even more extreme in individual floods. There was a flood with 38,000 W m^{-2} SP 4,900 years ago, and at Ranaghat, near the confluence of the Yarlung Tsangpo-Siang and the Brahmaputra, and the SP for a flood 3,600 years ago was 11,000 W m^{-2} (Borgohain et al. 2020). In the lowland reaches of the Brahmaputra, the average annual SP is <10 W m^{-2} (Kale 2003). The flood 3,600 years ago caused SP of 17 W m^{-1} and 46 W m^{-2} for a flood 7,000 years ago. During this flood, there would have been at least ten meters of water at the site of the modern town of Dibrugarh. These results show that most GPE dissipation by floods occurs in the Gorge and along the Siang stretch of the river.

Steep mountain landscapes result from rock uplift, valley incision, and landslides on valley sides. SP is, therefore, a critical variable and is a surrogate for landslide erosion. SP's relationship to uplift and relief can be expressed in its simplest form as follows:

$$\frac{\partial h}{\partial t} = U - \mathrm{SP} \tag{4}$$

where the rate of change of elevation (height) ($\partial h / \partial t$) is given by the difference between the uplift rate (U) and SP (Yuan et al. 2019). For relief to develop, U must be less than SP, a state that has been the case for the Himalaya for most of its history.

In the Tsangpo Gorge, the highest landslide erosion rate (4–21 millimeters per year) occurs where the exhumation rate (mostly erosion) is highest (2–9 per year), and the average annual SP is also highest. In this state, rock uplift and river incision are approximately balanced by the adjustment of landslide erosion rate on steep riverside hillslopes. Extreme floods aid the connectivity between hillslopes and the river. The antecedent Yarlung Tsangpo was steepened as the Namche Barwa Syntaxis was uplifted in the Late Miocene (approximately 23.03–5.333 million years ago) to the Pliocene (5.333–2.58 million years ago) and then again during the Pleistocene (2.58 million–11,700 years ago) when the Indian subcontinent's continued indentation of Eurasia caused between 12 and 21 kilometers of rock to be uplifted. This rock eroded at rates between 4 and 7 millimeters per year (Bracciali et al. 2015). But now the system appears to be in approximate balance, with high values of GPE dissipation through landslide erosion, aided by megafloods, and river incision occurring at rates of the same order of magnitude as uplift.

(Continued)

This concentration of GPE and its dissipation makes anthropogenic interventions on the Plateau's edges technologically difficult and inherently risky. The resulting dynamic landscape may underpin the cultural risk-limiting practices adopted by local peoples (see Chapter 6, 7 and Box 5b), which appear to have been ignored by technologists determined to tap the mountains' hydroelectric potential. Humans have only developed the technology to make limited use of the Highlands' GPE in the past century, with mixed results, and further developments will be complicated by climate change (Li et al. 2022). The existence of high GPE highlights the need for adaptation to the changing climate and landscape to manage risks better. This should include avoiding the construction of hydropower dams on rivers in favor of smaller, safer off-river pumped storage hydropower projects (see Chapter 9).

The mismatch between geomorphic and human timeframes has hampered human interventions in the Highlands. Human interventions seek decadal outcomes, and the Highlands' geomorphic processes operate over longer timescales, often with long periods between extreme floods and earthquakes. It could also be argued that the mere presence of the Highlands' beyond-human-scale GPE highlights the need for a worldview that includes humans within Earth systems rather than attempting modernist mastery over these systems.

BOX 3b DUST AND SAND: WIND ACTION ON THE TIBETAN PLATEAU

Most of the Tibetan Plateau is semi-arid to arid with low vegetation cover, erodible soils and sediments, and often strong winds. Eolian processes (erosion, transportation, and deposition by wind) across the Plateau produce sand and dust storms, dunes, and loess (wind-deposited dust deposits). On the Plateau, dust storms can either add to or deplete soil nutrients. The mobile sand they create can divert rivers, sterilize otherwise productive soils, and interfere with infrastructure. Depending on when and where they happen and how strong they are, dust storms can also impact human health (Zhang et al. 2016). But the Plateau's dust and sand transportation systems are part of larger regional and global patterns that underpin Earth systems. For example, transported dust and sand fertilize oceans by aiding phytoplankton growth. Growing phytoplankton draws down atmospheric carbon dioxide, changes the radiation balance and, therefore, has a net cooling effect on the atmosphere (Kok et al. 2023).

(Continued)

The Plateau's topography also influences dust transport in surrounding areas. Its high margins produce strong upslope winds that deposit dust that moves from the Indian–Pakistan Desert onto the Himalayan front and dust from the Taklamakan and other deserts onto the Plateau's north. The topography of the Plateau enhances wind erosion and dust emissions in the Hexi Corridor (north of the Qilian Mountains) and the Gobi Desert (also north of the Plateau and east of the Taklamakan Desert). These processes, in turn, lift more dust to high altitudes, and it is then transported far afield. The rates of dust production increased as the Plateau's height increased. They were much lower when the Plateau was half its current height (Tan et al. 2021).

Deposits of sand and dust provide insights into eolian processes and rates. They also have ecological significance. An outstanding example of the impact of dust production on ecosystems is the *Kobresia* meadow or pasture ecosystem. It covers about 18% of the Plateau, is the world's largest example of this vital ecosystem, and is the primary resource of the Plateau's economically, socially, and culturally defining pastoralism (see Chapter 7). It is built on dust that accumulated over the past 3,000 years at approximately the same rate as the modern rate of dust fallout (Zhang et al. 2022b).

The Yarlung Tsangpo and Dri Chu (Yangzi) river valleys contain dust deposits, which shows that there are more than riverine processes at work within them. There are 1,930 square kilometers of dunes in the wide parts of the Yarlung Tsangpo Valley. They have accumulated on river terraces high above the river, on sand bars in the river, and as climbing dunes on hillsides. Dust and sand movement within these valleys also creates loess (windblown silt and nutrient-rich dust) on hillsides and hilltops. The dunes and loess are derived locally, mainly from the erosion of river deposits that, in part, originated from outwash from past melting glaciers. The latest phase of dust deposition and dune formation started after the last Ice Age, about 13,000 BP (Dong et al. 2017).

In the Dri Chu's widest parts, there is a 4,500 square-kilometer dune field, and loess is widely distributed on river terraces, foot slopes, and hillsides. Loess fills small valleys in hilly terrain and lies between dunes. Like the Yarlung Tsangpo valley's dust and sand, the Dri Chu valley's dust and sand originate locally (Zhang and Jia 2019).

The local origin of these valleys' loess means the large deserts to the Plateau's north have contributed either none or very little dust to them over the long term. Moreover, their loess's chemistry shows that their contribution to the vast Loess Plateau and ocean further east has been minimal. However, other parts of the Plateau appear to export dust in quantities sufficient to have a global impact.

(*Continued*)

FIGURE 3B.1 *Top*, Sand dunes in the western Yarlung Tsangpo valley. *Bottom*, Climbing sand dunes in the eastern Yarlung Tsangpo valley. Photographs Ruth Gamble, 2018.

In recent decades, human activity has intensified some of these processes. In 2015, the Chinese Academy of Sciences warned that desertification was occurring on the Plateau. It was taking the form of soil erosion, permafrost melting, the loss of alpine and swamp meadows, and the loss of wetlands. Its scientists gave the causes of this desertification as unsustainable grazing practices, the spread of arable agriculture, mining, and urbanization. Wind erosion is also an important process of desertification, and there is some evidence that human activities are beginning to affect these processes.

The Asian Highlands produced 97.8 million tonnes of dust annually between 2000 and 2020, mainly from the Chang Thang (northwest Tibetan Plateau) and the Pamir Plateau. This is equivalent to an average surface lowering of 0.05 millimeters annually for the whole Plateau (Du et al. 2022). This rate was significantly higher than the average lowering rate between 1989 and 2015, which was 0.03 millimeters per year. During this period, there was an overall

(*Continued*)

decline in dust production but an increase in the northwest caused by climate change and land use (Teng et al. 2021).

Since the 1960s, the wind erosion of meadow soils in the Yarlung Tsangpo valley has been, on average, ~2.0 millimeters per year on grassland and ~2.4 millimeters per year on farmland (Ping et al. 2001). In the Dri Chu (upper Yangzi) Catchment, the average rate of wind erosion between 2005 and 2015 was 0.19 millimeters per year (Song et al. 2020). Wind erosion on farmland in the Gonghe Basin (in the Huang He catchment on the Plateau's northeastern margin) is three times greater than on grassland and, during cultivation, can increase by 5.7 to 8.8 times (Dong et al. 2000). Given the likely slow rates of soil formation in this harsh environment, some of these anthropogenic soil erosion rates threaten the long-term viability of agroecosystems, and these erosion rates are likely to increase as the climate warms (Chinese Academy of Sciences 2015).

Estimates of the amount and impact of dust emissions from the Plateau vary considerably. Mao et al. (2013) estimate that it contributes 10% of East Asia's dust deposits, and most of the remainder comes from the deserts to the Plateau's north (Dong et al. 2017). In contrast, Du et al. (2022) suggest that dust from the Plateau may contribute up to 19% of global dust. Wei et al. (2021) estimate that over millennia Plateau dust has contributed significantly to the glaciers of eastern Tibet, the Loess Plateau, the South China Sea, Japan, and as far afield as Greenland by means of the high-altitude westerly jet stream (see Figure 3b.2). It is likely that as the climate warms and land use on the Plateau intensifies the Plateau will contribute more dust, and this will have global impacts.

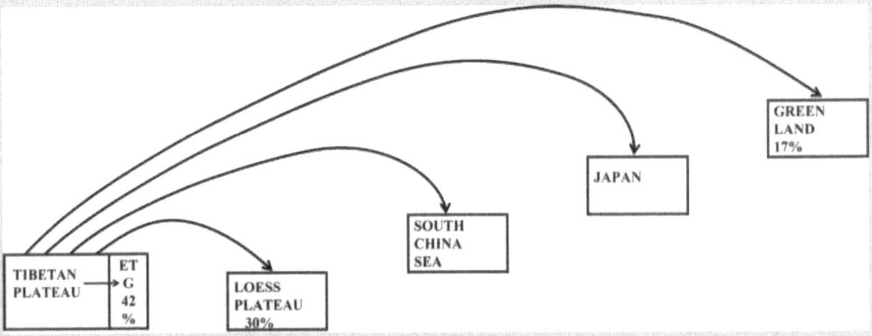

FIGURE 3B.2 Dispersal of modern dust from the Tibetan Plateau. The figures are the average contribution to the total dust fallout from the Plateau at each location, where known. ETG is Eastern Tibetan Glaciers. Data from Wei et al. (2021).

4

FROZEN RIVERS

The most common time markers in this chapter are BP ("before the present," i.e., before 1950), BCE (before the common era), and CE (common era).

Landscapes are rich and varied places with diverse living and non-living elements, plants, animals, soil, rocks, and water, set within unique topographies. Often, frozen landscapes—the cryosphere—are presented as an exception to this diversity, an expanse of empty whiteness. This is deceptive. Diversity thrives in the cryosphere, and the cryosphere plays a profound role in many of Earth's systems, underpinning climate and biological systems far beyond its icy expanses.

As much of Earth's cryosphere circles the poles, discussion about it tends to focus on the Arctic and Antarctic regions. But the Asian Highlands, often called "the Third Pole" because they are home to Earth's third largest ice store, are also an important part of this Earth system. Water in its solid form as ice covers a total area of 2,542,230 square kilometers (Ma et al. 2023) of the Highlands. This is equivalent to roughly five-sixths of India's land mass. This ice-bound area includes permafrost, snow cover, glaciers, frozen lakes, and frozen river waters. Climate and topography control its spatial and temporal distribution and sustain it through complex feedback systems.

Although it shares some characteristics with the other two poles, the Third Pole is unlike them in many ways. While they house vast polar ice sheets, the Third Pole comprises a heterogeneous mosaic, within which humans have

DOI: 10.4324/9781003392033-4

developed complex agrarian societies, cultures, and polities (see Chapters 6 and 7). In the past, the Third Pole's ice coverage and associated systems have defined the hydrological and geomorphic processes of the major rivers that flow from it. These have changed over long periods, through colder and warmer periods. Increasingly, however, human activities are overprinting these processes and responses, transforming patterns that have existed over much longer periods. This has never been a static environment. By its very nature, it was and remains finely tuned to climate variability and, therefore, particularly susceptible to climate change. However, anthropogenic changes are now destabilizing this fine balance, and this destabilization will increase in the future as these changes intensify.

Like the Asian Highlands' climate histories that we explored in the last chapter, we can discover much about the cryosphere's past through its self-creating geoarchives, including ice cores, tree rings, and sediments. These records not only tell us about past changes to the cryosphere but also reveal how it is responding to current anthropogenic climate change, and suggest what may happen in the future. This chapter uses these geoarchives to explore each of the significant components of the Asian Highland cryosphere through their *processes* and *responses* to natural variability and anthropogenically

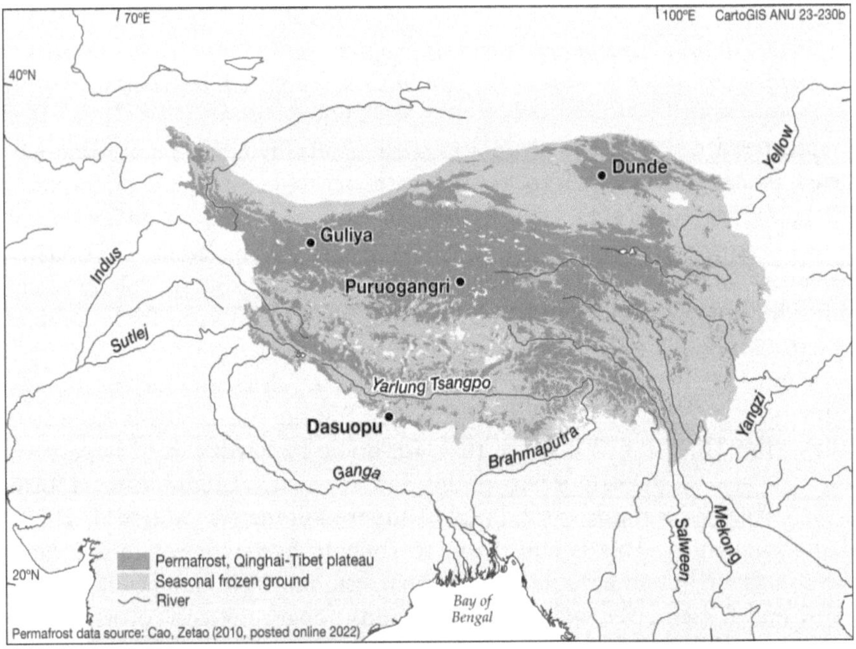

FIGURE 4.1 Distribution of glaciers, permafrost, ground ice, and ice-core retrieval sites in the Asian Highlands. Map produced by ANU's CartoGIS with data published as an addendum to Cao et al. 2023.

forced change. Examining the geoarchive and the current cryosphere at this time of accelerated change is crucially important, as global warming converts ice to liquid water.

The cryosphere constantly interacts with the other spheres. Over time, its influence shaped the geosphere, changing Earth's shape along with other geological agents. It transforms climate systems in the atmosphere, and water melts from it into the hydrosphere. Understanding this frozen landscape is a valuable prelude for navigating the wild rivers that discharge their waters from this region and support billions of people and water-dependent ecosystems in the biosphere (Chapter 5).

4.1 The Third Pole Cryosphere's Deep Time and Historical Past

From the deep past to the present, the Earth's climate has naturally cycled through cold and warm periods, which climatologists call glacials and inter-glacials. The extent and distribution of snow and ice across the Tibetan Plateau have waxed and waned in response to this climate variability. We know this because we can create paleo-reconstruction of past climate from our analysis of ice cores, lake sediments, stalagmites in caves, tree rings, and organic carbon. Ice-core data from the Khariya (Guliya) Ice Cap, situated at the northwest edge of the Tibetan Plateau, have been instrumental in providing reconstructions extending back 500,000 years (Thompson et al. 1997). Pollen analysis from lake sediments in the Central Himalaya and ice-core data from the Dunde Ice Cap on the Plateau's northeastern rim provide data at a finer granularity for the more recent Late Pleistocene (129,000–11,700 BP) and Holocene (11,700 BP to now) (Liu et al. 1998). The latter two data sets identify major periods of oscillating climatic conditions that expanded and shrank the Asian Highland's cryosphere. During the Late Pleistocene, warm, moist conditions prevailed at 30,000 BP. Then they continued for several thousand years before transitioning to the cold, arid conditions of the last Ice Age (or Glacial), reaching the most extreme conditions of the Last Glacial Maximum (LGM), at approximately 26,000–18,000 BP (Thompson et al. 1990; Kotlia et al. 2008).

The LGM was the most recent period of glaciation when the area of ice on Earth was at its greatest extent. As we discussed in Chapter 2, for many years, scientists operated under the understanding that an extensive sheet of ice covered the Tibetan Plateau at this time. Kuhle (1995) argued that this ice was 3 kilometers thick, covering even lower-lying areas like the Yarlung Tsangpo Valley. This view affected how people understood the peopling (human colonization) of the Plateau (Chapter 6) and its geological composition (Chapter 2). This view is no longer accepted. Owen et al. (2008) concluded that the best estimate of the extent of glaciation on the Plateau during the Last Glacial had been provided by Li et al. (1991), who made their estimate at around the same time as Kuhle but did not receive as much attention for their work. They argued that the glaciation area was only ~10% of the Tibetan Plateau.

Ice-core extraction, which began in the Highlands during the 1990s, and other geoarchive sampling techniques, such as lake sediment coring, have produced a much more precise picture of the Plateau's history of partial glaciations than early geomorphological models. The Khariya (Guliya) ice core, lake sediment cores in the northeastern Tibetan Plateau (Yan et al. 1999), and glacial moraines across the Himalaya (Owen et al. 2002) all suggest peak aridity and low temperatures occurred around 18,000 BP. However, its onset appears to have been regional, and it affected different parts of the Highlands dissimilarly. Although the timing and duration of these climate cycles across the Highlands are broadly synchronous with global oscillations, variations in the relative role of the monsoons and Westerlies, the steep north–south temperature gradient, and topographic effects all resulted in asynchronous glaciation and deglaciation (Owen and Dortch 2014). In the Himalaya, moraine studies indicate that there were multiple glacier advances that extended up to 14 kilometers beyond their current position (Sharma and Owen 1996; Owen et al. 2002). Rapid deglaciation followed due to increasing global temperatures heralding the inter-glacial period, the Holocene, in which we now live.

This spatial and temporal complexity was repeated during the Holocene. Changes in the strength of the monsoons—both the Indian Summer Monsoon and the Southeast Asian Summer Monsoon—and the degree of moisture delivery by the Westerlies have been linked to oscillating temperatures in the Highlands (Chen et al. 2020). Landscape responses to these oscillations included complex temporal changes in lake levels across the region. These differed between the southwest (wettest in the early Holocene) and northeast (wettest in the Mid to Late Holocene). Landscape responses also included changes in vegetation cover, the extent of forests, and changes in glacial activity (Chen et al. 2020). These patterns of climate that imposed changes within the Highlands' frozen landscape provide valuable contexts for current trends as anthropogenic activities overprint climate forcing (physical processes that affect climate through various driving factors such as changes in the sun's energy output or hydrological shifts).

However, as our primary sources of information on these events come solely from the geoarchives rather than archives produced by humans, our sense of what happened during these periods and our ability to relate to them are limited. If we want a more precise understanding of the effects of climate change in the Highlands—one that we can use as a comparator for current events—we need to investigate more recent climate changes that have occurred on the centennial rather than millennial scale. Two episodes of abrupt climate changes occurred within this scale: the Medieval Climate

Anomaly (950–1250 CE) and the Little Ice Age (1400–1900 CE). These two events created rapid responses from the Asian Highlands' frozen landscapes.

The Medieval Climate Anomaly was a warming–drying period with several decadal cold intervals in its latter stages. It occurred across the Northern Hemisphere and has been attributed to a coherent shift in ocean-atmospheric circulation patterns (Graham et al. 2011). Climatologists have identified two trends that occurred in the Highlands during this time. The first was a north–south mode by which each area oscillated in opposition between wet–dry conditions. The second was that the Westerlies dominated the region's moisture regime. This was represented by an east–west mode north of latitude 30°N, where each area also oscillated and was in the opposition between wet–dry conditions (Chen et al. 2015).

There were two particularly warm centuries in the Tibetan Plateau during this period: 890–990 CE and 1140–1250 CE. Their temperatures were 0.3°C above the mean annual temperatures between 900 and 1900 CE (Hao et al. 2020). During these periods, warming would have contributed to ice melt and cascading impacts on the landscape, including its glaciers, proglacial lakes (described later in this chapter), and downstream rivers. There is evidence of large floods and higher flood frequency during the Medieval Climate Anomaly in the Upper Ganga Catchment and that of its tributary, the Alaknanda River system (Srivastava et al. 2017; Wasson et al. 2013), which can be attributed to these linked processes. More work needs to be conducted to assess how these changes played out elsewhere in the Highlands.

As the name suggests, the Little Ice Age (LIA) was a period of cooler weather. It occurred between 1400 and 1900 CE and was characterized by a drop in mean annual temperature of 0.5°C. The LIA is well documented in archives, literature, and other historical sources throughout the Northern Hemisphere and in ice-core, tree-ring, speleothem, and moraine data. During this period, most of the Earth's glaciers advanced. In the Himalaya, glacier advance, thickening, and active moraine building occurred throughout this period, but particularly between 1400 and 1600 CE. This period was associated with cooler temperatures, a southward displacement of the monsoons, and an increase in Westerly winter precipitation. Together, these factors contributed to high-altitude glacier growth (Rowan 2017). Again, due to the complex topography of the region, as well as the relative influence of different climatic drivers, there was spatial and temporal asynchrony in the LIA's glacier maxima (greatest total extent and depth). The southern Tibetan Plateau reached glacier maxima in the late fourteenth century and the northwest and northeast in the early fourteenth century. The southern glaciers retreated from the early sixteenth to eighteenth centuries, the northwestern glaciers from the late fourteenth to early fifteenth centuries, and the northeastern from the early sixteenth century. There was another whole Plateau glacial advance from the late eighteenth century to the early nineteenth century and a period of retreat from the late nineteenth century onward (Xu and Yi 2014).

These recent climatic shocks left their mark on human and environmental history. Evidence shows that Chinggis Khan's (1162–1227) rise was partly helped by the Medieval Climate Anomaly, which brought more grass to the Mongolian steppe (Pederson et al. 2014). The transition from the Medieval Climate Anomaly to the Little Ice Age proved to be a time of climate uncertainty that contributed to famines in Tibet (Gamble et al. 2022) and China (Che and Lan 2021). There is also some evidence that the Little Ice Age imposed changes to human movement and health and coincided with episodes of severe drought and famine in China and India. Some have even suggested that the Little Ice Age led to the collapse of the Ming Dynasty (1368–1644)—an event that had political, economic, and social impacts in the Highlands—but the size and variety of climatic zones in China, India, and the Highlands mean that it is difficult to attribute large political, social, and economic changes to any one climatic anomaly. Instead, as we see happening now, climate change exacerbates eco-social fractures and political unrest. Understanding these past events through a multifaceted lens gives us frames to read current events with more nuance and, therefore, make better predictions about the cryosphere's possible futures. To rephrase a common Highland saying attributed to the Buddha, if we want to know what happened in the past, we should examine our present. And if we want to know where we will be in the future, we should examine our present.

4.2 The Third Pole's Precarious Present

Studying the components of the Asian Highlands' threatened cryosphere gives us insights into how it was formed and suggests ways we can protect its future. The primary components of the cryosphere are: (1) permafrost and thermokarst lakes (formed when permafrost thaws and creates surface depressions that fill with melted water); (2) snow; and (3) glaciers and glacial lakes. Each of the cryosphere's components has a unique history and related contemporary threats, but none operate in isolation from each other and broader Earth Systems. All three play a profound role in Highland river systems. The rivers flow through the cryosphere, albeit slowly, and changes in the cryosphere transform rivers' flow volumes, courses, and biodiversity.

4.2.1 Permafrost and Thermokarst Lakes

Permafrost comprises soil that maintains temperatures below freezing for more than two consecutive years. Its thickness ranges from a meter to over a kilometer, and it can occur as spatially continuous, discontinuous, sporadic, or, in isolated, discrete patches. Permafrost's thickness varies depending on its extent. Continuous permafrost can be 100–800 meters thick, while sporadic permafrost may only extend to a depth of 10–50 meters (Anisimov and Svetlana 2006; in Frey and McClelland 2009). Regardless of the overall thickness, permafrost has a surface layer that seasonally thaws and is called its "active layer."

Permafrost is formed when the water in soil's pore spaces (or within rocks' cracks and spaces) is frozen and acts as "cement," binding soil mineral grains or rock fragments. A temperature gradient with depth occurs within the permafrost, commonly represented by a "trumpet plot" (see Figure 4.2). Fluctuating seasonal temperatures are evident in the permafrosts' top few meters but diminish rapidly with depth. Below the point at which zero annual seasonal variation occurs, the temperature begins to increase due to geothermal heat. The permafrost base coincides with the point at which temperature exceeds zero degrees Celsius.

The active layer at and near the ground surface can support the growth of different forms of vegetation. The amount and type of vegetation it can sustain depends on the depth of the permafrost's freezing and thawing. A shallow active layer will enable the growth of grasses typical of tundra and alpine meadows, while thicker active layers will be able to provide suitable conditions for trees of boreal forests. Most permafrost in the Highlands supports grasslands, tundra, and alpine meadows.

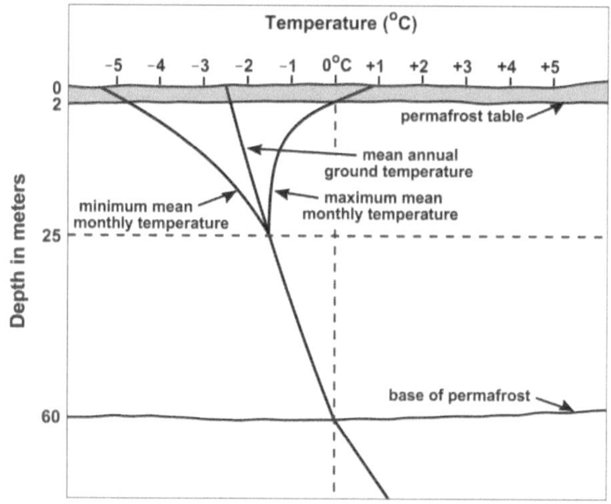

FIGURE 4.2 The theoretical temperature profile of permafrost (Devoie et al. 2019).

Permafrost ecosystems are fragile. They can be compromised by surface disturbance or temperature changes and only have access to restricted moisture. Permafrost imposes critical constraints on the local surface and groundwater hydrology as the frozen ground acts as an impermeable layer that impedes the groundwater flow and the volume of groundwater that can be stored within the soil or bedrock. Nevertheless, liquid water can still flow laterally along the permafrost base to provide essential groundwater discharge to streams.

Climate (temperature, solar radiation, wind) and topography (elevation and aspect) control permafrost occurrence and spatial distribution. Vegetation cover, water availability, snow cover, and substrate characteristics (grain size, water holding capacity) are also salient controlling factors. In contrast to snow- and ice-covered surfaces, the spatial distribution of permafrost is difficult to identify visually, but it has been estimated to cover approximately 1 million square kilometers of the Asian Highlands and two-thirds of the Tibetan Plateau (Gruber et al. 2017) (see Figure 4.2, Table 4.1). This is far greater than the distribution of glaciers and glacial ice and indicates that any thermal changes occurring in permafrost areas will have significant hydrological, geomorphic, ecological, and social impacts at a broad scale.

Permafrost results from past and present climates and has developed over tens of thousands of years. But under changing environmental conditions, increasing areas of Asian Highland permafrost are no longer in equilibrium. When it is no longer in equilibrium, permafrost's temperature profiles

TABLE 4.1 Comparative Data, Areas of Asian Highland Permafrost and Glaciers

Country	*Permafrost area (10^3 km^2)*	*Glacier area (10^3 km^2)*
Afghanistan	17.5	2.6 or 3.1
Pakistan	26.6	11.3 or 17.0
India	**40.1**	**11.8 or 19.6**
Myanmar	0.1	0.04 or 0.06
Nepal	11.1	4.1 or 4.8
Bhutan	**1.2**	**1.0 or 1.5**
China	**906**	**29.2 or 33.6**
Total	1,003 (516–1,538)	60.1/79.8

Source: Gruber et al. (2017).

The estimated permafrost area is based on Gruber's (2012) global model of permafrost extent with the range of uncertainty provided for the total area only. The two values for the glacier area are drawn from two studies (Bajracharya and Shretha 2011; Pfeffer et al. 2014, respectively, cited in Gruber et al. 2017). Countries that govern the Dri Chu, upper Yangzi, Yarlung Tsangpo, and upper Brahmaputra catchments are presented in bold.

FIGURE 4.3 *Left*, Permafrost, western Tibet, Yarlung Tsangpo headwaters. Photograph, Ruth Gamble 2018. *Right*, A melting permafrost landslide. WeChat video, 2017.

transform, initiating ecological, hydrological, geomorphic, and infrastructural responses at a range of spatial and temporal scales. The main causes for disequilibrium in contemporary Highland permafrost are global warming and the thermal changes that arise when black carbon, dust, and aerosols reduce snow cover, exposing the ground to warming by solar radiation. In both cases, conditions are increasingly ripe for permafrost thawing.

When permafrost thawing occurs, heat is transferred from the ground surface to the subsurface permafrost. The area near the surface of the permafrost heats first and relatively quickly. Heating at lower levels is delayed, creating a time lag before warming reaches greater depths. The temperature profile of permafrost changes under these thawing conditions (Figure 4.4), and a layer of *talik* (perennially unfrozen sediment—derived from the Russian word *tayat*, "to melt") develops at the base of the active layer.

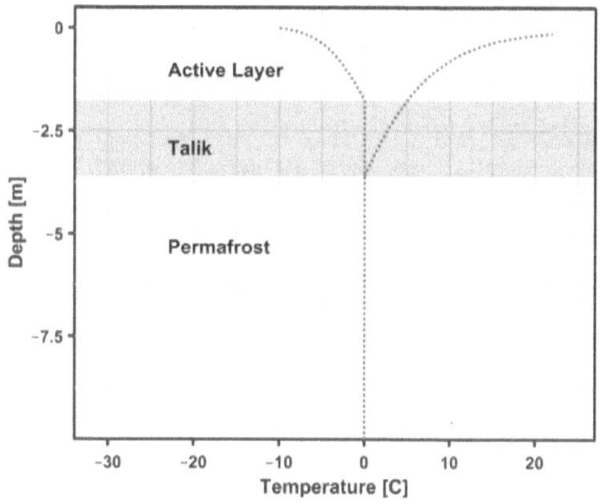

FIGURE 4.4 Measured temperature profile.

When frozen sediments thaw, the cohesion between mineral grains disappears, and the soil loses the structural "support" that a higher volume of ice provides. This causes land subsidence, which can result in the structural failure of built infrastructure, including roads, pipelines, airstrips, dams, and buildings. In addition, changes in the mechanical properties of permafrost can initiate slope instability as soil creep, landslides, and slumping or rockfalls, even on very gentle slopes. These processes of mass movement can be highly complex, changing from one type to another. Landslides, for example, can translate into avalanches or debris flows, and they can occur suddenly and in previously unaffected localities. Their effects then travel downstream through river systems, ultimately generating increased sediment loads down the catchment with additional risks to human populations.

Hydrologically, thawing permafrost with *talik* expansion allows liquid water to flow through the soil matrix. This can occur above and below the permafrost and within *taliks*, so the water creates shallow and deep flow paths. On the permafrost's surface, water ponding may be particularly evident in surface depressions, common where subsidence has occurred. Conversely, thawing causes a lowering of the water table and contraction of wetlands, marshy meadows, and grasslands that are important for local livelihoods (Cheng and Wu 2007; Chapter 7). This may explain the association between permafrost thawing, localized desertification, and an increased incidence of dust storms (Wang 2002b; Zhao et al. 2005; Cheng and Wu 2007).

When permafrost thaws and *talik* forms, this often leads to thermokarst landscapes, characterized by land subsidence and thermokarst lakes. Any process that induces permafrost thawing can be associated with thermokarst development, including climate change, but also includes anthropogenic disturbances to the ground surface, such as roads, railways (including the Qinghai-Tibet Highway), pipelines, and other engineering works (Yang et al. 2016b). Once these lakes form, they conduct warmth from the lake waters into the surrounding permafrost, and *talik* expands further laterally and vertically (Hinkel et al. 2012). Thermokarst lakes are, therefore, an expression of, and contributor to, permafrost degradation.

Studies of thermokarst lakes in the Beserpama (Beiluhe) Basin in the Plateau's north indicate that up to 60% of the region's lake inflows are sourced from permafrost (Yang et al. 2016; Yin et al. 2017). The changing climate has affected this region, and the Qinghai-Tibetan Railway crosses it. Although the contraction and expansion of lakes and wetlands seem contradictory, these different responses are due to the extent and direction of *talik* development, including its vertical growth and intersection with the water table (see Figure 4.5). Under changing climate conditions, a mosaic of hydrological and ecological responses will be expressed on the surface with associated implications for livelihoods and river systems down the catchment.

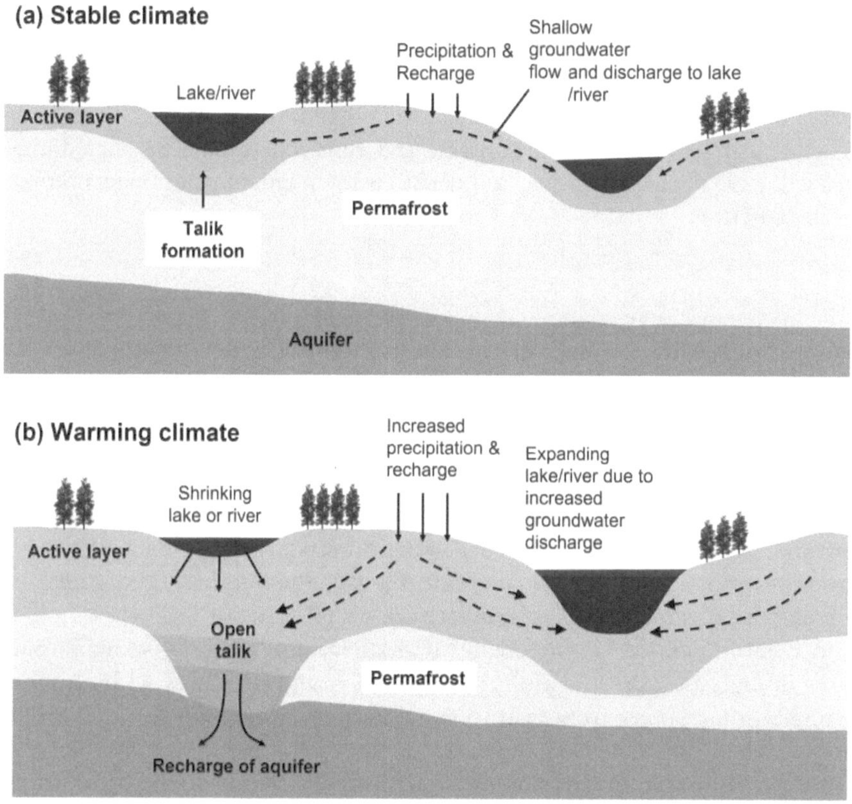

FIGURE 4.5 Hydrological impacts of permafrost thawing and *talik* development under a changing climate. Adapted from Walvoord and Kurylyk (2016).

Regardless of the pathways water takes as the permafrost melts, hydrological connectivity increases across catchments in tandem with higher volumes and flow rates (Hinzman et al. 2005). Over time, with thawing extending deeper into the permafrost, the volume of water being released from storage can be expected to taper off due to less ice being present in the first instance and the buffering of heat transfer with distance from the ground surface. As a result, melting permafrost may cause a temporary increase in water flows through otherwise arid regions, but eventually this flow will subside, and the regions will become even more arid. Scientists are still determining how long it will take for this stored water to flow out of the permafrost regions. It will vary according to climate, topography, and permafrost depth.

The flows of permafrost melt will not only affect the areas that are currently permafrost regions. Flows of water and various sediments will also transform the broader Highland river catchments. They will dissolve and mobilize different types of organic matter and elements that are key chemical components of this landscape's mineral soils and peats. Mineralized meltwater rich in dissolved carbon, nutrients, salts, and metals will change the geochemical signature of meltwaters, and researchers have predicted that this will have neutral, positive, and negative down-catchment impacts (Loiko et al. 2017).

4.2.2 Snow

Snow forms within clouds in the atmosphere when temperatures are at or below freezing, and both sufficient moisture and very fine ice crystals (<75 μm) are present. Warm air can hold more water vapor than cold air, so heavy snowfalls are generally associated with *relatively* warm air (around −9°C to 0°C). Very cold, dry regions may not experience snowfall because the air above them is too cold to hold moisture. Dry polar valleys in high latitudes are evidence of this. When snow does fall, it will either melt if the ground is above freezing or accumulate if it is below freezing. The conditions for snow to form, fall, and accumulate are complex.

A combination of factors creates permanent and seasonal snowfall patterns in high altitudes. As ambient air temperature decreases with altitude, precipitation changes from rain to snow at high altitudes. The exact elevation at which this occurs changes seasonally and is influenced by the solar radiation a site receives. At sites at higher latitudes, closer to the poles, snow will occur at lower altitudes and for a more extended period throughout the year. In the Northern Hemisphere, north-facing slopes tend to be colder than south-facing slopes. Despite all these differences, there is a point in high-altitude regions such as the Asian Highlands above which snow coverage lasts most of the year, called the "snow line." Given the Asian Highlands' position in the temperate zone at higher latitudes than most of Earth's cryosphere, this snowline is relatively high, 6,000–6,500 meters above sea level.

Snow is a general term for different types of frozen precipitation. These include powder snow (new, dry), new and old snow, *névé* (a Swiss-French word meaning fresh, granular snow that has gone through cycles of partial melting, freezing, and compaction), *firn* (a German word meaning snow that is never greater than one year old), seasonal, and perennial snow. Each type of snow will vary according to the presence or absence of original ice crystal structures, the presence or absence of freezing and melting cycles, its density, water content, and persistence.

The physical characteristics of snow when it is lying on the ground and the area it covers depend on the rate of snowfall and deposition, heat fluxes, and wind, both during and after snowfall. A region's climate determines the persistence of its snowpacks and its cycles of deposition, evaporation, and melting. A snowpack changes its structures as its snow accumulates. Much of fallen snow's volume is comprised of air, and as it accumulates, it compacts, pushing this air out. One way to measure the compactness of snowpacks is to consider their snow water equivalent (SWE), which refers to the ratio of snow composed of frozen water rather than air. This is a handy measure but does not describe many other snowpack variables. For example, it does not account for the fact that water in each snowpack can vary up to three orders of magnitude depending on weather conditions and snow density. Importantly, SWE is only one factor that determines the volume of meltwater generated during its thawing, making it difficult to assess how much meltwater any snowpack produces. Despite its complexity, the correct measurement of potential meltwater is important. It has profound implications for river flows and the associated hydrological and geomorphic processes that impact down-catchment human populations and ecosystems.

Snow plays a significant role in energy and water interactions between the atmosphere, lithosphere (of which the riverine environment is a component), and the biosphere at local, regional, and global scales. One example of these interactions is the thermal buffer that snow provides to its underlying soil. Depending on its depth and the timing and duration of melting, snow cover also influences soil moisture content and the processes of freezing and thawing in permafrost. Another surprising attribute of snow is its color. White snow reflects a large proportion of radiant energy compared to other, darker surfaces, and this imposes an atmospheric cooling effect that can be local or more widespread, depending on the areal extent of snow cover and type of snow. Fresh snow has the highest albedo, reflecting 80–90% of light. Glacial ice not covered in fresh snow reflects 20–60% (Gray and Male 1981; Ming et al. 2015; Wang et al. 2020c; Alessandri et al. 2021). Unfortunately, the reverse of this effect is also true. Snow with lower albedo, such as old snow and perennial snow covered with dust or black and brown carbon, will absorb more heat energy and exhibit higher surface temperatures.

Changes in snow albedo are one of the primary reasons that the deposition of natural and anthropogenically sourced aerosols—dust and black carbon—on the Third Pole's snow and ice is such a concerning trend. Scientists hold this deposition at least partially responsible for the Highlands' accelerated warming, which occurs at 0.32 K per decade compared to the global rate of 0.1–0.15 K per decade (Liu and Chen 2000). Researchers regard the positive feedback mechanism of increased surface temperatures due to reduced snow albedo as a significant contributor to snow cover variability, including losses of SWE (Ji et al. 2016; Zhang et al. 2018), the rapid melting of glaciers (Qu et al. 2014; Zhang et al. 2018; Kang et al. 2019), and a reduction in the seasonal duration of snow cover (Qu et al. 2014).

As aerosols are fed by specific sources and distributed by the region's multiple climate systems, their effect on albedo and climate varies across the Highlands. The Westerlies transport dust across Central and Western Asia onto the ice. Although this changes the snowpack's albedo, it is—with the caveat that anthropogenic global warming is modifying the Westerlies system—a long-standing and non-anthropogenic process. The monsoons' transportation of black carbon from densely populated South and East Asia is, by contrast, entirely anthropogenic. It is primarily caused by the burning of biomass and combustion of fossil fuels in South and East Asia. Growing human populations are expected to increase these emissions, causing accelerated snowmelt in the Highlands (Zhang et al. 2018).

Temperature and precipitation trends across the Highlands have increased in the past 50 years (Deng et al. 2017). This should mean a decrease in snowfall and an increase in rain. But the snowfall patterns are complicated by significant regional and temporal differences with relatively high inter-annual variability. Temperatures have played a role, but so has their relationship with the region's major synoptic systems, such as the Westerlies and the monsoons. Weakening Indian Monsoons have reduced snowfall in the eastern and northern Highlands, but strengthening Westerlies have increased snowfall in the western Highlands (Yao et al. 2013; Deng et al. 2017). These regional and annual synoptic systems are, furthermore, driven by key modes of inter-annual variability such as the North Atlantic Oscillation and the El Niño Southern Oscillation (You et al. 2020; Yuan et al. 2012).

Along with aerosols and climate systems, altitude is another key factor in snow distribution across the Highlands. The relative importance of precipitation and temperature in generating snowfall varies according to elevation (You et al. 2020). Over time, high-low areas experience increasing–decreasing trends, respectively (Deng et al. 2017; Li et al. 2018). Recent studies indicate that there has been a significant decrease in snow depth and the duration of snow cover in the Highlands over the past 40 years (Ma et al. 2023), but the sites of this decrease are strongly correlated with their elevation. For example, the date at which snow begins to fall is delayed by 1.1 days, and the date at which snow starts to melt begins earlier by 1.8 days for every 100-meter increase in elevation. This means that the period over which snow can accumulate is shortening and that the magnitude of this change is greater with increasing elevation. This, in turn, results in an elevation-dependent reduction of snow depth (Ma et al. 2023).

So far, the Asian Highlands' complex and dramatic topography has severely limited our understanding of snow cover change. Remote sensing is challenging as it is difficult to differentiate clouds and snow cover. Still, numerous studies have shown that climatic and anthropogenic forcing has generated quantifiable change in snowfall and snow cover across the Highlands. Broad-scale predictions of change in Northern Hemisphere snowfall suggest a decrease of 5 centimeters per decade in spring and autumn and an increase of 4 centimeters per decade in winter this century (Krasting et al. 2013). With strong anthropogenic forcing, the Highlands snowfall could be reduced by 30–50% in the Indus Catchment, 50–60% in the Ganges Catchment, and 50–70% in the Brahmaputra Catchment (Viste and Sorteberg 2015), and the elevation at which rain becomes snow will increase by hundreds of meters (Viste and Sorteberg 2015). What is more, climate change was already decreasing the mean annual snow cover during an early period, 1966–2001, by around 1% per year (Rikiishi and Nakasato 2006; Viste and Sorteberg 2015).

All this data suggests that snowfall and snow cover are responding in complex ways across the Highlands. The effects of these phenomena will compound with other changes in the region's frozen landscape. Kraaijenbrink et al. (2021) captured some of this complexity through hydrographs that identify four distinct snowmelt regimes in the Highlands. Their hydrographs (see Figure 4.6) show short (three-month) and long (six-month) meltwater seasons in the Highlands' northwest and central regions, with early and late peak meltwater seasons in the east (Huang He (Yellow) and Dri Chu (Yangzi)

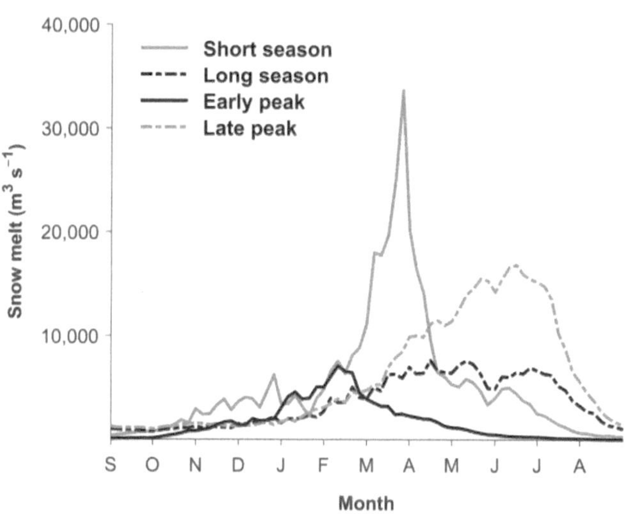

FIGURE 4.6 Hydrographs of several rivers. Their meltwaters and their timing and spatial variability is called "snowmelt." The following examples of rivers with the relevant meltwater timings: *short season*—rivers of the Hindu Kush Mountains (e.g., Helmand); *long season*—rivers of the Karakoram Mountains and the Tibetan Plateau; *late peak*—Ganga, Yarlung Tsangpo-Brahmaputra, western Gyelmo Ngul Chu (Salween), Da Chu (Mekong; early peak—Dri Chu (Yangzi), Ma Chu (Yellow).

rivers) and south (including the Indus, Ganga, Brahmaputra, Ayeyarwady, Da Chu (Mekong), and western Gyelmo Ngul Chu (Salween) rivers).

Critically, however, future projections for snowmelt indicate strongly *negative* trends for all river catchments, ranging from approximately a best-case warming scenario of less than 8% to a worst-case warming scenario of 41% (Kraaijenbrink et al. 2021).[1] This indicates that a future decline in Asian Highland snowmelt will have enormous significance for meltwater supply to Asia's major rivers.

4.2.3 Glaciers and Glacial Lakes

Glaciers are perennially frozen rivers. They consist of accumulated ice, snow, rock (as moraine), and sediment that move very slowly downhill under the influence of their own mass and the force of gravity. They are classified according to their size and thermal regime (either cold, polythermal, or temperate) and accumulate snow and ice over millennia.

Not all areas of a glacier are the same. Glaciers usually descend from high-altitude regions to lower-altitude regions. Their high-altitude sections tend to have a positive annual net balance of snow accumulation and are called accumulation zones. Their lower-altitude regions tend to lose deposited snow through melting, generate a negative mass balance, and are, therefore, called ablation zones. The boundary between these areas is called an equilibrium line altitude (ELA). At the ELA, accumulation and melting rates are equal. The ELA's position is sensitive to variations in winter precipitation, summer temperature, and the transport of snow by the wind. This means it changes according to climatic conditions. Glaciers (and the meltwaters they produce) are critical for ecosystems, hydrological cycling, and populations dependent on river flows discharging from the cryosphere. There is nowhere else on Earth where glacial discharge supports as many people and ecosystems as it does in the Highland rivers' Lowland catchments.

The Asian Highlands house approximately 36,800 glaciers, totaling an area of 49,873 square kilometers and 4,561 cubic kilometers of ice (Yao et al. 2012a). Climate controls the glaciers' spatial distribution and their type. There are continental (or polar-type) glaciers on the northwestern and central Tibetan Plateau and many marine or temperate glaciers in the Highlands' southeast. There are many glaciers in the southeast Highlands, and specifically in the Eastern Himalaya and Hengduan. These were formed by the monsoons' high precipitation rates. There are fewer continental-type glaciers on the Plateau's north and west because this area is drier and dominated by the Westerlies (Yao et al. 2012b).

Glaciers are not only an essential part of the cryosphere, but they also provide the most extensive geoarchive of its past climates. In 1987, Lonnie

FIGURE 4.7 *Left*, Marine or temperate glacier, Khawa Karpo Glacier, southeastern Tibet. *Right*, Continental glacier, Karo La Glacier. Photographs, Ruth Gamble, 2008.

Thompson and Yao Tandong were the first to drill an ice core at Dunde Glacier on the northeastern edge of the Tibetan Plateau. During the following 30 years, they extracted another three ice cores from the Highlands, two at Khariya (Guliya) (1992 and 2015) in the northwest and Dachuphu (Dasuopu) Glacier on the slopes of Mount Shishabangma in the Central Himalaya. Other ice cores have also been extracted from East Rongphu (Rongbuk) Glacier on the slopes of Chomolungma (Everest) and Purok Khangri (Puruogangri) in the Plateau's center, among others.

Extensive studies of these ice cores have produced evidence for the Highlands' climate variability across millennia and, more recently, of anthropogenically forced climate change (Thompson et al. 1998; Liu et al. 1998; Thompson et al. 2006; Chen et al. 2020). Glacier change is often described in terms of retreat or advance. Most studies of the Highlands' glaciers speak of their broad-scale, accelerated retreat since the mid-twentieth century. Comparative analyses indicate that the rate of glacial retreat is significantly faster there than in other frozen landscapes. The Highlands' glaciers were still advancing in the first half of the twentieth century. They began retreating in the second half of the twentieth century before a hiatus during the 1970s (Shi et al. 2006; Yao et al. 2007, 2012b), followed by stable conditions for the next few decades. Then, toward the end of the 1990s, they began retreating again at a more accelerated rate.

Although their retreat rate is a useful indicator of climate variability and change, it is also instructive to look at other parameters. One of the most sensitive indicators of glacier change is the calculation of their mass balance, that is, the difference between gains (accumulation) and losses (melting). A negative mass balance indicates a reduced ice mass. If sustained over time, this leads to glacial retreat, but also the less well-known phenomenon of glacial thinning.

Shrinking and thinning can also be accompanied by other factors that portend glacial mass loss, including decreases in glacial volume due to disintegration and collapse (An et al. 2022) and changes in glacier surface elevation, which are an indicator of reduced thickness (Wang et al. 2013c). Each of these processes is associated with increased meltwater production and has implications for downstream hydrology, including increased runoff, the development and expansion of glacial lakes, and increased river flows.

Glaciologists have noted a high degree of spatial variability in glacial mass balance across the Tibetan Plateau, with the most negative mass balances occurring along the Himalaya and the least negative in the interior (Yao et al. 2012b). Climate scientists attribute these variations to factors of both climate and topography, and they are similar to the causes of melting permafrost. First, the Indian Monsoon is weakening, and the Westerlies are strengthening. Second, the rate of warming in this region tends to be greatest at high elevations (4,800–6,200 meters ASL), which is the ablation zone for most of the southeastern glaciers (Yao et al. 2012b). Along with these Highland-wide trends, glaciologists have also found that local melt moderators and accelerants affect individual glaciers (Carrivick and Tweed 2013; Wang et al. 2023b). Melt moderators include debris cover, and accelerants include proglacial lakes. Proglacial lakes form at the front or side of glaciers when moraines, ice, or landslide debris block the ablating waters' flow away from the glacier (see Figure 4.8). The deposition of light-absorbing impurities—including organic carbon, black carbon, and mineral dust—can

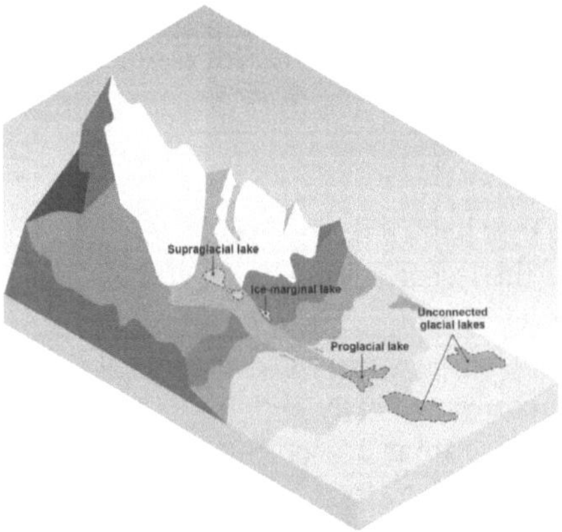

FIGURE 4.8 Supraglacial, ice-marginal, proglacial, and unconnected lakes. Based on Dou et al. (2023).

also affect melt rates by reducing the surface albedo of glacier ice and accelerating melting (Kaspari et al. 2014; Qu et al. 2014; Li et al. 2017b). As glacial lake water and other runoff have a lower albedo than glacial ice, the creation of lakes can also create a critical feedback process that encourages melting.

The retreat and thinning of glaciers support the formation and expansion of glacial lakes more generally. These bodies of water are classified as supraglacial, proglacial, ice-marginal, or unconnected lakes (Dou et al. 2021). In the Highlands, supraglacial and proglacial lakes predominate. Supraglacial lakes are generally associated with glacial thinning and form when meltwater accumulates within depressions on the surface of a glacier; they expand when adjoining ponds coalesce. By contrast, proglacial lakes form at the retreating terminus of a glacier, where meltwater is dammed and accumulates behind moraines (rock and sediment) left behind by the retreating glacier.

Glacial lake mass has expanded during the past few decades. As this expansion has occurred, scientists have observed two trends in relation to Highland glacial lakes' total surface area. There has been an expansion of pre-existing lakes and an increase in newly formed lakes. The expansion of pre-existing lakes is the dominant change process. In the Central Himalaya, 80% of net lake expansion has occurred through the growth of existing lakes. In the Tibetan Plateau's southeastern corner, the increased surface area and volume of existing lakes account for 67% of net expansion (Wang et al. 2011, cited in Song et al. 2016).

Scientists have used remote sensing data to track changes in the glacial lake mass across three periods: 1990–1999, 2000–2012, and 2013–2019 (Dou et al. 2021). Despite the linear temporal trend, there is a high spatial variability in the rate and, in some locations, even the direction of change. While significant increases in total lake area have occurred in the central Tibetan Plateau, rates have been lower in the Qilian Mountains, the Central Himalaya, the Nyenchen Thanglha Mountains, and the Khangtise (Gandese) Range. The Western Pamirs and Eastern Hindu Kush to the west and north of the Tibetan Plateau have experienced glacial lake shrinkage (Dou et al. 2021). Like changes to permafrost cover, this spatial variability should be attributed to climatic factors, type of lake, and positioning in relation to glaciers.

Lake expansion happens because of several reasons: (1) ice melts along the water line, (2) there is sub-aerial (at or near the Earth's surface) melting of ice-cliffs that surround lakes, and (3) ice blocks calve (break off) into glacial lakes (Gardelle et al. 2011). There is also a close relationship between glacier retreat and proglacial lake expansion. As meltwater accumulates in a lake, the slightly higher temperatures of liquid water and the air above it change the thermal conditions of the glacier's face, which causes the accelerated collapse or calving of glacial ice (Song et al. 2016, 2017). This is why proglacial lakes expand upstream into the glacier rather than laterally or downstream (Song et al. 2016). The connection between glacial retreat and proglacial lake expansion leads to faster glacial melt rates and intensifies the threat of glacial lake outburst floods (GLOFs) and debris flows (see Box 4a at the end of this chapter). GLOFs occur when the volume of glacial meltwater exceeds the lake's capacity or the strength of the moraine damming that blocks the water's flow.

4.3 Conclusion

The Asian Highlands are a critical part of Earth's cryosphere, comprising permafrost, seasonal and perennial snow and ice-covered landscapes, and glaciers. They are the only region of the cryosphere near large and growing populations that depend on them for meltwater and climate stabilization. Over their long histories, Highland communities and ecosystems have evolved to respond to a system in equilibrium. Human inhabitants could rely on relatively even levels of meltwater entering their river systems from the cryosphere.

They experienced some climate and cryosphere change as the Highlands warmed during the Medieval Climate Anomaly and cooled during the Little Ice Age, but nothing as dramatic as the anthropogenic forcing that has accelerated cryosphere transformation in the past few decades. The uneven changes brought about by anthropogenic climate change, construction work, and increased aerosol contamination have destabilized this balance. Increased meltwater has entered the system, creating larger lakes and threatening GLOFs. These accelerating processes are forcing socio-ecological systems to change as quickly (or hopefully more quickly than) the climate. As we learned in Chapter 1 and will return to in later chapters, snow and ice also hold cultural and social significance for the communities that live near the cryosphere.

BOX 4a LARGE FLOODS, PRECIPITATION, GLOFS, AND LLOFS

Some of the planet's largest floods occur in the rivers that drain the Asian Highlands. In historical records and contemporary conversations, many Tibetans and other Highlanders attribute these floods to a combination of human and non-human factors. The human causes of floods include the residents' negative karma, ignorance of geomancy, and insufficient river

(Continued)

management. Non-human causes for floods could include an insulted local *lha* (god), *lu* (subterraneous and subaqueous spirit), or another supernatural being seeking revenge (Romain 2021). In recent years, many local communities have approached these threats through preventive religious rituals and hazard reduction (Box 5b).

According to contemporary science, the largest and most destructive of these events are caused by extreme precipitation, GLOFs, from the retreat of glaciers and bursting of glacial ice dams, and landslide lake outbursts (LLOFs). Large floods and landslides expend some of the large stores of gravitational potential energy (GPE) that exist at the Tibetan Plateau's margins (Box 3a). This GPE results from the formation of high topography as the Indian and Eurasian tectonic plates collide (Chapter 2). High elevations, steep topography, uplifted monsoonal clouds that induce high rainfall, and earthquakes on giant faults are all consequences of this collision. These factors predispose the region to glacial formation and landslides, which, in turn, are conducive to GLOFs and LLOFs.

Enormous floods set and change the form of rivers and valleys and erode large amounts of rock and soil in river gorges. They may impoverish riverine and aquatic ecosystems, and the ecosystems can take decades to recover from these events. Large flood events can also destroy human settlements and infrastructure, such as bridges and hydroelectric plants (Liu et al. 2019; Dahlquist and West 2021; Li et al. 2022). The people of the Asian Highlands have the highest human vulnerability and exposure to GLOFs on Earth (Taylor et al. 2023), and the threat of GLOFs will multiply as temperatures rise and glacial ice disappears, which in turn will increase the volume of glacial lakes (Shukla and Sen 2021). As permafrost melts, more landslides occur, and therefore, LLOFs are likely (Patton et al. 2019).

To better understand landscape change in this environment and to undertake flood risk assessment and mitigation planning, we need to understand the histories of large floods. In keeping with the historical emphasis of this book, it is important to note that risk assessments for dams and other infrastructure are based on short records of river flows that are unlikely to include the largest floods known from other sources of information, mostly from deep time Earth science archives.

Figure 4a.1 illustrates the importance of these long records. It shows the gauged flows, historical floods, and paleofloods of the Han River, a major lower Yangzi tributary that flows through Shanxi and Hubei provinces. It shows that much larger flows are recorded over millennia than are captured by a century of gauging records. The Danjiangkou reservoir, part of the middle route of China's South-to-North Water Diversion Project, is in this catchment, and the

(*Continued*)

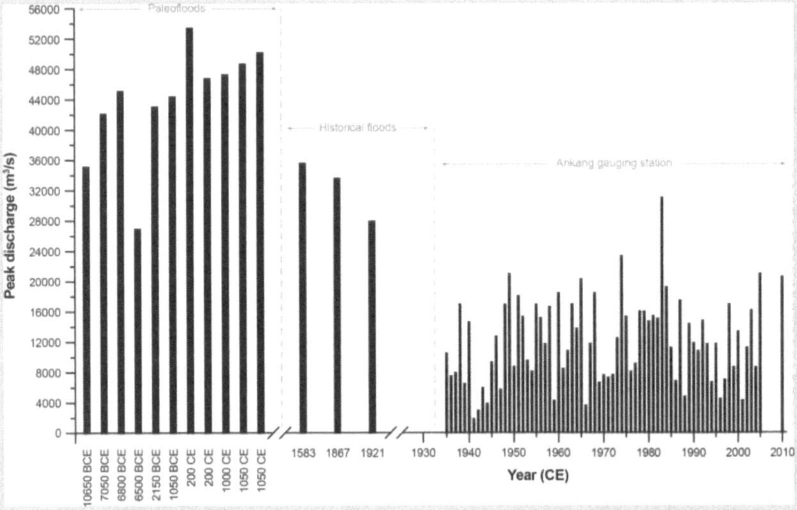

FIGURE 4A.1 Gauged peak flows, historical floods, and paleoflood reconstructions from the upper Han River over 13,000 years. Based on data in Guo et al. (2015).

best possible flood frequency understanding is clearly of great value for managing this reservoir. Over about 8,000 years, the largest peak flood of 51,000 cubic meters per second was in about 200 CE (estimated from paleoflood sediments), the largest historical flood of 36,000 cubic meters per second was in 1583 CE (based on documentary records), and the largest gauged flood was 31,000 cubic meters per second in 1983. This means the largest flood in 200 CE was 1.7 times larger than the largest gauged flood in 1983. With knowledge of deep-time floods, those overseeing the Danjiangkou reservoir should be able to manage it to withstand the largest floods in the area rather than the largest recent floods.

Another example comes from the Yangzi River in its mid-reaches near Chongqing, where flood levels recorded on stone inscriptions from 1153 to 1870 CE were up to 1.6 times larger than gauged floods between 1882 and 2012 CE. Knowledge of this historical data led to improvements in the design of the Three Gorges Dam (Li et al. 2020a).

The examples of floods from the Han and Yangzi rivers resulted from monsoonal precipitation. The naturally occurring dams that produce GLOFs and LLOFs in the Asian Highlands amplify floods by storing large quantities of water that can be released quickly and travel tens of kilometers downstream. Not all GLOFs and LLOFs are larger than monsoon floods; their magnitude depends on the size of a dam and its constituent material. However, large GLOFs and LLOFs

(Continued)

pose a significant threat to those living below them, and the risks associated with them need to be well understood.

Many hydroelectric projects built in the Himalaya recently (see Chapter 9) have not considered long-term flood magnitude estimates in their design. Instead, they are built to withstand "design floods" calculated from short flow records and modeling. Schwanghart et al. (2016) showed that, at most, 33% of estimated GLOF peak flows exceed design floods by up to tenfold.

Arguably, the section of the Asian Highlands most susceptible to very large floods is the Yarlung Tsangpo Gorge, through which the Yarlung Tsangpo descends at the eastern edge of the Himalaya. When it enters this Gorge, the river flows around 3,000 meters above sea level between two high snow and glacier-covered peaks, Namche Barwa (7,782 meters) and Gyela Pelri (7,293 meters). When it exits the Gorge around 200 kilometers later, it flows around 600 meters above sea level. The disputed border between India and China crosses the river at this point (see Chapter 8), and it enters the lands of the Adi people, who call this reach of the river Siang. The Yarlung Tsangpo-Siang's steep descent through the gorge means it dissipates some of Earth's largest GPE. It is south facing and receives the brunt of the South Asian Monsoon. Monsoonal rains swell the waters that pass through the gorge from the Plateau, and this flow is supplemented by glacial meltwater from the surrounding mountains.

Geomorphic and human archives record floods with peak flows between 30,000 and 40,000 cubic meters per second through the gorge (Mao et al. 2023). In 2000 CE, a LLOF descended from the Yigong Tsangpo, a Yarlung Tsangpo tributary that enters it just above the gorge. It peaked at 125,000 cubic meters per second before it reached the Brahmaputra's floodplain in upper Assam (Shang et al. 2003). In about 9500 BCE, a flood of a million cubic meters per second cascaded from a glacial lake burst in the Yarlung Tsangpo's middle reach in what is now the Lhasa region. Even earlier, around 2060 BCE, a flood of uncertain cause peaked at around 300,000 cubic meters per second in the same region (Yang et al. 2022a). Montgomery et al. (2004) documented paleofloods of 1 million cubic meters per second between 360 and 800 CE (see Chapter 6 for possible semi-mythical descriptions of this flood) and 5 million cubic meters per second between 9200 and 8000 BCE. They were all most probably a result of GLOFs or glacier bursts.

Borgohain et al. (2020) provide evidence for five very large floods between 12500 BCE and 100 CE on the Siang, probably all resulting from GLOFs, LLOFs, or both. These peak flows were between 130 and 30 times greater than the largest peaks gauged at Pasighat (where the Siang enters the plain) between 1978 and 2004 CE. The peak flow of the flood 7,000 years ago would have submerged the city of Dibrugarh, which now sits on the banks of the Brahmaputra,

(*Continued*)

under 10 meters of water. The flood 3,500 years ago would have covered it with 5 meters of water.

The frequency of these very large floods is low; therefore, they tend to pass from human memory. But their number is likely to increase as the climate changes. Along with the region's earthquakes, such as the 1950 CE event with a magnitude of 8.6, the very large floods are a real threat to dams, hydroelectric facilities, and villages in the Yarlung Tsangpo Gorge and below it. Perhaps humans need more humility in their interactions with the elemental powers in this region.

Over generations, local communities have developed long-standing, localized solutions to avoid damage and death from large floods. These have included locating buildings in safe areas, relocating their homes and villages after large floods, using porous—non-concrete—flood embankments, creating positive karma, and propitiating the *lha-lu*. These memories and responses help residents to survive seasonal flooding events and are, therefore, important to maintain. However, generational knowledge does not protect as effectively against millennium-scale events beyond human—but not geological—memory. Given that large floods are likely to increase in frequency and magnitude as the climate changes, probably beyond the range of current residents' written and oral histories and memories, it is unclear whether seasonal mitigation approaches will effectively cope with the most severe events. As "once-in-a-millennium" floods become much more common, residents in particularly flood-prone regions must combine all their generational knowledge with cutting-edge geomorphic, climatological, and other forms of monitoring and disaster response to continue living in their ancestral lands.

Note

1 Kraaijenbrink et al. (2021) use the term SSP (shared socioeconomic pathways) to indicate this change. They say that SSP 1–1.9 is the most optimistic scenario where global CO_2 emissions are reduced to zero by 2050. SSP 5–8.5 is the worst-case scenario where current CO_2 emissions double by 2050 and temperatures increase by 4.4°C.

5

LIVING RIVERS

Asian Highland rivers not only channel power and water but also cultivate life and support diverse habitats. These rivers serve as vital lifelines, nurturing rich biodiversity. Like the processes that brought about the Asian Highlands' geomorphology, hydrology, and cryosphere, life forms that inhabit the Highlands' river systems took millions of years to evolve. They adapted to changes in altitude, changes in temperature and changes in water supply as Earth's systems evolved and took advantage of the multifarious Highland climes. Because all living things depend to some extent on water, rivers—as collections of flowing fresh water—play a crucial role in supporting many forms of human and non-human life (Best 2019).

One of the most influential life forms to make the Highlands home are, of course, humans and over the millennia they have lived in these environments, they have developed various ways of understanding their habitats. As we discussed in Chapter 1, many Indigenous and pre-scientific lifeways understood the environment, its plants and animals relationally, viewing them as the abodes of an array of beings they called *lha-lu* (see Box 5b after this chapter, which describes the *lu*, a class of being that inhabit wet places.) Another way of assessing the riverine environments and the life that flourishes within them is through the lens of biophysical science research, environmental values associated with this research, and the ecosystem services paradigm. Although it is open to critique (see Chapter 10), way of looking at the biosphere is

DOI: 10.4324/9781003392033-5

globally influential and often determines the river management decisions that we will explore in later chapters. Therefore, in this chapter, we will describe how these integrated biophysical approaches explain the importance of rivers. We will do this by focusing on several eastern Asian Highland rivers, the two that are the focus of this book, the Dri Chu (upper Yangzi) and Yarlung Tsangpo (upper Brahmaputra), and the Highland reaches of the Mekong, which is known as the Da Chu in Tibetic languages.

Before we begin describing and assessing the biodiversity of these catchments, we must note that biophysical studies of the region have been patchy, and there are gaps in our knowledge about it. This means our assessment of the living rivers' biological health has had to rely on a subset of iconic environmental attributes. Primary among those we have used are data sets that speak to the region's fish fauna conservation and high-altitude wetlands. The picture these data sets paint of the rivers' health—and lack thereof—elucidates the importance of conserving these freshwater ecosystems. In addition, as we outline in the chapter, this conservation merely fulfills the obligations the catchments' governments have agreed to under several international treaties on conservation, including the Convention on Biological Diversity (CBD), the Ramsar Convention on Wetlands, and the maintenance of UNESCO World Heritage Sites.

As we will see, the rivers in the eastern Asian Highlands play a unique but threatened role within the biosphere. As a broader category, rivers are already rich in resources and are vital corridors through the landscape for human and non-human animals (Pittock and Thieme 2018). Freshwater biota is more diverse per unit area within rivers than other biomes (Dudgeon 2019). This diversity is intensified in the eastern Asian Highlands by altitudes, which created the multiple ecological niches that the multitude of species developed within. These niches have since become essential for regional livelihoods and economies (Allen 2010). But rivers in general—and these rivers in particular—are imperiled by multiple threats from warming climate and anthropogenic changes to rivers (Dudgeon 2019). The stakes in their conservation could not be higher.

5.1 The Ecological Significance of Eastern Highland Rivers

The rivers of the Eastern Himalaya span the greatest elevational range in the world and form the most consistent gradient in a mountainous landscape before reaching the coastal plains (see Table 5.1).

Like the plants and animals of other freshwater ecosystems, the distribution of biota within these catchments is linked to temperatures, precipitation, and seasonal conditions (Arthington et al. 2018). Within these catchments, however, the rugged and—in geological terms—rapidly evolving landscape provides a tremendous diversity of substrates, aspects, geomorphological

TABLE 5.1 Characteristics of the Dri Chu, Da Chu, and Yarlung Tsangpo

River	*Length (km) (Gupta 2008)*	*Annual volume of discharge (mm per year) (Wohl 2008)*	*Atmospheric river contribution to Q10 high flow (%) (Best 2019)*	*Catchment area (km2) (Wohl 2008)*	*Elevational range (m)*	*Annual total suspended and dissolved sediment load (106 tons) (Best 2019)*	*Human population (millions)*
Dri Chu (Yangzi)	6,300 (second globally)	870,000 (third globally)	14.12	1,940,000	5,342	650 (second globally)	326
Da Chu (Mekong)	4,880 (fifth globally)	667,000 (fourth globally)	14.83	811,000	5,224	215 (tenth globally)	76
Yarlung Tsangpo (Brahmaputra)	2,900 (fifteenth globally)	612,500 (sixth globally)	12.16	580,000	5,210	603 (fourth globally)	93

features, and barriers that support species' evolution and diversity. Glacial action, rapid erosion, earthquakes, and landslides that mobilize large sediment loads all contribute to terrestrial diversity.

These hydro-geological activities also transport nutrients downstream, where deep and nutrient-rich soils accrue on coastal plains, supporting further flora and fauna diversity, including various fish, birds, amphibians, reptiles, and mammals. Among the notable fish species in the region is the golden mahseer (*Tor putitora*), a large freshwater fish found in the Himalayan rivers of the Indian subcontinent, known for its striking color, large size, and sporting value. This fish is important to the rivers' ecology for multiple reasons. It sits at the top of the fish food chain and is crucial for maintaining a balanced food web. It may even eat small invasive species (Bhatt 2016). The riverine corridors also provide critical habitats for iconic mammal species such as the Indian rhinoceros (*Rhinoceros unicornis*), the Asian elephant (*Elephas maximus*), and the Bengal tiger (*Panthera tigris tigris*) and serve as essential stopovers for migratory birds. The Chinese softshell turtle (*Pelodiscus sinensis*), an endangered species, is among the aquatic animals found in the region.

Within the rivers, the historic—pre-damming—size of their flow supported many migratory fish species, perhaps as many as 520 (Allen 2010), which makes them some of the most biologically significant concentrations of fish species in the world (Abell et al. 2008) (see Table 5.1). The rivers' entire catchments form part of the Indo-Burma and Himalaya biodiversity hotspots, which sustain more than 1,500 endemic plant species, 175 species of mammals, and 500 species of birds (Allen 2010; Manish et al. 2017).

The slow onset climate change that transformed these regions in the past (see Chapters 3 and 4) allowed terrestrial and aquatic flora and fauna to move up or down the mountains' altitudes to find suitable new niches. The eastern Asian Highlands is notable for its high levels of endemism, and numerous species are found exclusively in this area. The range of habitats in the Eastern Himalaya, from the high-altitude alpine meadows to the subtropical forests and Lowlands, creates a variety of ecological niches that support a broad range of flora and fauna. The region's diverse ecosystems and habitats are formed by the world's most consistent gradient in a mountainous landscape before the mountainous areas meet the plains. The biodiversity of the Eastern Himalaya plays a critical role in maintaining essential ecosystem services such as water regulation, soil conservation, and nutrient cycling. The Eastern Himalaya is also a significant food, medicine, and cultural value source for local communities. Therefore, the conservation and sustainable use of the region's biodiversity are essential for its ecological, social, and economic well-being.

As we write, however, there are multiple threats to the rivers' biodiversity and general health. Globally, all riverine habitats are threatened by anthropogenic activities—primarily population growth and climate change (Sharma et al. 2019), but also by unsustainable development. The golden mahseer (*Tor putitora*) is one example of these threatened species. Its populations have been declining due to the indiscriminate fishing of brooders and juveniles and the adverse effects of dams, which interrupt its upstream migration during pre-monsoon and monsoon seasons for spawning (Bhatt and Pandit 2016). As Chapters 3 and 4 explained, the Asian Highlands are warming at one of the Earth's fastest rates (Doblas-Reyes et al. 2021). Conservation scientists warn that as these trends continue, the region's terrestrial and aquatic ecological zones will be transformed rapidly.

Another example of this anthropogenic enforced change is the upward movement of species seeking to escape rising temperatures and regional environmental pressures. According to some models, Himalayan endemic trees will undergo upslope migration of about 140 meters by the close of the twenty-first century in the Eastern Himalaya, which will result in the loss of approximately 20% of existing habitats for endemic flora (Wang et al. 2022a). In the catchments, about 16% of endemic angiosperm (flowering plant) species in the Sikkim Himalaya will lose their habitats by 2050, and 3% of the current geographic spread of meadows will transform into shrubland by 2070 (Sharma et al. 2019). The white-winged wood duck (*Asarcornis scutulata*) is an endangered species, with fewer than 800 individuals estimated to be left in the wild, out of which 450 live in Assam, Arunachal Pradesh, and Manipur (Sharma et al. 2020). Its status is set to become even more precarious as the duck's current highly suitable habitat is projected to decline from 5,123 square kilometers in 2020 to 4,544 square kilometers by 2070.

Glacial melt will increase river flows in the short term, which may increase the size of wetland areas on the Tibetan Plateau (Xu et al. 2019). But long term, as we discussed in the previous chapter, this glacial melt is expected to change the rivers' seasonal flows and exacerbate floods and droughts (Rees and Collins 2006; Shrestha et al. 2015). Floods and droughts significantly impact the hydrology and ecology of wetlands in the Eastern Himalayan region. These wetlands act as natural water storage systems during floods and provide crucial habitat for aquatic and terrestrial species. However, excessive flooding can lead to erosion, sedimentation, and altered hydrological regimes, which will negatively affect biodiversity and wetland functions (Junk et al. 2013). Droughts can cause water scarcity and alter the wetland's hydroperiod, decreasing productivity and biodiversity (Junk et al. 2013).

These changes in river flows will, in turn, impact fauna, particularly wildlife that are adapted to previous conditions. In addition, glacial retreat leads to the creation of lakes formed by unstable moraines (see Box 4a). These unstable lakes, the glacial lake outburst floods (GLOFs) that occur when they break, and the increase in other floods and landslides downstream create sediment loads detrimental to wildlife species that depend on clear water. For example, the Eurasian and smooth-coated otter can be found in the Eastern Himalayan regions. Otters rely on clear water to hunt for food and breed. Decreasing water quality can hinder their ability to locate and catch food (Khatiwara and Bhutia 2020). Mountain ecosystems worldwide are susceptible to ongoing changes in climate, but the effects are particularly pronounced in the Himalaya due to their faster warming rates and glacier melting. Among the vulnerable species, Himalayan cold-water species are of particular concern due to their limited thermal tolerance (Sharma et al. 2021). The common snow trout (*Schizothorax richardsonii*), which thrives in cold water environments in the Himalaya (see Box 6a), is expected to shift its range upslope to high-altitude streams while experiencing habitat loss at its current lower elevation range. This shift will likely result in a net loss of habitat, estimated to range under different climate change scenarios from 7.41% to 16.29% by 2050 and from 9.46% to 26.56% by 2070 (Sharma et al. 2021).

Climate change is also altering freshwater habitats and impacting their biodiversity in multiple ways (Finlayson and Pittock 2018; Woodward et al. 2010). These effects will be exaggerated or mitigated by the orientation of rivers. In sections of the rivers that flow from higher to lower altitudes, and north to south, biota may adapt to climate change impacts by moving to higher and cooler niches within the mountainous river systems. Flatter river sections (like those on the Plateau) provide fewer opportunities for biota to move to high elevations with cooler environments. This is particularly true of river systems that flow west-to-east (or east-to-west) along similar longitudes, like the highest reaches of the Dri Chu and the Yarlung Tsangpo, because their orientation limits the northward movement of their biota as the climate heats. At most risk within the river systems on the Plateau are its significant peatlands, including the Earth's largest alpine wetland, the Dzoge (Songpan) Peatland, which occupies around 2,600 square kilometers between the Dri Chu and Ma Chu (Huang He) (see Box 5a). These ecosystems are crucial to the Highland–Lowland hydrological systems and support low-season flow to rivers; they are also at significant risk of desiccation and destruction. As they become drier and perhaps even catch fire, they release their large stores of carbon dioxide and contribute to heating the local environment (Cao et al. 2019).

5.2 Societal Significance

As accumulations of fresh, flowing water, rivers are foci for human societies and generate many ecosystem services (Yeakley et al. 2016). Ecosystem services is a term that conservation scientists and development specialists use to refer to the benefits that ecosystems bring to humans as goods and services, such as clean air, water, and food (see Box 9a and Chapter 10 for a discussion of this term and its uses). People need access to the freshwater that rivers provide. Riverine floodplains comprise the flattest land with the deepest and most fertile soils that support the best pastures and croplands. Globally, inland rivers are often vital sources of nutritious foods, especially as sources of proteins and micronutrients from fish. In mountainous regions like the Himalaya, river valleys usually provide the most logical communication and land transport routes. Before the advent of railways and trucks, shipping on Lowland rivers was the most reliable, fastest, and highest-capacity means of communication and transport. For these reasons, human cultures and political entities have formed near and in interaction with rivers (Best 2019). For good and for ill, riverine corridors have allowed cultural influences, goods, and militaries to traverse landscapes (see Chapter 6).

Highland communities adapted to their precipitous environment. They built homes and farms in river valleys to avoid the extreme weather on mountain ridges and to access water and flat lands for their crops and livestock. Although they chose to live in river valleys, humans established their habitation at a safe distance from main river channels to avoid flood hazards. They have also developed communal memories of other hazards, such as earthquakes, avalanches, and landslides, and the environmental scouring these events can cause. These memories often direct their habitation choices.

However, anthropogenic activity is increasing the threats to human settlements. Climate change has increased the potential danger from GLOFs (see Chapter 4 and Box 4a) that can severely scour valleys (Veh et al. 2020). The expanding number of dams in the mountains exacerbates this threat with the risk that such GLOFs might destroy structures with cascading impacts in lower reaches of valleys. This kind of climate-induced threat was evident in Uttarakhand, India, in February 2021 when part of a glacier is believed to have collapsed, causing a flood that impacted the Tapovan hydropower project and killed more than 170 people (Biswas 2021), and in Sikkim in October 2023 (see Box 9b).

Riverine environmental values and local societies are threatened by the direct impacts of climate change and the adverse effects of societal choices in response to climate change (Pittock 2015; Pittock et al. 2017). As India, China, and Bhutan seek to decarbonize, they have turned to hydropower, leading to a proliferation of dams across the eastern Highlands. These dams have significant local impacts and limit human and non-human adaptation opportunities as the climate changes. The construction of hundreds of dams in these river systems (see Chapters 8 and 9) is fragmenting the movement of wildlife and the flows of nutrients and sediments. The dams turn high-energy rivers into chains of still reservoirs, and they inundate ecologically crucial riparian ecosystems. Hydropower generation drives river flows and changes flow patterns. The alteration of river flows through these kinds of dams only has the potential for small ecological benefits through climate change mitigation (Pittock 2021), but its negative impacts can be quite profound.

There are better ways to build and manage dams (Hydropower Sustainability Secretariat 2023; Pittock 2021). If dams could be well-built and well-managed—something that has rarely happened—this would ameliorate many of their negative ecological impacts and even help mitigate some of the effects of climate change. A well-built dam would be situated in an area where its construction would have minimal impact on the surrounding ecosystem and communities. It should incorporate features such as fish ladders and bypasses to allow for the passage of aquatic animals and sediment bypass systems to maintain natural sediment flows downstream. To further mitigate climate change, pumped-storage hydropower dams could be built off rivers (Gilfillan and Pittock 2022a, 2022b). These would act like batteries for other renewable energy generators, such as solar and wind, to complement their variable electricity output. A well-managed dam would be operated in a way that balances the needs of the ecosystem, downstream communities, and power generation. For example, multi-level off-take towers could maintain pre-climate change temperatures and release water from reservoirs. Large reservoirs could be managed to mimic yearly water cycles based on historic flow patterns in the form of "environmental flows." Until these and other new building and management systems are introduced, however, the potential benefits of these dams will be overshadowed by their negative impacts on the environment and people. The combination of the dominance of traditional engineering thinking, powerful financial interests, and the political appeal of large infrastructure projects—dubbed the hydraulic bureaucracy—have frustrated efforts to adopt more sustainable practices (Molle et al. 2009).

A further threat looms because climate change may exacerbate water scarcity in downstream countries already experiencing water stress due to growing populations and economies (Yang et al. 2016a). The rapidly expanding populations and mismanagement of groundwater resources in the north Indian plain have led to a particularly acute water crisis, but more generally, nearly 2 billion people in China and India are experiencing water scarcity (Mekonnen and Hoekstra 2016). This makes further exploitation of the water resources of Highland rivers very likely.

Governments will probably respond to these shortages by building more reservoirs to store water in the wet season for use in the dry season, which will disrupt natural river flows. The construction of reservoirs will have concomitant ecological impacts, including changes in water temperature and dissolved oxygen levels, along with trapping of nutrients and sediments. These changes can, in turn, impact fish and other aquatic species or cause extensive impacts on the whole ecosystem, such as from eutrophication. Along with increased reservoir development, Chinese and Indian governments have plans for inter-catchment water transfer (Feng et al. 2019). China has already built two of the "three lines" of the "south to north water transfer scheme," which moves water from the central and lower Yangzi Catchment to north China (Ma et al. 2006). A few Chinese engineering advocates have also proposed a western water transfer line that would move water from the Yarlung Tsangpo to the Yangzi Catchment and then on to north China, but the central government has not yet accepted this plan (Barnett et al. 2015; Webber et al. 2017). India's "National Interlinking of Rivers Project" has been a more ad hoc process, part of which proposes an extensive diversion from the Brahmaputra River in Assam to the Ganga (Ganges) River catchment (Best 2019).

5.3 The Environmental Values of the Upper Brahmaputra, Da Chu, and Dri Chu

Along with the specific local ecological and social value of these Highland rivers, their diversity of fish species, wetland habitats, and the multiple niche areas within them have been identified as priorities for conservation under international treaties. It is important to note that the assessment of their environmental values is constructed through scientific assessments of the region—counting fish species, habitat varieties, and so forth—and the structures of international treaties rather than traditional and local knowledge. As we have explored elsewhere in this book, local communities may well value other elements and characteristics of their environment than those on which conservation science focuses (Pradhan et al. 2021).

While fish are not the only species associated with the rivers and wetlands within the Highlands, they are obligate freshwater taxa (species confined to freshwater) and are more consistently monitored than most other biotas. The diversity of fish species in these large river systems is globally significant, and there is broad agreement that such species should be conserved under the CBD and Ramsar Conventions (Abell et al. 2008). International compacts also stipulate that wetland ecosystems should be conserved. The Ramsar Convention (Ramsar 2018) describes nine criteria by which wetlands can be designated sites of "international importance." Generally, national governments choose these in an *ad hoc* manner if the sites meet at least one of Ramsar's nine criteria. Several sites within the upper river valleys are also deemed to be UNESCO World Heritage areas of "outstanding" cultural or natural value, which are assessed by ten cultural and natural criteria (World Heritage Convention 2022). All those sites already recognized are listed in Table 5.2, but it is likely that other sites fulfilling the Ramsar and World Heritage criteria could also be found in these rivers' catchments. Table 5.2 lists government-established protected areas to preserve freshwater biodiversity in each river catchment. Under the CBD, the world's governments adopted targets for biodiversity conservation of 30% of terrestrial and freshwater ecosystems in designated protected areas by 2030 (CBD 2022). The use of protected areas to conserve freshwater ecosystems has been limited and poorly managed (Pittock et al. 2015b). In this context, the large number of freshwater-protected areas in the Dri Chu (Yangzi) Catchment is unique globally.

5.4 Fish Biodiversity

The Dri Chu (Yangzi) is the world's fifth most diverse fish habitat after the Platte, Amazon, Orinoco, and Congo River catchments (Abell et al. 2008; Best 2019). In total, around 520 species are estimated to live in these rivers. Seventy are categorized as threatened, forty-six as near threatened, fifteen as endangered, and five as critically endangered (Allen 2010). The endangered and critically endangered species live in the Ganga-Brahmaputra River system and the Chindwin catchment in Myanmar (Allen 2010). The "Freshwater Ecoregions of the World" mapping project, which uses biogeographic analysis to identify ecoregions for conservation (WWF/TNC 2019; Abell et al. 2008; Allen 2010), determines the following freshwater ecosystems within the three catchments:

1. Upper Brahmaputra: Upper Brahmaputra, Middle Brahmaputra, Ganges Delta and Plain;
2. Da Chu (Mekong): Upper Lancang, Lower Lancang, Khorat Plateau, Kratie-Stung Treng (Mekong), Mekong Delta; and
3. Dri Chu (Yangzi): Upper Yangzi, Middle Yangzi, Lower Yangzi.

TABLE 5.2 Selected examples of the Dri Chu, Da Chu, and Yarlung Tsangpo's environmental values

River	*Fish species (Best 2019)*	*World Heritage Sites*[a] *(World Heritage Convention 2022)*	*Ramsar wetlands of international importance (Ramsar Convention 2021)*	*Other wetland habitats (Ministry of Ecology and Environment (MEE) 2016; People's Government of the Xizang Autonomous Region 2013)*	*Riverine protected areas (on the mainstream and tributaries)*
Dri Chu (Yangzi)	322 (fifth globally)	Three Parallel Rivers (Yunnan) Protected Areas (China)	Nakpak Tso (Napahai) (China) Pota Tso (Pudacuo or Bitahai) (China) Lashihai (China) And 12 other sites east of the Tibetan Plateau (China)	Sanjiangyuan National Park (established in 2019) covers 123,100 square kilometers of the headwaters of the Dri Chu, Da Chu and Ma Chu) (part World Heritage) Hong Hu Lake (China National Nature Reserve) Jiduansha Wetland (China National Nature Reserve) Danjiangkou Reservoir Wetland (China Provincial Nature Reserve)	14 (1/13)
Da Chu (Mekong)	244 (eighth globally)	Three parallel rivers Yunnan protected areas (China)	8 sites in the Lowland portions of the river catchment	Zhengyuan Wanghe (China County Nature Reserve) Mengsuo Longtan (China County Nature Reserve)	5 (1/4)
Yarlung Tsangpo (Brahma-putra)[b]	126 (seventeenth globally)	Kangchenjunga National Park (India) (Also known as Kangchendzonga National Park) Kaziranga National Park (India) Sundarbans Mangrove Wetland (Bangladesh & India)	Mapam Tso (China) Mitikha (Maidika) (China) Bumdeling (Bhutan) Gangtey-Phobji (Bhutan) Khotokha (Bhutan)	Lhalu Wetland (China National Nature Reserve) Lake Namuchu (China Provincial Nature Reserve) Yarlung Zangbo Grand Canyon (China National Nature Reserve for Forests) Wetlands of Arunachal Pradesh (1,534 wetlands) Mouling National Park (India) Namdapha National Park (India) Dibang Wildlife Sanctuary (India) Mehao Wildlife Sanctuary (India) Kamlang Wildlife Sanctuary (India) Yordi-Rabe Supe Wildlife Sanctuary (India)	None

[a] Excluding sites listed for cultural values only; UNESCO World Heritage Centre (2024).
[b] Xu and Pittock (2018, 2020a).

These divisions are significant in defining biogeographic regions, also called ecoregions. Conservation institutions under the CBD and Ramsar Conventions require reservation and conservation management of a representative selection of the ecoregions as a surrogate for the habitat of species. These regions help to define, for example, how national governments should implement the CBD protected area target for at least 30% reservation of habitats by 2030 (CBD 2022). All these rivers form complex networks and sustain multiple ecoregions with high biodiversity. The three river catchments have (Abell et al. 2008):

- Medium to high fish species diversity;
- Low endemic species diversity;
- Low to medium proportions of endemic fish species; and
- A medium to high number of species per unit area.

Fish species diversity is lower in the harsher and ecologically newer post-glacial freshwater environments of the Tibetan Plateau (headwater and upper reaches). Most of these species have responded to the low water temperature and low food availability by evolving to be cold-adaptive, slow-growing, long-lived, and later-maturing (Huo et al. 2014; Liu et al. 2021a). Fish diversity is much higher in the middle and lower reaches and is highest in the large and nutrient-rich Lowland river systems (Abell et al. 2008). Fish species in the upper-middle and middle reaches are from a wide range of taxonomic categories, with a higher proportion of endemic species. For example, the Dri Chu (upper Yangzi) and middle Yangzi contain 94 endemic fish species accounting for 58.4% of the Yangzi's total number of fish species (Liu et al. 2020b). Most of these Highland fish have adapted to the rapid current and inhabit the river bottom. Some fish, like those in the catfish (*Sisoridae*) families, not only live in the river bottom but have also evolved to adhere to and climb on riverbed stones (Huang and Li 2016; Xu and Pittock 2019). Fish in the middle-lower and lower Yangzi reaches are abundant, but there is a lower proportion of endemic species (Huang and Li 2016), and more migratory fish species live there, including some euryhaline marine fishes (Huang and Li 2016; Xu and Pittock 2019).

Fish diversity in these rivers' Lowland reaches depends substantially on water flows, sediments, and nutrients generated in the upper catchments. These three rivers have the following high levels of diversity:

- The Da Chu has 22 native fish species upstream and 142 downstream (Xu and Pittock 2019).
- The Dri Chu has 14 native fish species in its Plateau headwaters and 357 in its middle and lower reaches (Huang and Li 2016; Xu and Pittock 2020a).
- The Yarlung Tsangpo has 28 native fish species in its Plateau headwaters, and 141 were recorded in its middle and lower reaches (Bhattacharjya et al. 2017; Liu et al. 2021a).

Human activities have severely impacted these rivers' fish biodiversity, extirpating four fish species in the Da Chu's upper reaches and 49 in its downstream reaches (Zhang et al. 2019a). Within the Dri Chu, 46 native and 20 endemic species have been extirpated, and 37 species are listed on the International Union for Conservation of Nature's (IUCN) Red List (Xu and Pittock 2020a). Much of this population decline has been caused by habitat loss. The Chinese sturgeon (*Acipenser sinensis*) lost 99% of its natural spawning habitat when the Bezhouba and Three Gorges dams were constructed (Xu and Pittock 2020b). There is less information about human impacts on fish biodiversity in the Brahmaputra system (Bhattacharjya et al. 2017; Liu et al. 2021a). But we do know that between 20 and 30 of its species are endangered or critically endangered, including *Amblyceps arunachalense* (a catfish species), *Badis tuivaiei* (sharped-spined, small fish), *Schizothorax integrilabiatus* (a snow trout species), *Pterocryptis barakensis* (another catfish species), and *Lepidocephalichthys arunachalensis* (a carp species) (Allen 2010; Bhattacharjya et al. 2017). Fourteen of fifteen endangered species and all five critically endangered species are endemic in the Ganga-Brahmaputra catchment and the nearby Chindwin catchment in Myanmar (Allen 2010). The loss of these species is a blow to the conservation of endemic species in the Eastern Himalayan reaches of the Brahmaputra catchment.

Large fish species and other fauna have been even more profoundly impacted, primarily by habitat fragmentation (Prajapati 2021; Xu and Pittock 2020b), than smaller fish species. Iconic river dolphins used to live throughout the middle and lower reaches of all three river systems. The Ganges River dolphin (*Platanista gangetica gangetica*) still lives in the Ganga, Brahmaputra, and Karnaphu River systems (Smith et al. 1998). But the construction of 20 irrigation barrages in these systems between 1886 and 1971 shrank its population by 80% (Braulik et al. 2014; Smith et al. 2001). Water infrastructure expansion in the past 20 years has further reduced their habitat (Smith et al. 2001; Braulik et al. 2014), and now only around 2,500–3,000 dolphins are left in these rivers (WWF 2023). The Yangzi River dolphin (*Lipotes vexillifer* also known as Baiji) has probably been extinct since 2007 (Braulik et al. 2014).

5.5 Wetland Biodiversity

Along with fish biodiversity, another way to understand the biodiversity import of these rivers is to learn about their wetlands. All three river systems—the Dri Chu (Yangzi), Yarlung Tsangpo (Brahmaputra), and Da Chu (Mekong)—house high-altitude wetlands, which the conservation protocols developed in global agreements such as the CBD and the Ramsar Convention on wetlands consider to be very important.

The Ramsar Convention's definition of wetlands is broad. It includes all lakes and rivers, underground aquifers, swamps and marshes, wet grasslands, peatlands, springs, estuaries, deltas and tidal flats, mangroves and other coastal areas, coral reefs, and all human-made sites such as fishponds, rice paddies, reservoirs, and salt pans (Ramsar 2023). A common misconception holds that the Convention focuses only on conserving waterbirds. In reality, it "provides the framework for national action and international cooperation for the conservation and wise use of wetlands and their resources" (Ramsar 2023). The 1971 Ramsar Convention is one of the world's earliest environmental treaties. Through a joint work plan, the CBD endorses the Ramsar Convention to implement the conservation of "inland water" biodiversity (CBD and Ramsar 2011).

It is a legal requirement under the Ramsar Convention that national authorities "make wise use and maintain [the] ecological character of all wetlands" (Pittock et al. 2010). "Ecological character" is a legal concept defined as the "combination of [the] ecosystem components, processes and benefits/services that characterize the wetland at a given point in time" (Ramsar Convention 2021). The Ramsar Convention recognizes that wetlands are cultural landscapes, that many wetlands are formed by people, and that wetlands are valued in part due to the ecosystem services that they generate for humans. This is important to remember in the Highlands, where many wetlands—particularly peatlands—are socio-ecological spaces (Gamble 2021; Hayes 2013; Yeh 2009).

Under the Ramsar Convention, national authorities are encouraged to conserve wetlands through such measures as:

- National inventories applying the Convention's wetland typology;
- Designation of wetlands meeting one or more of Ramsar's nine criteria as wetlands of "international importance" (Ramsar sites);
- Establishing national committees for government and non-government stakeholders to collaborate for wetland conservation; and
- Encouraging international cooperation for the conservation of transboundary wetlands.

The Highlands wetlands formed following the last glacial period (10,000–7,500 BP), when the Tibetan Plateau's ice sheet retreated to mountain peaks. There are now multiple types of high-altitude wetlands on the Plateau that all have important environmental and socioeconomic values. The Plateau has an area of approximately 131,900 square kilometers of wetlands and has the largest area of swamp wetlands in China (Zhao et al. 2014). These wetlands are also threated by anthropogenic activities. Around 2.2% of wetlands were lost between 1990 and 2006.

Peat wetlands are common on the Plateau. This wetland type provides habitats for many plants and animals, especially waterbirds. The Plateau has some of the world's largest alpine peatlands (Chen et al. 2014a; Zhao et al. 2014), such as the aforementioned Dzoge (Songpan) Peatlands, which sequester large carbon stores but are very vulnerable to drying out as permafrost melts. Palynologists can assess the layers of peat (and sediments in lakes) to identify different kinds of pollen, animal remains, sediments, and their ages to determine environmental changes on the Plateau.

Springs are another type of wetland. They provide permanent water sources in arid areas of the Plateau, as they often underpin irrigation systems and have great cultural significance (see Box 5b at the end of this chapter). They are also important as a source of base flow for rivers.

Glaciers are, arguably, wetlands, although this has not been confirmed under the Ramsar Convention. As discussed in the last chapter, glaciers are important sources for Himalayan rivers, providing them with a base flow. Melting glaciers caused by global warming increased the area of lakes on the Plateau by 27.3% between 1976 and 2009 (Zhao et al. 2014). Approximately 66% of wetlands on the Plateau were degraded between 1978 and 1990, but this was offset by the 6,000 square kilometers of new wetlands created after the melting of covered glaciers between 1978 and 2008 (Zhao and Shi 2020). This climate change dividend is unlikely to last.

Rivers and lakes can also be classified as wetlands. They provide habitats for many plants and animals, especially fish and waterbirds. Human irrigation systems create artificial wetlands within river systems where migratory birds nest. For the most part, however, humans have a negative effect on wetlands. For example, both urbanization and hydropower projects fragment and destroy wetland habitats. Dam reservoirs create poor habitats for wildlife as they tend to be too deep to sustain high levels of biological activity.

Despite operating in very different governance models, Bhutan, China, and India are each contracting parties to the Ramsar Convention, and they outline their activities for the conservation of wetlands every three years in national reports. To give a sense of how these nation-states view their contributions to the bioconservation of wetlands, we have summarized their reports below.

5.5.1 Bhutan

Forty-three percent of Bhutanese territory is above 3,000 meters elevation. Conservation inventories have recorded 3,027 non-riverine wetlands that occupy 10,200 hectares or 0.61% of this high-altitude region (WWF Bhutan 2011). In the main, these wetlands are different forms of lakes and marshes, and three of them have been designated as Ramsar sites (see Table 5.2).

In 2018, Bhutan's Royal Government announced plans to develop a national wetlands strategy, which included management plans for wetlands sites, enforcement of conservation regulations, and enhanced cross-agency collaboration to conserve wetlands (Royal Government of Bhutan 2018b).

In a 2022 report to the Ramsar Convention, the government of Bhutan indicated that it had adopted a strategy to accomplish this, but further detail is not publicly available. In the same announcement in 2018, the government said its water governance systems "treat wetlands as natural water infrastructure integral to water resource management at the scale of river basins" (s. 9) and that "traditional knowledge and management practices relevant for the wise use of wetlands have been documented and their application encouraged" (s. 10). In 2018, there were no institutions in place to manage invasive species, a national wetlands committee, or reports on transboundary wetlands. But the government did state that environmental assessment is an essential prerequisite for future policies and proposed developments.

5.5.2 China

The Tibetan Plateau's wetlands represent 30–40% of China's natural wetlands and constitute the largest area of alpine wetlands globally (Wei et al. 2020). Around 6% of the Plateau is wetland, dominated by peatlands—including the large Dzoge (Songpan) Peatlands—and lakes (Mao et al. 2020). The extent of vegetated wetlands is reported to have declined from the 1960s to the 1990s and is now stable or expanding, forming an extensive carbon sink (Wei et al. 2020).

In 2021, the Government of the People's Republic of China announced an extensive program to reform and improve the conservation of wetlands (DWMNFGA 2021). This included supporting research into the adoption of a series of laws and policies that would be focused on the Dri Chu-Yangzi River catchment, including the "Yangzi River Protection Law (Changjiang baohu fa)," the "Ecological Protection and Restoration Plan for the Yangzi River Economic Belt (Changjiang jingji dai shengtai huangjing baohu guihua)," the "Scheme for Returning Agricultural Land to Wetlands in the Yangzi River Economic Belt (Changjiang jingji dai tuigeng huan shi xiufu gongcheng)," and the "Action Plan for the Uphill Battles for the Conservation and Protection of the Yangzi River (Changjiang baohu xiufu gongjianzhan xingdong jihua)."

In 2020, the powerful National Development and Reform Commission and the Ministry of Natural Resources released the "Master Plan for the Major Program of Protecting and Restoring Ecosystems of National Importance (2021–2035) (Quanguo zhaongyao shengtai xitong baohu he xiufu zhongda gongcheng)," which prioritizes wetland conservation and restoration. Its report accompanying this plan noted that the wetland area had expanded through restoration and that more than half of the area of wetlands has legal protection. In 2020, fishing was prohibited in 332 aquatic protected areas, and in 2021 this ban was extended for ten years throughout the Yangzi Catchment.

The Chinese government reported that its priorities included developing a new wetlands conservation law, redoubling its efforts to restore better and manage wetlands, and improving international cooperation (DWMNFGA 2021). The central government noted that many provincial governments already have wetland conservation laws. The report noted that the government had undertaken extensive work to deliver the environmental flows required to maintain major wetlands, including research on coordinated environmental flows on the cascade of hydropower stations on the upper reaches of the Yangzi River. Further, from 2019, water allocation plans were approved for 41 transboundary rivers.

The government reported that 76% of Chinese households were connected to sewerage systems, and another 25% had improved septic systems. Improperly designed or maintained sewerage systems can discharge untreated or partially treated wastewater into wetlands, resulting in water pollution, loss of biodiversity, and changes in wetland ecosystem structure and function. Conversely, well-designed and properly managed sewerage systems can provide clean water to wetlands, supporting their health and vitality. Treated wastewater can serve to supplement water supplies and provide nutrients to plants. Constructed wetlands and other natural processes can be employed in wastewater treatment to remove contaminants and improve water quality while enhancing wetland ecosystem health. Official sources say that institutions are also in place to manage invasive species. The government reported that the condition of wetlands generally had improved over the preceding three years.

The Chinese government claims it has a national wetlands committee and conducts environmental assessments of policies and proposed developments. The authorities do not acknowledge international cooperation in the management of transboundary wetlands. The People's Republic of China (PRC) has pursued water diplomacy through the Lancang-Mekong Cooperation, and it has signed a Memorandum of Understanding with India to share limited hydrological data about the Yarlung Tsangpo-Brahmaputra in the monsoonal flood season. However, broader international water governance issues are not addressed (Liu 2015b, see Chapter 9).

In all, the Chinese government's own reports suggest it is making a significant investment to develop a technical understanding of wetlands and river systems. It has committed to sustainable management—from a biophysical sciences perspective at least—and significant investments in wetland conservation. It has also indicated vastly greater river and wetland governance activity than Bhutan or India. This is partly because of China's more significant economic and governance resources and partially because the government has recognized how important wetlands are to the country's water and food security. But as we will discuss later in the book, many criticisms can be made of their technocratic approach to conservation.

5.5.3 India

Two Indian states, Arunachal Pradesh and Sikkim, sit within the Yarlung Tsangpo-Brahmaputra catchment at high altitudes (ISRO 2009). Their government agencies have identified some 2,206 lakes that cover 15,188 hectares within them. These represent 0.74% of the high-altitude lands in these two regions (ISRO 2012). The National Wetland Atlas also identifies riverine and other kinds of wetlands at lower altitudes between the mountains and the sea.

The Republic of India's Ministry of Environment Forests and Climate Change 2021 report to Ramsar stated that there were more than 1,000 wetlands across the country, and "wetlands rejuvenation" was a national priority. Still, it also recognized a need to improve information dissemination, capacity building, and management. As the report explained, India's wetland management is subsumed within its National Wildlife Action Plan's 17 priority areas, namely, the conservation of inland aquatic ecosystems. Within the report, the authors also claim to have conducted partial assessments of wetland-maintenance environmental flows; that there are patchy institutions in place to manage invasive species; that nearly all households are linked to sewage systems; that the wetlands have not changed in the previous three years; that it has a cross-sectoral national wetlands committee; and that it undertakes the environmental assessment of wetland impacts before implementing policies and proposed developments. It also reports that the government engages in limited international cooperation in managing transboundary wetlands.

The report also contained details about the Indian government's understanding of wetlands. It states that authorities have undertaken an environmental assessment of policies and proposed developments impacting wetlands. Limited international cooperation is reported for the management of transboundary wetlands.

Within these reports, similar characteristics in the management of high-altitude wetlands stand out. First, neither the Bhutanese nor Indian governments classify rivers as wetlands, despite the Ramsar Convention's definition of wetlands. Moreover, the maintenance of the wetlands they do recognize consists primarily of inventory and management. By contrast, China has a triage approach to river management. It has offset its extensive exploitation of main stem river resources by conserving and restoring selected headwaters and tributary rivers (Xu and Pittock 2020a). Unsurprisingly, neither Bhutan nor India reports providing adequate environmental flows to sustain wetlands as part of their water management systems. China has only recently begun this practice.

The other thing to note about these reports is the importance and geographic concentration of non-riverine high-altitude wetlands. These areas, comprising thousands of lakes, marshes, and peatlands, occupy less than 1% of the high-altitude mountains and around 6% of the Tibetan Plateau. Despite occupying such a small portion of the region, their sustainable management is essential for conserving biodiversity and maintaining the ecosystem services they provide.

5.6 Conclusion

The river systems originating in the eastern Asian Highlands are among the highest and longest in the world and are one of the three regions with the greatest diversity of freshwater biota (as indicated by fish species). The ecosystem services generated by these rivers are crucial to the societies that have emerged along them.

But these river systems, the biota they support, and the ecosystem services they provide are experiencing profound and rapid change. The conservation response to this change has been, so far, inadequate. Except for a few stem rivers in China, all three governments have focused on conserving lakes and peatlands rather than rivers. Human activities and climate change already stress these rivers, and the stresses will only intensify in the coming years. As the Asian Highlands rapidly warm and precipitation changes, a cascade of impacts will unfold along the rivers. The region is already experiencing more severe weather, heat, droughts, and floods (see Chapter 3). Glaciers and permafrost are melting, and landslides, outbreak floods, and avalanches are increasing (see Box 4a). Peatlands are degrading, which impacts pastoralists' livelihoods and river inflows, and this threatens to exacerbate greenhouse gas emissions. These events by themselves are a cause for alarm. But it is also important to remember that such changes funnel into the rivers from their catchments and affect the freshwater species whose lifecycles are tied to specific river flows, along with the people who live along the rivers.

A changing climate also directly impacts the countries through which these rivers flow. As national governments seek to generate low-emission electricity, hydropower seems appealing. They also need to supply water to nearly 3 billion people in the Lowlands, whose water supply alternates between shortages and crises. These impacts compound, as graphically illustrated by the climate change-induced floods in Uttarakhand (2021) and Sikkim (2023), which both destroyed hydropower projects. All of these challenges make regulating the rivers' flows challenging, and we will return to this topic in later chapters of the book.

BOX 5a DEFINING WETLANDS

Can we define wetlands, and does it matter if we cannot? It is a fiendishly difficult task but helpful in a book about rivers, so we will attempt it here for our readers' sake. Wetlands are freshwater ecosystems that are defined and demarcated by multiple synonyms and overlapping names in a variety of scientific and legal contexts (Burton and Tiner 2009; Gerbeaux et al. 2016). These descriptions and demarcations are used to construct and implement international and domestic laws and the conservation programs and institutions that flow from them. That means these laws, institutions, and programs do not operate with an agreed definition of wetlands. What we can do here, then, is provide an overview of how some of these international conservation regimes understand wetlands.

Asian Highland wetlands are all freshwater ecosystems on or under land between the atmosphere and the sea. They are legally defined as wetlands because all Asian Highland nations—China, India and Bhutan, Nepal, and Pakistan—are contracting parties to the Ramsar Convention on Wetlands. By signing this global treaty states commit to conserving these ecosystems and using them wisely. The treaty defines wetlands as:

> Areas of marsh, fen, peatland, or water, whether natural or artificial, permanent or temporary, with water that is static or flowing, fresh, brackish, or salt, including areas of marine water the depth of which at low tide does not exceed six meters.
>
> (UNESCO 1994: Art. 1.1)

It also states that wetlands "may incorporate riparian and coastal zone(s) adjacent to wetlands, and islands or bodies of marine water deeper than 6 meters at low tide lying within the wetlands" (UNESCO 1994: Art. 2.1). This broad definition of a wetland includes rivers and groundwater-dependent ecosystems within its purview. Hence, it defines all freshwater ecosystems as wetlands. The Convention has been silent on the status of glaciers, but their definition seems open to their recognition as wetlands.

The UN CBD demarcates wetlands differently. In its programmatic work, it distinguishes between "inland waters" and "marine and coastal" biodiversity. Despite this difference from the Ramsar Convention, it maintains a joint program of work conserving inland waters with it (CBD 2010). This program largely delegates the conservation of "inland waters"—terrestrial wetlands—to the Ramsar Convention.

Unlike the Ramsar Convention and CBD, individual governments and scientists tend to define wetlands very narrowly to exclude rivers and lakes. For

(*Continued*)

example, the United States Government defines wetlands as: "Lands transitional between terrestrial and aquatic systems where the water table is usually at or near the surface or the land is covered by shallow water" (Burton and Tiner 2009, 508).

International law is another system of codes that impacts wetlands governance—particularly the international compacts concerning rivers divided by borders. Several Asian Highland states contest these laws, which were notoriously difficult to establish. The UN Convention on the Law of Non-Navigational Uses of International Watercourses (UNWC) was only agreed upon in 1997, after 27 years of negotiation. It was not until 2014, another 17 years later, that enough countries had ratified the treaty to bring it into force. This accord focuses on the conservation and sustainable use of watercourse systems, namely internationally shared rivers and connected water bodies (such as aquifers). China voted against the adoption of the UNWC, and India abstained. These nations are part of a group dubbed "hydro-hegemonic" states (Zeitoun and Warner 2006). They are likely resisting international river treaties because of the concern they may bestow greater legal rights on downstream nations and curtail their aspirations to fully exploit the rivers within their borders (Mirumachi et al. 2013). If the treaties were implemented, they would likely benefit smaller nations. These benefits include the greater recognition of their right to be consulted about developments on rivers in neighboring countries that may impact them and the capacity to invoke formal, graduated dispute resolution procedures. Surprisingly then, in South and Southeast Asia, only Vietnam ratified the treaty, whereas other downstream nations such as Bangladesh have not. China and India only make bilateral agreements with neighboring states to manage shared rivers. As powerful upstream nations, this confers advantages on them that they are unlikely to relinquish without a substantial—and probably decades-long—international negotiation process.

Given these multiple and contested terminologies, we use the Ramsar Convention's broader definition of wetlands in this publication. Further, we use the term "catchment" applied to aquifers, lakes, and rivers. This is synonymous with the terms "watershed," as used in North America, and "basin," as used in many other countries (McCracken and Wolf 2019). These definitions and governance institutions are vital because they influence the extent of cooperative management of wetlands and basins by countries and any commitment to doing no significant harm to their neighbors.

BOX 5b *LU* AND WATER

The oldest stories in the Tibetan cultural sphere describe the *lu* as one of the Highlands' earliest inhabitants. In these stories, which are preserved in 1,200-year-old manuscripts and later collections of purportedly ancient oral traditions, the *lu* perform a multiplicity of roles, including the creation and sustenance of the universe. Their important role in Tibetan cultural life was reduced as Buddhism was introduced between the eighth and twelfth centuries. Within Buddhist cosmology, they were aligned with Indian *naga* (shape-shifting creatures with human heads and snake bodies), and their domain shrunk from the cosmos to the Highlands' subaqueous realms. They became associated with all kinds of water and moisture, springs, the edges of rivers and streams, river confluences, lakes, wells, wetlands, and trees that extracted moisture from underground. Despite or perhaps because of the contraction of their realm, they are known to react capriciously to the intentional or unintentional desecration of moist spaces.

5b.1 The *Lu* of the *Lu Bum*

Much of what we know about the roles *lu* played in pre-Buddhist Tibetan culture comes from several collections of *lu* lore that are known jointly as the *Lu Compendium*. This collective name refers to several smaller compendiums, including the *Collections of White, Black and Multicolored Lu* and the *Collection of the Pure and Precious Lu from the Perfect Collection*. The *Collections of White, Black and Multicolored Lu* were created by Bon practitioners at some point before the thirteenth century, from which time an extant manuscript exists (Helman-Ważney and Ramble 2020, 74), but some of the narratives and rituals within it are much older. Within the collated stories, disconnected and sometimes contradictory narratives are placed next to each other. This suggests that the authors' motive was to combine as many oral traditions about *lu* as possible rather than determine a singular, authoritative story.

In the *Compendium*'s older stories, the *lu* bring the world and themselves into existence (see Box 2a). The four elements of the cosmos—earth, water, wind, and fire—arise, and then the *lu* and the world arise from these elements. As the *Lu Compendium* says, the *lu* are "the origin of everything; they created the universe with the elements and are also able to destroy it" (Bon Canon 1999, vol. 139, 40).

There are two *lu* progenitors, the male Sipai Gyelpo (Lord of Existence, who is also known as Zangpo Bumtri (Pure One of a Thousand Thrones) and the female Chucham Gyelmo (Aquatic Queen Consort). Many of the couple's manifestations are aqueous creations, conch-colored frogs (or perhaps turtles), and

(*Continued*)

turquoise-colored fish, but they also appear in many other non-aqueous forms, such as geographical features. In one *lu* origin story, the couple lay eggs from which *lu* with various shapes and supernatural abilities appear. These offspring *lu* change the environment, producing wind with their wings and evaporating seas with their heat (Zeren 2021, 188, 191, 196). In another *lu* origin story, Zangpo and Chucham live on a continent called Kalpa, where they manifest as frogs and fish but produce three oxen as offspring. These *lu*-oxen can control three elements—water, fire, and air—and with this power, they shape the world (Zeren 2021, 188). Elsewhere, Zangpo and Chucham appear as a golden mountain and a turquoise lake and give birth to *lu* of various forms. In yet other stories, *lu* operate as *sadak* (earth lords) or *nyen* and produce deities from eggs. In some stories, the *lu* offspring of Sangpo and Chucham appear as marmots, in others as snakes, frogs, fishes, turtles, cattle, and most often oxen. Occasionally they take human form. Sangpo Bumtri's eldest son has a human form but is made of white, radiant light. At other times, the *lu* take a hybrid form, with a human body and the head of an ox, bird, horse, snake, or tiger (Zeren 2021).

Along with the role they play in these cosmic origin stories, *lu* continue to appear in the stories and histories of the early Tibetan kings, whose installation as rulers draws from pre-Buddhist traditions. In these stories, the kings depend on their association with *lu*. As Charles Ramble noted, these stories suggest a particular connection between *lu* and fish in underwater realms and the *khyung* and birds in the sky. These two, *lu* and *khyung*, were understood as opposites, often enemies, but also as mirror images of each other. The story of Drigum Tsenpo—a semi-mythical Tibetan king who was kidnapped by a *lu*, held underwater near the entrance to the Yarlung Tsangpo Gorge, and only released when a child with bird eyes was exchanged for him—points to this *lu-khyung* connection and enmity (Ramble 2013; see also Chapter 6).

5b.2 *Lu* and *Naga*

When Buddhism was introduced to the Highlands between the seventh and eleventh centuries, it was accompanied by an extensive translation project, through which large canons of Buddhist teachings were translated into Tibetan. Within this project, Tibetan translators chose the word *lu* to render the Indic word *naga*. In Indian culture, *naga* are serpentine beings that reside in a netherworld. They can take human or part-human forms and are associated with types of trees, snakes, and some human communities. It is uncertain whether they are gods, demigods, or a subclass of animals, but there are multiple *naga* narratives in the Buddhist canon. *Naga* guard Sumeru (see Chapter 1), protecting the deities that live there from demigod attack. One *naga* king, Mucilinda, used his seven snake heads to shelter the meditating Buddha from the rain

(*Continued*)

(Johns and Nag 2021). Traditional Tibetan histories also claim that the legendary third-century Indian philosopher Nagarjuna traveled to the *naga* realm to retrieve the Mahayana scriptures from the *naga* king who was minding it. This story reflected another trope in *naga-lu* lore, their role in guarding and preserving treasure (Woodhouse et al. 2015).

The *Mahavyutpatti*, an eighth-century Sanskrit-Tibetan dictionary upon which Tibetan translators relied, not only evidences the link the translators made between *naga* and *lu* but also provides a list of eight great Indian *naga* kings, who came to play a significant role in Tibetan culture. The list of *lu-naga* kings' names changed over the centuries, and various versions of the list exist (Zeren 2021, 159), but there were always eight of them. One king who is sometimes on the list and sometimes not is Tsukna Rinchen, who is a protagonist in the *Epic of King Gesar*. A *naga* king, he gave his daughter and the *Lu Compendium* to Padmasambhava after Padmasambhava freed his land from disease (Kornman et al. 2012).

Over time, the conception of Indian *naga* and Tibetan *lu* cross-pollinated, and the Tibetan understanding of these beings included a range of possibilities stretching from local, pre-Buddhist spirits to the *naga* in Buddhist stories (Nebesky-Wojkowitz 1996, 290). As *lu-naga* lived in specific sites and interfered with human health and wellbeing, they also came to play a significant role in the *mentsi* (medicine-astrology-geomancy) knowledge lineages, and norms from both these lineages made their way into *lu-naga* lore.

Both Buddhist and *mentsi* influences are evident in the later layers of the *Lu Bum*. In these, the *lu* are classified in accordance with Indian social classes and manifest animal heads that align with these classes. Royal *lu* have horse heads, Brahmins have chicken heads, nobles have duck heads, commoners have ox heads, and outcasts have bear heads. Their color, type of throne, direction, character, and the realm in which they dwell also reflect their social class. Royal *lu* are white, their king sits on a central golden throne, and they live in the Desire Realm heavens. Noble *lu* are yellow, their king sits on a turquoise throne in the east, and they live in the demigod realm. Common *lu* are purple, their king sits on a bronze throne in the south, and they live in the continents and subcontinents of the human realm. They particularly like water and warmth. Brahmin *lu* are red, and their king sits on an iron throne in the west. They prefer wet ground. Servant *lu* live in the north, their king sits on a bronze throne, and they live underground, underwater, and anywhere there is humidity and warmth. They can also live in trees, stones, and rocks (Zeren 2021).

In the *Lu Compendium*, the *lu*'s color reflects the harm they may deliver. White *lu* generally bring good fortune to humans, but as they are also responsible for rainfall and the ripening of harvests, they may cause thunder or crop burning if angered. Yellow *lu* are ambivalent. Their actions toward humans

(*Continued*)

depend on human behavior. Purple *lu* bring sudden disease, red *lu* cause chronic afflictions, and black *lu* cause serious illnesses like leprosy. Although there are males and females within all these color-coded categories, female *lu* are sometimes divided into separate classes of white, black, and *lusin*. *Lusin* dwell in specific sites like rocks and springs and are especially virulent. They are associated with destroyed ground and crops, often due to the introduction of worms and insects (Zeren 2021).

According to the geomantic tradition, *lu* may also become annoyed by non-anthropogenic events. For example, if solstice sunrays hit human-like rocks and reflect onto them, noble *lu* may become angry and destroy northern villages. If a solstice breeze moves water on a northern lake, they may consider the lake poisoned and ravage settlements in the south. If they identify a tree growing in the west as a *khyung* bird demon—which are often equated with Indian *garuda* and cast as the enemy of *lu*—they may eradicate a settlement in the east (Sangs rgyas rgya mtsho 1997, *smad cha*, 248.2–6; for a German translation, see Maurer 2009, 188–189, and Maurer 2024).

5b.3 *Lu* in Contemporary Highland Life

In the contemporary Asian Highlands, peoples' understanding of *lu* reflects a combination of local *lu* lore, particularly propitiated by Bon priests (Tashi 2023), Buddhist ideas of *lu-naga*, and the descriptions of *lu* found in the *mentsi* literature. *Lu* are still viewed as powerful actors able to cause disease and disaster when they are insulted or polluted. They live across the Plateau in large numbers, and individual villages can be home to up to 40 *lu* (Coggins 2019). As the textual tradition suggests, they live in moist places, including springs, wetlands, and the edges of rivers, and are particularly associated with the headwaters of rivers and springs (see Chapter 7). But there are also many more *lu* trees across the Highlands than the textual tradition recognizes (Kocurek 2013).

Lu can sometimes live in larger *lukhang* (*lu* houses), to which village and city inhabitants make offerings. Some of these *lukhang* are quite large, and the Dalai Lama's *lukhang* which sits below the Potala Palace in Lhasa on a lake, is very well-known (see Figure 5b.1).

Lu bring water-associated disasters such as floods and droughts to those who defile their residences, but, more regularly, arthritis and skin diseases such as leprosy, acne, and psoriasis. Communities often recognize *lu* sites after one of their members falls ill with these diseases (Kocurek 2013). Specific human activities that annoy *lu* and cause them to inflict these illnesses include anything that pollutes water, digging in the ground, constructing buildings, and neglecting *lu* rituals. The presence of *lu* acts as a strong deterrent to spring pollution (Gaerrang 2017).

(*Continued*)

FIGURE 5B.1 *Left*, Lu image from Lhasa Lu Khang. *Right*, Lu Khang in Lhasa. Photographs, Ruth Gamble, 2018.

Families and communities perform rituals for individual *lu* according to local traditions, or in the case of specific illnesses or other negative events, through the advice of a Buddhist teacher or *mentsi* specialist. Rituals to appease *lu* can be as simple as offering cooking vessels at the bottom of a *lu* tree or spring (Coggins 2019) or may involve complicated rituals that are performed over the course of days (Kocurek 2013). *Lu* lore also influences and moderates pastoralists and agriculturalists' use of water and their approach to water resources (see Chapter 7 for more details).

6

HUMAN RIVERS

9,650 BP——1800s CE

The most common time markers are BP (before the present, i.e., 1950) and CE (common era).

The first humans who entered the Asian Highlands walked up river valleys. They settled on their banks in the Highlands' vertiginous margins and depended on their waters in the high-altitude, arid zones. As humans found spaces to thrive within the Highlands' hydrological systems, rivers and humans became not only interdependent but also interpenetrated. Humans drew drinking water from springs, wells, and rivers. They grew rain- and irrigation-fed crops. Their animals drank from streams, fed on rain-fed grass, and then gifted their humans water-dependent dairy and meat. Water cleaned human bodies, animals, clothes, and buildings. Then, after the humans had used as much water as needed, the waters continued their journey downstream.

Just as the rivers passed through humans and their animals, human societies became enmeshed in the rivers' flows. They altered Highland glaciers, springs, and rain-fed rivers. They diverted the rivers' flows into fields to grow crops or away from towns to alleviate seasonal and outburst floods. They used the water's power to grind their grains. More than any other species that settled in the highlands, humans found ways to bend the rivers' systems to their needs, but the extent to which they bent the rivers depended on their understanding of and relationship with the river.

DOI: 10.4324/9781003392033-6

Until late-modern Lowland empires transformed Highland governance, humans lived within Asian Highland river systems for millennia, viewing their waters as an endowment, working with their seasonal flows to produce crops and feed their animals. This chapter traces the *longue durée* of these pre-modern human–river relations in the Highlands. It begins with the initial peopling of Asian Highland river valleys around 30,000 to 15,000 BP, then outlines the uses of water during the Tibetan Empire (seventh to ninth centuries CE) and its various successor states up to the introduction of modernist river regulation in the eighteenth century (Chapter 8 will take up this story). It focuses on the markers of the human–river relationship: fishing, pastoral watering, irrigation, boats and ferries, bridges, drinking water, and sewerage systems. The long view of human–river entanglements highlights patterns that a more detailed analysis of a singular topic could miss. The long view tracks changes and continuities in community and political understandings of river systems. It occurs in the same time frame as the Highlands' climate geoarchives (see Chapter 3) and interacts intimately with all Earth Systems.

It highlights differences between the Highlanders' river relationships and those of their neighbors in the Chinese and Indian Lowlands. Despite Tibet's aridity, the states that developed on the Plateau did not become water despots (Wittfogel 1957), nor did Highland rivers repeatedly consume state resources as some lowland Chinese rivers did (Zhang 2016; Mostern 2021). But Highlanders did operate small state-run river-management projects, valley-wide schemes, and single-family irrigation ditches like those found on the pre-modern Chinese plains (Perdue 1987, 171). Rivers did not become central figures in Highlander cosmologies as they did in India (Sen 2019). But human–river relationships were negotiated through supernatural proxies, and cosmopolitics pervaded the hydrological system, from the snow and ice at the rivers' sources, which were viewed as the gods' homes, through rivers, to springs and underground water resources, the home of the *lu* (see Box 5b).

The rivers' hydrological tentacles permeated Highlanders' lives, below ground, on the ground, in their homes, and into their bodies. They were the stuff of everyday life and the cause of some deaths. The rivers were perennial, allowing human settlements to flourish on their banks, but they were also constantly changing. They fluctuated through annual seasons and more significant climatic shifts such as the Last Glacial Maximum (26,000–18,000 BP), the Medieval Climate Anomaly (950–1250 CE), and the Plateau's extended Little Ice Age (1400–1900 CE) (see Chapter 3 and Rowan 2017, 292–308).

Given all this historical climate and environmental change, it may be easy to assume that what is happening now merely continues historical trends. This would be a mistake. It is important to keep pre-twentieth-century human interventions in perspective. Certainly, Highland hydrosocieites served humans, transformed *more-than-human* ecologies, and made these changes within uneven power structures that benefited elites and disadvantaged the poor and marginalized. But these pre-modern hydrosocial systems were distinct from those imposed on the Highlands by Lowland governments in the past century and a half (see Chapter 8). As they relied on human and animal metabolisms to change the environment, they did so slowly, extracting and diverting water seasonally. They did not block rivers' flow completely or ameliorate the rivers' seasonality. All the human interventions in river systems that this chapter outlines were minor compared to humans' transformation of the catchments during the Great Acceleration (1945–). If anthropogenic environmental change unfolded in a hockey stick graph (Mann 2012), the *longue durée* this chapter outlines is the shaft, not the blade.

6.1 Peopling Rivers (Pre-sixth Century)

Although the timescales of the Plateau's peopling are still debated, it is most likely that humans arrived before the Last Glacial Maximum (26,000–18,000 BP) (see Chapter 4) and hunted, fished, and foraged around moraine-blocked lakes in forested river valleys (d'Alpoim Guedes and Aldenderfer 2020; Miehe et al. 2006). There is increasing evidence that the humans who first inhabited the Tibetan Plateau and Himalaya were *Denisovan hominins* rather than *Homo sapiens* (Chen et al. 2019; Chen 2022; Wang 2023a) and that interbreeding between *Denisovans*—or a *Denisovan*-like species—and *Homo sapiens* provided some of the physiological adaptations that have enabled Highlanders to thrive at altitude (Zhang et al. 2022a). This genetic admixture may have also helped humans survive on the Plateau through the Last Glacial Maximum when glaciers and ice fields increased, but lower riverine and lakeside environments remained ice free (Miehe et al. 2006).

Our understanding of the Highlands' peopling process is more apparent from the period after the Last Glacial Maximum, during the Holocene. This settlement unfolded in stages. It began with limited lower-altitude human settlements in the southeast, whose residents lived sedentary lives, hunting, fowling, fishing, raising pigs, and growing millet (Wang et al. 2023a; Li 2007), before these settlements expanded to Central Tibet, along the Kyi Chu (Lhasa River), near present-day Lhasa (Meyer et al. 2017). When the climate became colder and drier around 3,800 BP, it was more challenging to hunt and forage, so settlers herded bovines (yak and cows) and caprines (goats and sheep), fished (Wang et al. 2021d), and grew grains (Ren 2000). As the humans settled, they deforested much of the juniper (*Juniperus convallium* and *tibetica*), cypress (*Cupressus gigantea*), peach (*Prunus mira*), and butterfly-bush (*Buddleja crispa*) that had grown in these ecological niches (Miehe et al. 2006) and replaced these forests with crops and grasslands for pastoralism.

Humans, plants, and animals climbed into the Highlands from all sides. *Homo sapiens* made their way to the Plateau from North and East Asia, the Central Asian steppe, and the Southeast Asian Highlands (Wang et al. 2023a). The barley (*Hordeum vulgare, Hordeum volgare var. nudum*) they sowed was first cultivated in Anatolia, and traces of crops genetically related to it have been found in 9,000-year-old archeological sites in East Asia, Harappa, Kashmir, and Nepal (Spate et al. 2017). Its short growing season and high tolerance to low temperatures, and—for a cereal—the ability to withstand aridity made it eminently suitable for the Plateau. At lower altitudes in the Highlands' southeast, communities grew rice that arrived from the lower Brahmaputra and Ayeyarwady Catchments (Wang et al. 2021c).

We are not yet certain about the journey yak (*Bos grunniens*) and hybrid *dzo* (*rdzo, B. grunniens* × *B. primigenius taurus*) took into the Highlands, but we do know the crucial role they have played in Asian Highland life, providing humans with milk, meat, dung, waterproof hide, and wool, which enabled them to spread further across the Plateau (Rhode et al. 2007). Yak dung fertilized fields, supplemented river clay in earthenware, and became an alternate fuel source for wood. Yak tails were used for tethers and fly whisks.

We know more about the route goat and sheep took into the mountains. Most Highland species of domesticated sheep and goats were probably imported from East Asia, where there was a significant increase in population around 4,000 BP (Cai et al. 2007), and from Central Asia, where they played a fundamental role in pastoral life (Taylor et al. 2021b).

The last stage of the Highlands peopling, which began around 700 BCE, was the development of the agropastoral systems that are sustained today (see Chapter 7). This system has several components. In the highest, drier regions where agriculture was impossible, *drokpa* (nomads) practiced pastoralism. Highlander *sanampa* (agriculturalists) cultivate barley in *zhing* (arable land) in high-altitude river valleys, using snow-melt, spring, and river-fed irrigation. *Samadrok* ("neither farmers nor nomads," agropastoralists) combined agriculture and pastoralism across these altitudes. In steep, lower-altitude river valleys, agriculturalists maintain terraces for millet and rice in paddy and *jhum* (shifting agriculture) fields.

Until the introduction of Buddhism, many people hunted, and a few groups continued hunting into the Buddhist era (Huber 2012). Fishing became much less prevalent (see Box 6a). Farmers' houses were made of either earth, clay, cow dung, or wood, if available. Their roofs were generally flat and used for drying food. Many pastoralists lived in black yak-wool tents. The separation of labor between pastoralists and farmers led to trade between them and may have, in turn, encouraged the production of the quintessential Highlander food, *tsampa* (roasted ground barley), which can easily be transported.

Farmers in the drier areas (and those in lower altitudes who grew rice) developed rudimentary, gravity-driven irrigation systems from an early period. Some of these systems directed water from melting glaciers, springs, and seepage at valley tops onto fields and then out into rivers. Others redirected water from the rivers' main channels onto their fields via irrigation canals. In a few places, like the Lhasa Plain, irrigation in the early spring season had to be combined with drainage in the wet summer season.

Highlanders lived in *chulung* (river valleys), within which their *chudrenpa* (irrigation) systems included *chuyur* (canals), *chuka* (ditches), *churak* (dams and embankments), and *chudzo* (reservoirs) and were overseen by a *chupon* (water chief). *Chukor* (water-powered wheels) milled grain and turned prayer wheels. Clocks and hours were both called *chutso*, "water measure," because early Tibetan clocks measured water.

Villages arranged hydropolitics themselves, appointing a *chupon* (water chief) who worked with their *ponpo* (valley chief). The multi-leveled, patronage-based Highland polities' key role in water governance was mediating water disputes. All irrigation, agricultural, and pastoral systems operated seasonally. Farmers began plowing fields and preparing irrigation channels in the highland valleys when the spring meltwaters started in the second Tibetan month (March–April). In the third month (April–May), they spread manure or river silt on the fields and then sowed them. In the summer, they prayed for rain, weeded, re-wet their crops with irrigation water if the rains failed. In late summer, they worried about hailstorms and early frosts. Migratory birds visited them from the south in summer, including bar-headed geese (*Anser indicus*), ruddy-shell duck (*Tadorna ferruginea*), and the treasured black-neck cranes (*Grus nigricollis*), whose arrival was taken as a good omen (Prosser et al. 2011; Dorji et al. 2021, see Box 10a). Fields next to the rivers were less likely to experience frost and were highly prized despite their flood risk. Farmers grew one cereal crop yearly in the high valleys, usually barley. At lower altitudes, they could follow rice crops with a second crop, such as mustard or peas. While cultivators protected their crops, pastoralists moved their herds to higher pastures as the weather warmed (see Chapter 7). Winter was everyone's rest season.

6.2 Empires and Irrigation (Seventh to Ninth Centuries)

The Highlands' residents did not record their history until the seventh century. They began to write them during a time of significant change as a local ruling family, the Pugyel, who were based in the eastern and central sections of the Yarlung Tsangpo Valley, gradually subjugated its neighbors and formed the Tibetan Empire (seventh to ninth centuries). One of the Empire's significant innovations was the introduction of a written script, the Indic-inspired Tibetan script, which was adopted across the region and remained a unifying cultural phenomenon (Van Schaik 2011).

Our written sources for the Tibetan Empire include several traditional histories that were written centuries after the Empire ended but drew from non-extant texts and a stash of manuscripts that were preserved in a cave near Dunhuang, an oasis on the other side of the Taklamakan Desert that was under Tibetan imperial rule. By comparing the later histories, these Dunhuang documents, and the geoarchives of ice cores, we can piece together many details about life in the imperial period. All these sources suggests that the Tibetan Empire was an intensely fluvial state.

One of the most remarkable stories this interactive data tells is that of the paleo-lakes that existed in Kongpo. According to the geological record of the past 60,000 years, Kongpo—a name that means "concave" or "bowl"—has been inundated multiple times by large and small paleo-lakes. The largest was a saline lake that lasted for tens of thousands of years, from around 51,000 BP to 13,000 BP, covering the entire Kongpo region (Xu et al. 2020c; see Chapter 2). This means it existed past the Last Glacial Maximum into the time the Plateau was settled. Another smaller but significant lake inundated Kongpo in the first millennium, around 200 to 800 CE. It was caused by a glacial blockage and stretched from the top of the Yarlung Tsangpo Gorge through the Nyang Chu Catchment and the Yarlung Tsangpo's main channel near Menling. The water emptied when it began warming around the seventh century (Montgomery et al. 2004; Zhu et al. 2013).

There are multiple aquatic stories about Kongpo from the Imperial-Era sources, but they do not describe this lake directly. The clearest description of an inundation comes from a later, fourteenth-century history of Tibet, Sonam Gyeltsen's (1312–1375) *Mirror Illuminating Royal Genealogies*, which contains the following description of a pre-imperial Kongpo flood:

> It is said that, at that time, the mountains became forests, and the valleys filled with water. Kong Chu's arm [reaches?] opened, and all the water went there. [Eventually,] the waters in Kong Chu's arm dissolved, the plains became fields, and towns were built.
>
> *(Bsod nams rgyal mtshan 1973, 116–117, based on Sørensen's 1994 translation, 132–133)*

The term "Kong Chu" in this passage could refer to either the Nyang Chu river that flows through Kongpo or Kongpo's waters, more generally, including the sections of the Yarlung Tsangpo upstream and downstream from its confluence with the Nyang Chu. The paleo-lake would have covered the riverside fields in all three reaches, or perhaps "arms."

The lines that follow this description in the *Mirror Illuminating Royal Genealogies* explain that the waters' dispersal facilitated the coming of the first Pugyel King, Nyatri Tsenpo. Many versions of this tale describe Nyatri Tsenpo descending from heaven to Mount Lhari Gyangtho in Kongpo, where he discovers the land *chuwo kunkyi chenpo go* (at the head of all great rivers). Dunhuang manuscripts *Pelliot tibétain 1286*, also known as the

Chronicle, probably composed around 800 CE (Dotson 2013, 13), describes Nyatri Tsenpo's descent like this.

> The mountains bowed their heads, trees bent for him, the wind blew their branches, springs bubbled clear water, and hard boulders bent like elastic cartilage. This is how he became lord of Tibet's six gorges ... [Looking around, he saw] a place under the center of the sky, in the middle of the earth, the heart of the continent, encircled by snow, *the head of all rivers*, in high mountains, with pure earth, fertile land, its people, wise and righteous, and swift horses that reproduced themselves quickly.
>
> *(PT 1286, emphasis added, collated in Imaeda et al. 2007, 197–199)*

Subsequent histories and declarations repeat the same imagery, including the stone pillar that memorialized the 821 CE Pugyel-Tang Dynasty treaty and still stands outside the Jokhang Temple in Lhasa. It describes the Pugyel kings as "gods in heaven who became lords of men [in the] *center of high snow mountains, at the head of great rivers*, in a high country, with pure earth" (author's translation based on Richardson 1985, 108–109, emphasis added).

Seven generations after Nyatri Tsenpo, the river and Kongpo played key roles in another king's story, the tragic—and well-studied—tale of Drigum Tsenpo (PT 1286; see also—among others—Haarh 1969, 401–406; Karmay 1998; Dotson 2011). By this stage, the Pugyel kings lived in the Yarlung Chu catchment, south of the Yarlung Tsangpo in U. Drigum Tsenpo was the eighth Pugyel King, but the omens were misread when he was born, and he was given an inauspicious name. He sealed this unpromising fate by picking a fight with his guard, the renowned warrior Longam, who killed him and then placed his body in copper pots, which he threw in the Yarlung Tsangpo. Longam then took control of the kingdom and banished Drigum Tsenpo's two sons, Shay Kyi and Kya Kyi, into exile in Kongpo. Meanwhile, a *lu* named Long Oede Bede captured the erstwhile king's corpse and held it captive underwater, also in Kongpo. A few years later, one of Drigum Tsenpo's subjects, Ngarlekye, discovered Drigum Tsenpo's situation and made a deal with Oede Bedé; if Ngarlekye brought him a human whose eyes closed upward like a bird, Oede Bedé would give him Drigum Tsenpo's body. Ngarlake found such a child and made the exchange. This allowed Drigum Tsenpo's sons to bury their father and Sha Kyi to retake his father's throne. As King, Sha Kyi was known as Pude Gunggyel.

Drigung Tsenpo was the first of the Pugyel kings to be buried in the Yarlung Chu Valley, and his tomb has been dated to around 600 CE (Hazod 2007). His reign marked the start of the Pugyel Kingdom's rise. It also marked the beginning of a warmer climate cycle on the Plateau, which increased pasture sizes, enabling them to spare grass for horses and build armies (Hou et al. 2023).

(Similarly, good weather aided Chinggis Khan's rise around the turn of the thirteenth century CE, Pederson et al. 2014.) It is unlikely—as Hou et al. (2023) claim—that this was the primary reason for the Tibetan Empire's rise, but it certainly aided it.

Rather than crediting the climate for the Empire's rise, Tibetan histories tend to credit the rulers' personal innovations, many of which involved water. Another later history that includes information from non-extant sources is the sixteenth-century text, *A Feast for the Fortunate,* by Tsuklak Trengwa (1504–1566). It credits Pude Gunggyel with the following set of innovations.

> By charring wood, he made charcoal. By smelting ore with charcoal, he extracted gold, silver, copper, and iron. By drilling holes in wood, he made plows and yokes. *By digging in the ground, he directed upper-valley water into irrigation canals.* By fastening oxen in pairs, he turned meadows and plains into cultivable fields. *Where he couldn't pass, he built bridges across rivers.* And [during his reign] plowed harvests appeared for the first time.
>
> *(Gtsug lag 'phreng ba 1986, 134, emphasis added, cited in Haarh 1969, 122)*

The type of irrigation that Tsuklak Trengwa credits Pude Gunggyel with inventing is widespread across the Highlands. It involves creating irrigation canals at the point where the groundwater coagulates near the base of a hill, the *chugo* (water head), using gravity to direct it down through a central channel and then gates to re-direct it onto specific fields through minor channels. Any excess water continues to flow into the river system. The first field Pude Gunggyel harvested, Zorthang, is still a pilgrimage destination today. Farmers often scoop up dirt from the area to return to their fields to bring them luck (see Figure 6.2, *top left*). Further west, where there was less rain and more snow melt used to water the fields, ice reservoirs were created at the *chugo* in winter, which provided meltwater in early spring (Angchok and Singh 2006). Along the braided sections of the Yarlung Tsangpo, the riverbed itself was cultivated, which enabled the fields to be inundated easily (see Figure 6.2, *top right*).

Over centuries, the Pugyel Kingdom's rulers used allegiances and war to gain control over most of U, Kongpo, and Tsang. Then, in the early seventh century, a later king, Songtsen Gampo (c. 605–650) extended their realm, conquering the Zhangzhung Kingdom in western Tibet, forcing the Kathmandu Valley Kingdoms and the Azha of the upper Huang Catchment into vassalage, and marching on the Chinese capital at Xi'an. During his reign, Songtsen Gampo also moved his capital to the Lhasa Plain, built a palace on its Marpo Ri (Red Hill), and drained one of its wetlands to build the Jokhang Temple (Sørenson and Hazod 2007, 451–456).

A hundred years later, Songtsen Gampo's descendant Tri Songdetsen (r. 754–c799) expanded the empire further, briefly occupying Xi'an and

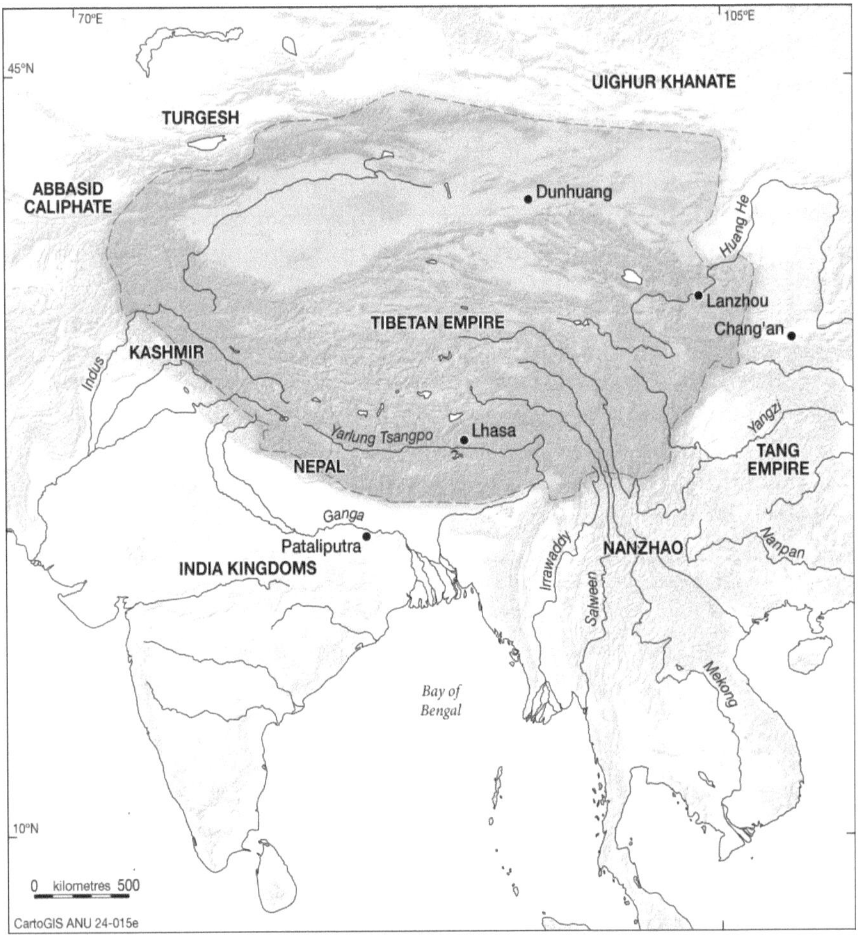

FIGURE 6.1 The Tibetan Empire at its Greatest Extent in 790 CE.

conquering Dunhuang and other oases north of the Taklamakan. But in Tibetan histories, Tri Songdetsen is best remembered for inviting numerous Indian Buddhist teachers to Tibet and building its first monastery, Samye, on the banks of the Yarlung Tsangpo across the river from the Yarlung Chu Valley. His grandson, Tri Ralpachen (d. 836), was even more devout than his grandfather. According to the traditional histories, this devotion to Buddhism caused his brother and successor, Langdarma (d. 842 CE), to instigate an anti-Buddhist purge when he took the throne. Only four years into his reign, a Buddhist monk assassinated him.

Following Langdarma's assassination, the Tibetan Empire fragmented, but its legacy continued to shape the Highlands, including its visions of rivers. Manuscripts from this period, including the *Chronicle*, represent ideal rivers, which flow from snow-capped peaks through fertile territory down to the Yarlung Tsangpo or the Lowlands. The Kyi Chu, a *chuzang* (good river), flows from the sacred snow mountain Nyenchen Thangla. The Yarlung Chu, a good river, flows from the equally sacred mountain Yarlha Shampo, which sits at the head of the Yarlung Valley. The Yarlung Tsangpo, a good, powerful, pure river, flows from Khangtise's snow (see Box 7a). The gods who occupy these mountains are presented as the Tibetan rulers' ancestors (see Chapter 1), and they bless their descendants with flowing, clean, life-giving water. As Brandon Dotson (2008) has shown in his analysis of Dunhuang healing texts, the Yarlung Tsangpo also acts as a proxy for life's journey. Stories explaining the onset of illness track downstream toward *chabkyi mashug* (the river's tail) in Kongpo, which represents death. Stories of recovery track upstream, toward the river's source, Khangtise, and healing.

Within the inter-valley competition that underpinned the Pugyel Empire's allegiance networks, the good rivers and their ecologies were even presented in competition. For example, an old Tibetan text boasts of the Yarlung Chu, which flows through the homeland of the Pugyel Emperors, as "the bluest and deepest of all rivers" (PT 1060, cited in Dotson 2008). And after describing the Pugyel Kingdom's conquest of Zhangzhung, the same text denigrates the kingdom's sacred lake, Mapang Tso (Lake Mansarovar). "Before last year and the year before that," it says, "geese and ducks flew over Mapang Tso's waters, flocking at river headwaters. These days, geese and ducks do not stay at the holy lake, the rivers' high-altitude source. If geese and ducks stay too long, Mapang's waters bury them" (PT 1060, collated in Imaeda et al. 2007, 83–87).

6.3 Bridge Building (Tenth to Sixteenth Centuries)

After the Tibetan Empire's demise, various centers of power headquartered in *dzong* (fortified strongholds) rose and fell within the Asian Highlands' river valleys. Local lords built *dzong* in Purang and Guge on the Langchen Tsangpo (Sutlej) and Sakya, Zhigatse, Gyantse, and Rinpung along the Yarlung Tsangpo in Tsang province. Lhasa remained a center of influence in U, and forts were built in the northern Kyi Chu Valley at Pudong and Redreng. Tsethang, at the base of the Yarlung Chu Valley, and Kongpo were also power centers.

While we still have few written records about the Dri Chu, Nyak Chu, and Eastern Himalayan river systems from this period, they are mentioned more regularly in the writings of wandering Buddhist teachers. Buddhism was introduced to the Tibetan court during the imperial period, but it was in this post-imperial setting that it became a mainstay of the Plateau's culture, society, and politics. Tibetan hierarchs used their resources to send Tibetans to study in India and sponsored Indian teachers' travel to Tibet. This exchange led to several new Buddhist lineages in Tibet, collectively known as the Ngak Sarma (New Tantras).

As the Ngak Sarma developed, practitioners of the old school, the Nyingma, began revitalizing their tradition. This revitalization was based on reimagining the Tibetan Empire as a Buddhist golden age, during which Bodhisattva Chogyel (Dharma Kings) ruled with the guidance of Indian gurus. In this period, the traditional stories of Songtsen Gampo, Tri Songdetsen, and Ralpachen were written, and they began to be called the Chogyel Sum (Three Dharma Kings). However, even their stories were eclipsed by the tales that developed during this time about Padmasambhava. He was granted a special place in Highland culture and society for his role in taming Tibet's *lha-lu* and convincing them to support Buddhism (see Chapter 1 and Box 2a).

Among the stories about Padmasambhava, several involve rivers and water. He and one of his consorts, Mandarava, are said to have transformed their funeral pyre into Lake Rewalwar in the Western Himalaya. Along the upper reaches of the Nyamchang Chu catchment, he is said to have created the 108 waterfalls that now sit on the Line of Actual Control (the defacto border) between India and China (Tenpa 2018, 52–53). Even more intriguingly, there is tantalizing evidence that his legend grew at least in part from his abilities as an irrigator. In the eleventh-century retelling of Padmasambhava's tale in the *Testament of Ba*, he is presented as an engineer whose importation of water management techniques to Tibet was so successful it threatened Tibet's *chabsi* (its hierarchical hydraulic political order) (Wangdu and Diemberger 2000, 14). The Tibetan ministers were so threatened by his innovations that they forced him to leave the highlands before his *lha-lu* taming mission was complete, gifting later *tantrikas* opportunities to continue his work.

The era of fragmented kingdoms ended in the thirteenth century when the Mongol Empire (1206–1368) annexed most of the Highlands. Mongol rule changed Tibet's soft (governance) and hard (technology) infrastructure and, therefore, its hydrosocial systems. The Mongols instituted postal services, censuses, and trade routes (Van Spengen 2000). Mongol officials encouraged new forms of horticulture in river valleys, including more extensive tree plantations in the Yarlung Tsangpo Valley (Byang chub rgyal mtshan 1989, 46), and they allowed and encouraged more bridge building (Rang byung rdo rje 2006, 404). The Mongol Empire also had other, less direct, and less positive impacts on Highland rivers. Written records from the time note the strains intermittent Mongol army invasions placed on resources and the effect their horses had on the environment, including its water resources, and there were occasional outbreaks of famine and revolt, at least in part as a consequence of this (Gamble 2018, 234–236).

More importantly for the Dri Chu Catchment, Mongol Empire trade routes spurred migration into eastern Tibet. This migration may also have been aided by the southward movement of refugees from the Western Xia (or Tangut) Kingdom (1038–1227)—located between Mongol and Tibetan lands—which they sacked in 1227 (Sperling 1987; 1994). This demographic and trade shift would have consequences in the coming centuries.

The Mongol Empire in China fell in 1368, when its last Emperor, Toghun Temur (1320–1370), fled his summer capital, Xanadu, and returned to the steppe. Mongol influence across the Tibetan Plateau decreased around the same time, and local states rose to fill the political vacuum. Once again, these states were based in riverside *dzong* (forts) that could control movement into and out of valleys, collect taxes, and guard against invasions and bandits. But the post-Mongol states did not operate in the same manner as earlier Highland states. They adopted new governance technologies such as legal codes (Pirie 2013) and improved hard infrastructure such as irrigation systems and grain stores. The grain stores lasted longer than one season because of the Plateau's cool, dry climate (*Rgya bod yig tshang* 1979; Wangdu and Diemberger 1996) and enabled the communities to survive the droughts that occurred as the Medieval Climate Anomaly (950–1250 CE) gave way to the Little Ice Age (1400–1900) (Gamble et al. 2022). The rulers of this period also changed human–river relations by improving riverside roads and adding more bridges (Gyatso 1980). However, the most famous development in human–river relationships during this period came not from governments but from one Buddhist guru, Thangtong Gyelpo (1361–1485).

Thangtong Gyelpo, "the King of the Empty Plain," is credited with the invention of chain link bridges, improved boats and ferries, and—in his spare time—Tibetan opera (Lo chen 'gyur med bde chen 1982; Gyatso 1980). Although many of the stories about Thangtong Gyelpo are probably apocryphal, even those few that are verifiable make him an intriguing character. There is little doubt, for example, that he worked with miners and blacksmiths in Kongpo to produce 25–35 centimeter chain links that he exported across the Highlands to construct iron-chain bridges that replaced earlier yak-hair rope, bamboo, or wooden bridges, all of which had been prone to rotting. Some of the bridges Thangtong Gyelpo is claimed to have built are still functioning today, including the Tachogang Bridge in Bhutan (see Figure 6.2, *center right*). The ruins of two of his other bridges—one that used to cross the Yarlung Tsangpo near the Kyi Chu confluence at Chushul (see Figure 6.2, *center left*) and another that crossed the Yarlung Tsangpo between Tsethang and Samye—are now pilgrimage sites.

Thangtong Gyelpo's alleged interventions in Highland ferry and boat design are even less historically verifiable than his bridge building, but Highland boat-going communities hold him in very high regard. These specialist communities already existed by the fourteenth and fifteenth centuries, and Tibetans had been using animal skin boats to navigate the flatter sections of their rivers for centuries. The technology improved slowly.

The simplest boat forms were inflated animal skins used as either paddle boards or strapped with wood planks to make a raft (Lange 2014). But Highlanders also used larger, triangular coracles made from animal skins—mainly yak—and wattle or saltbush (see Figure 6.2, *center right*). These coracles could be taken downstream relatively long distances and then returned on boatmen's backs as they walked them over mountain passes. Some boatmen from fishing villages, like Jon, used them for fishing (see Box 6a), and they were increasingly used as ferries to cross heavily trafficked trade and pilgrimage routes near Lhasa, Samye, and Derge in Kham (Rockhill 1891, 198, 226). Thangtong Gyelpo's exact relationship to these boats is not known, but each boat boasted a special triangular piece of skin on their broadside that they called "Thangtong Gyelpo's hat" and credited him with protecting them from *lu* and disasters (Lange 2014).

As river traffic increased, boatmen began to build larger wooden ferries known as "horse-head ferries" because of their distinctive horse-head prows (Figures 6.2, *center left*). These vessels could hold more passengers but were more challenging to navigate and only worked at large ferry crossings, such as the one next to Thangtong Gyelpo's Tsethang-Samye and Chushul Bridges, and those that serviced Lhasa and Zhigatse (Hedin 1909, 294).

6.4 River Politics in Tibet, Bhutan, and Kham (Seventeenth to Mid-nineteenth Centuries)

Sporadic infighting between central Tibet's fort-based clans intensified during the seventeenth century when it became sectarian, and competing Mongol nations offered military support to either side of the sectarian divide. The eventual outcome of these upheavals was the establishment of the Ganden Podrang (1642–1959) government by the fifth Dalai Lama, Ngawang Lozang Gyatso (1617–1682), a reincarnate teacher and leader of the reformist Geluk school of Tibetan Buddhism. The Dalai Lama's government was initially backed by the Khoshut Mongols (1642–1717) and later the Manchu—another Steppe nation—Qing Empire (1636–1911). The influence of both these external powers ebbed and flowed over the next three centuries. Like the Pugyel Empire, the Ganden Podrang survived by cultivating allegiances with monasteries and clans rather than governing directly as an autocratic state. By this stage, monasteries had become more powerful than most clans. They owned large swathes of irrigated land and developed large trade networks that relied on ferries and other boats (Van Spengen 2000). The Ganden Podrang let them raise their own taxes and accepted tribute from them.

Agriculturalists and agropastoralist communities worked on land tenanted by local monasteries and other landholders. They paid taxes through goods and services or *corvée labor*. Goods taxes were paid as grain, meat,

FIGURE 6.2 *Clockwise from top left*, Yarlung Valley. Zorethang is on the riverbank at the back of this photo. The building at the front, the Yumbu Lhakhang, was probably built by Sha Kyi-Punde Gunggyel's descendants at the turn of the eighth century. Photograph, Ruth Gamble 2018. Irrigated Yarlung Tsangpo Flood Plain near Lhasa, using dzo to plow a small embankment with a drainage system to irrigate. Photograph Ernst Schaeffer, 1936–1937. Tachogang Bridge, Bhutan, construction attributed to Thangtong Gyelpo. Photograph Ruth Gamble, 2023. Coracle at Chushul. Photograph by Bruno Beger, 1936–37, courtesy Bundesarchiv. Horsehead Ferry, near Lhasa. Photograph by Ernst Schaeffer, 1936–1937, courtesy Bundesarchiv. The remains of Thangtong Gyelpo's bridge at Chushul, Central Tibet. Photograph Edmund Chandler, 1904.

and timber. Part of the calculation for how much tax they paid was how much water they received (Gelek 1986). *Corvée* taxes could involve the construction of significant public works, including levees and large irrigation canals, and providing transport services, including ferries and boats. Pastoralists were concentrated in eastern and northern Tibet and operated somewhat outside the Ganden Podrang's rule. However, they also supported the Ganden Podrang by negotiating with government officials and Buddhist leaders to exchange pastoral products for religious services and protection (Lange 2023).

The Ganden Podrang was as dependent on the region's riverine systems as previous governments had been, and it, too, developed river and water usage. In 1645, when its first leader, the fifth Dalai Lama, began constructing the imposing Potala Palace on Marpo Ri (Red Hill), the site of one of Songtsen Gampo's palaces, he used 300 hide boats to ferry stone, wooden rafters, and whole stone pillars to the site (Sørensen and Hazod 2007, 476). As Lhasa's population grew, the Ganden Podrang installed more water utilities. There were, for example, several wells drilled within the city, but most observers complained about their waters' dubious quality (Shakabpa 2009, 53), and a new well had to be created for the Dalai Lama's private use (Bell 1992, 128). Backed by the Ganden Podrang, city officials also established a basic sewerage system, an open drain that ran through the city (Shakabpa 2009, 54).

As Lhasa's population increased during the eighteenth century, the government slowly added more infrastructure. In 1721, they built a dual-use irrigation and flood control canal that collected water from the *chugo* (waterhead) on the Lhasa plain's northeast, diverted this water to fields north and south of the city through an aqueduct, then out into the Kyi Chu. This canal was only needed in the rainy season. It was completely dry in winter (Hummel and Vogliotti 2000). Throughout the Ganden Podrang period, they also built and repaired levees and canals to control flooding on the plain. Along with other aristocratic families, the Ganden Podrang also managed the Lhalu Wetlands, which sits behind the Potala Palace. Unlike other wetlands on the Lhasa Plane, this area was never drained but instead managed as a source of marsh reed fodder for animals and peat turf for heating (Yeh 2009, 110–111).

The rise of the Ganden Podrang was accompanied by the development of a series of less influential but sometimes more resilient states along the Dri Chu and across the Eastern Himalaya in the Brahmaputra tributaries. These included Nangchen, Derge, Lithang, and Bathang on the Dri Chu, and Bhutan, Sikkim, Monyul, and Poyul in the Eastern Himalaya. All these kingdoms depended on irrigated agriculture and pastoralism but practiced varied mixes depending on their altitudes and ecologies. In Kham's steep gorges, for example, farmers used hollowed tree trunks as *tronpa* (irrigation gutters). According to Hummel and Vogliotti, these gutters extended "over a distance of several kilometers and across ravines, supported by scaffolding" (2000, 9). In the middle-altitude regions of Sikkim and Bhutan, farmers grew rice in terraces fed by the region's heavy monsoonal rains and snowmelt-fed streams. Precipitous Sikkim became so famous for its terraced paddies that the Tibetans called it Drenjong (land of rice).

Almost all these small kingdoms had tribute relationships with the Ganden Podrang, the Qing Empire, or both and maintained religious links with large Tibetan religious institutions. Bhutan, the exception, was established in resistance to the Ganden Podrang and fought several wars to maintain its independence. Its interactions with Lhasa were limited to trade, technology, and cultural exchange. But much of Bhutan's approach to water governance resembled those of other highland states. Bhutan's founder, Zhaptrung Ngawang Namgyel (1594–1651), is praised in much the same way as early Tibetan kings for bringing "water to places where there was no water, increasing water where it was scarce, making infertile fields fertile, and bridging bridgeless rivers" (Ardussi 1977b, 240).

6.5 Conclusions

Highland human and river histories have been thoroughly entwined since humans arrived on the Plateau. Humans used rivers to access the Highlands and depended on their waters. They hunted, fished, foraged near them, and then gradually settled beside them. Eventually, they developed complex irrigation systems that used gravity to utilize water seeping out of mountainsides, gurgling through springs, and flowing through rivers.

Multiple polities developed in Highland river valleys, but most were conquered by the Highlands' only Indigenous Empire, the Tibetan Empire (seventh to ninth centuries). From the ninth century onward, the Plateau was ruled by a combination of local governments and larger states that engaged each other through tribute networks. The Dalai Lamas' Ganden Podrang (1642–1959) operated a state in conjunction with monasteries and aristocratic estates and relied on patronage networks that stretched into the Dri Chu Valley, Sikkim, and the Eastern Himalaya. In turn, it had a patronage relationship with the Manchu-Qing Empire.

As the states rose and fell, the Highlanders gradually developed more complicated irrigation systems and used boats, ferries, and some bridges to facilitate trade and pilgrimage. By the turn of the mid-eighteenth century, they had long traditions of irrigated agriculture, water transport, and trade but resisted pressure to open their lands to outsiders. The European colonial project had taken its time to penetrate the Plateau's socio-ecology, but it was about to bring the Highlands into a new global, geopolitical, and technological order that would irreversibly transform its society and hydrology.

BOX 6a HUMAN–FISH RELATIONS IN THE HIGHLANDS

Many fish in the Dri Chu and upper Brahmaputra River systems are endemic to the region, and they have evolved in response to the river systems' extremes. Those living on the Tibetan Plateau face intensely cold waters with limited dissolved oxygen and food sources. Those in the precipitous reaches on the Plateau's edge live in small ecological niches between quickly changing water temperatures and altitudes.

One stress the fish did not have to face historically was overfishing. Unlike most human communities, Highlanders did not eat fish regularly, and most Tibetans avoided and denigrated fishing and fish eating. Highland attitudes toward fishing and fish eating have only begun to change in recent decades as Lowland migrants moved into the Highlands and the Chinese government encouraged aquaculture and fish eating. In response to this influx of fish culture, some Tibetans began to eat more fish, and others refrained entirely from fish eating as an affirmation of their identity. Along with this abstinence, many Tibetans and other Buddhists are increasingly practicing fish release or *tsetar* (life release). Unfortunately, when this practice is not managed correctly—when invasive fish are released—this causes more harm to fish and riverine ecosystems than overfishing. While the two worlds of Highland fish and humans used to be entirely separate, they are now becoming more insolvably entwined.

6a.1 Asian Highland Fish Species

More Tibetan Plateau fish live in lakes than rivers, and more live in the Yarlung Tsangpo and the other Himalayan Brahmaputra tributaries than in the Dri Chu Catchment. The Dri Chu has less diversity because of its flow velocity, altitude, and low dissolved oxygen levels (Yan et al. 2022). Many of the Highland fish that live in these waters do not live anywhere else, and most belong to four classes: carp (*Cyprinadae*), loach (*Nemacheilidae*), catfish (*Siluriformes*), and introduced trout (*Salmonidae*). Eighty percent of eastern Highland fish are carp and loach (Weng 2012). The fish's categorization can be confusing because of the common tendency to call the Highlands' largest family of carp "snow trout," even though they are not trout. The family's scientific name is *Schizothoracinae*.

Schizothoracinae is the only family of carp adapted to the eastern Highlands, and 76 of its species live in Highland rivers. As an adaptation to the Highlands' extreme cold and altitude, they grow slowly and have low fecundity and late sexual maturation (Chen and Cao 2000). As there is little connectivity between the Highlands' rivers and there are geographic boundaries within them—like

(Continued)

waterfalls and rapids—each catchment hosts its own fish species (see Chapters 2 and 5). Many species are endemic to the Highlands and the individual river or lake systems in which they live.

The most common *Schizothoracinae* in the Yarlung Tsangpo is *Schizopygopsis younghusbandi*, which grows to about 50 centimeters and lives in swift-flowing rivers with rocky substrates (Zhu et al. 2019). Others include the 30-centimeter-long *Schizopygopsis pylzovi*, *Schizopygopsis oconnori*, the threatened *Schizopygopsis macropogon*, and *Schizopygopsis molesworthi*. The Yarlung Tsangpo Gorge acts as a boundary to fish migration, and the *Schizopygopsis waltoni* and *Schizopygopsis curilabiatus* live below it (Liu 2021a). The *Schizopygopsis younghusbandi* and *Schizopygopsis oconnori* are the favorite food of another endemic carp *Oxygymnocypris stewartia*, which is slow-growing and late maturing, inhabits fast-flowing rivers with rocky bottoms and grows up to 67 centimeters and three kilograms. All three of these fish species, *Schizopygopsis younghusbandi, Schizopygopsis oconnori*, and *Oxygymnocypris stewartia*, have been commonly fished in the last few decades.

Schizothoracinae also live in the Dri Chu, but they are different species. The Dri Chu's endemic fish include *Schizothorax kozlovi*, *Schizothorax davidi*, *Schizopygopsis wangchiachii*, *Schizopygopsis dolichonema*, and *Schizopygopsis malacanthus* (He et al. 2009; Yan et al. 2022). *Schizopygopsis wangchiachii* and *Schizopygopsis malacanthus* are the river system's dominant snow trout species. The Dri Chu's fish need to be even hardier than those in the Yarlung Tsangpo. One species of *Schizothoracinae, Herzensteninia microcephalus,* was found swimming in the Dri Chu's headwaters at 5,350 meters above sea level, making it the highest cyprinid ever recorded (Zhu 2021b).

The Himalayan rivers of the upper Brahmaputra system are home to another carp genus known locally as mahseer (*Tor*). Mahseers live in rivers and lakes, are omnivores, and ascend rocky-bottomed rapids to breed. The largest of these, the Golden Mahseer (*Tor putitora*), can grow up to 2.75 meters and weigh up to 54 kilograms. Mahseer are commercially important game fish that fetch high market prices. Their numbers and size are declining in the wild (see Chapter 5), and they are now being restocked to encourage recreational fishing.

Catfish and loaches live on the bottom of rivers and lakes. They tend to be smaller than other Highland species and are fished even less than snow trout. Loaches are an old fish species more broadly distributed across Highland rivers than snow trout. Their existence in multiple now disconnected rivers suggests there used to be a fluvial connection between these rivers that enabled the fish to swim between them. *Triplophysa tibetana*, for example, lives in the Yarlung Tsangpo and Sengge Tsangpo (Indus) catchments, and *Triplophysa stenura* lives in the Dri Chu, Da Chu (Mekong), western Gyelmo Ngul Chu (Salween), and Yarlung Tsangpo catchments (Yan et al. 2022).

(*Continued*)

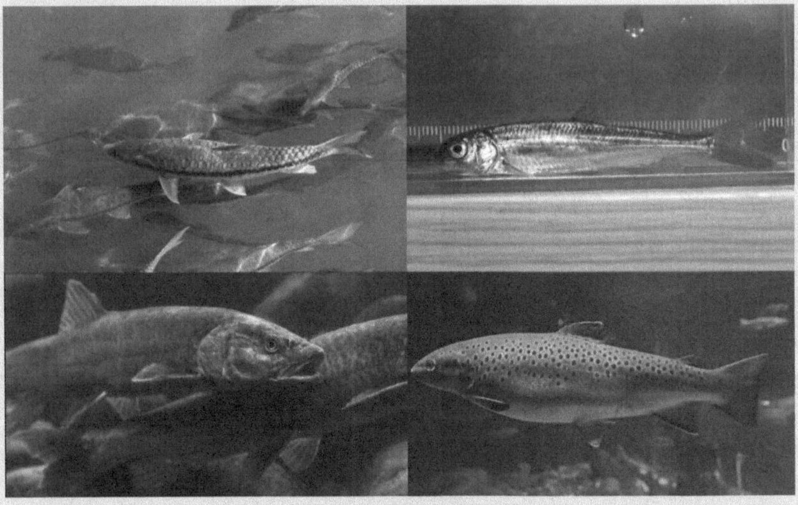

FIGURE 6A.1 *Top right*, Golden mahseer. *Top left*, Brown Trout. *Bottom left*, *Oxygymnocypris stewartia*. *Bottom right*, *Schizopygopsis younghusbandi*.

6a.2 Fish–Human Relations

Tibetan society developed an aversion to fish sometime between agricultural settlement 15,000 years ago and the Imperial Period (sixth to ninth centuries). Archeological evidence suggests that some of the Plateau's earliest inhabitants ate fish (Wang et al. 2021b), but documents from the Imperial Period show social proscriptions against fish consumption (see the main text of this chapter).

There were multiple biophysical reasons for Highlanders to grow wary of fish over the intervening millennia, and these probably acted symbiotically with socio-cultural and religious fish-eating proscriptions to marginalize fish eating. Biologically, the most plentiful Highland fish family, snow trout, can be poisonous, and their stocks do not replenish quickly, making them a dangerous and unsustainable food source. Economically, Highlanders have plenty of yak, sheep, and goat meat from the animals they herd and do not need to eat fish meat. Culturally, Tibetans have associated waters with *lu* (see Box 5b), the underworld, and water burial practices. And religiously, Vajrayana Buddhism does not prescribe meat eating but asks its followers to limit killing as much as possible. Killing one yak will feed a family for many months. Killing one fish may not provide it with one meal. Therefore, Vajrayana Buddhists would prefer to kill one yak than many fish. These factors combined to push fish-eating to the Highlands' social and geographic fringes. The association between fringe dwellers and fish further discouraged fishing as a significant food source. There

(Continued)

were communities and individuals who, for various purposes, resisted these pressures and continued to eat fish, but they were either geographically marginal, in Southern Bhutan, for example, or socially marginal, like the residents of Jon village near Lhasa (Altner 2009, 2010).

As humans generally only developed knowledge of animals they use, and the Tibetans had little use of fish, one outcome of Highlanders' widespread aversion to fish was that they did not develop the same widely distributed knowledge traditions about fish as they did about yak, horses, and other domestic and wild animals (see Chapter 7). It was left to outsiders, Lowlanders traveling in the Highlands, to develop an in-depth knowledge of Highland fish.

The first Lowlanders to conduct extensive surveys of Highland fish were colonial British officers in the late nineteenth and early twentieth centuries. Their surveys were part of their larger colonial scientific program, which produced and used knowledge to govern the region. After multiple attempts, the British also introduced trout species to the Highlands. The Scottish carpet entrepreneur Frank Mitchell introduced brown trout (*Salmo trutta*) to Kashmir in 1899 (Mitchell 1930). A few years later, he sold ova to some of the British officers occupying northern Sikkim and the Dromo (Chumbi) Valley (see Chapter 8), and they released them in the Khangbu Chu (Torsa) (Crawford and Muir 2008). Several years later, they introduced rainbow trout (*Oncorhynchus mykiss*) to the same system and the Teesta. The brown trout settled into their new environment, becoming naturalized within decades. The rainbow trout have been more problematic across the Eastern Himalaya, pushing endemic snow trout out of their habitats (McGlade et al. 2022).

During the 1950s, as two new states, the People's Republic of China and the Republic of India established their administrations in the Highlands, they engaged in an initial flurry of fish studies. In the early 1950s, Chunlin Zhang led surveys of the Tibetan Plateau's fish species that identified nine (Zhao et al. 2008), including two new species of *Schizothorax*. Between 1959 and 1961, the same team identified nine further fish species belonging to six genera, three families, and two orders. Chinese fish surveys stopped soon after this, however, and did not begin again until the 1990s, when Chinese scientists started working with international teams to conduct extensive surveys. From this point onward, the Chinese state has supported comprehensive studies of the Highlands' fish. These have been primarily instigated by local governments and conducted to investigate possibilities for aquaculture and, therefore, economic development (Cai 1992).

Indian scientists have also worked closely with governments to establish aquaculture in the Highlands. During the 1960s and 1970s, they worked with the governments of the North East Frontier Agency (NEFA) and independent Bhutan and Sikkim to establish fish farms. After NEFA became Arunachal Pradesh

(*Continued*)

and Sikkim became an Indian state in the 1970s, scientists and governments further increased the new states' aquaculture reserves (Choudhury 1996; Chettri 2021), and Bhutan continued to develop its fish farms independently (Royal Government of Bhutan 1986). Since the 1990s, all three governments have intensified aquaculture production in their territories, and it is now a mainstay of the Bhutanese economy (Royal Government of Bhutan 2018a).

6a.3 *Tsetar* (Life Release)

Like most rivers worldwide, the Highland waterways have experienced an increase in invasive species entering their ecologies through fish seeding for aquaculture and sport fishing and the dumping of pet species and live fish from restaurants in local waterways (Feng et al. 2023). Invasive species such as pond loach (*Misgurnus anguillicaudatus*), stone moroko (*Pseudorasbora parva*), and the notorious *Carassius auratus*, more commonly known as goldfish, now threaten the Highlands' endemic species, and may have entered Highland waterways through these means. But fish invasions have also come into the Highlands' waterways through another unusual source, the Buddhist religious practice of *tsetar* (life release). Life release is a ritual in which Buddhists ransom an animal that would otherwise be slaughtered and release it while saying prayers (Tan 2016). For most of *tsetar*'s long history, its Highland practitioners released terrestrial animals such as yak, sheep, goats, and sometimes deer (Barstow 2017). The release of fish and aquatic animals was usually practiced in associated Buddhist rituals downstream in East and Southeast Asian communities (McCarthy 2022). In recent years, however, the popularity of fish *tsetar* has grown exponentially in the Highlands.

Multiple factors have encouraged this trend. One factor is the efficient merit creation that fish release grants *tsetar* practitioners. Fish are endowed with the same consciousness as other animals, so they are equally in need of release. As they are smaller than other animals, more can be bought cheaply and released quickly. Another factor is the recent rise in fish trade on the Plateau. Aquaculture encourages faster-growing invasive species, and transporting fish from the Lowlands is easier. Unfortunately, these non-native species are often cheaper and more available for *tsetar* practitioners than the region's endemic fish. This creates a cycle where their availability makes them more likely to be released into local streams, and then they become more readily available for purchase.

Provincial and national governments have not always helped the situation. Government officials in Tibet have used fish release to symbolize their environmental credentials, linking it to the broader Chinese government policy to promote an "ecological civilization" (McCarthy 2022). Government-sanctioned fish releases on the Tibetan Plateau are sometimes carefully matched to local

(*Continued*)

FIGURE 6A.2 *Left*, Official Fish Release at Zangmu Power Station, Yarlung Tsangpo. *News.cn*, 2017. *Right*, Fish bagged and waiting for release, Zangmu Power Station, Yarlung Tsangpo. *Xinhua*, 2019.

ecologies. At other times, the government has ignored the region's endemic specificity and allowed any Highland fish to be released in the waterways. One example of this was the 2016 release of naked carp (*Gymnocypris przewalskii*)—a fish species endemic to Tso Ngonpo in Qinghai—into the Yarlung Tsangpo at the behest of the hydropower company running the Zangmu Hydropower Dam (see Figure 6a.2).

6a.4 New Fish Politics

The multiple threats to fish populations in recent years and concerns about diminishing fish populations have led governments to engage in various conservation programs, but they have tended to respond to some threats while ignoring others. The multiple problems facing the region's fishes include the dams that fragment habitats and impede migration, dangers arising from dredging and other riverside construction, irrigation and eutrophication, and the introduction of invasive species (see Chapter 9).

The governments' aquaculture programs and the migration of Han Chinese into Tibetan cities have also led to a proliferation of fish and seafood restaurants in Lhasa, Zhigatse, and Nyingtri. At first, these restaurants only served migrants. Now, they serve more and more Tibetans. Tibetans report mixed reasons for patronizing these restaurants. Some feel coerced to patronize the restaurants to conform to Han Chinese norms. Others like fish and want to eat it (Anonymous 2015). Whether they choose to eat fish or not is not only a dietary choice for them but also a marker of identity. Tibetans make a statement about their modernity, affiliation with the government, or individualism by eating fish. By not eating fish, they state their adherence to Tibetan markers of belonging.

7

AGRICULTURAL AND PASTORAL RIVERS

(Co-written by Kelzang T. Tashi)

1980s–Now

Rivers play an integral role in Asian Highlanders' everyday lives. This chapter explores these everyday interactions by focusing on daily water regimes in two communities: nomadic pastoralists who live in the upper Dri Chu (Yangzi) catchment on the Plateau's northeastern edge in a region they know as Kham, and agriculturalists of central Zhemgang county, Bhutan, who live along the Mangde Chu, within the upper Brahmaputra catchment.

The Khampa pastoralists live at an average elevation of 4,000 meters above sea level, occupying the grasslands that cross the Qinghai-Sichuan border in the Yushu and Ganzi Tibetan Autonomous Prefectures. Their black tents dot the river plains between the Dri Chu and its nearby Dza Chu (Upper Yalong) tributary. They herd animals, primarily yak, across these high-altitude pastures, which are watered by snow melt and monsoonal rains.

The Bhutanese agriculturalists live in a series of villages along the Mangde Chu just north of its confluence with its tributary, the Chamkhar Chu, in south-central Bhutan's Zhemgang county. The region's largest town is Tingtibi, on the Mangde Chu's western bank, and there are several villages, Goleng, Tagma, and Shobleng on the river's eastern bank. A further series of villages are perched on the mountain ridges that create the Mangde Chu's narrow valley. These villages are in Bhutan's middle hills, at much lower altitudes than the Khampas' pastures. They have access to snow melt and a much stronger monsoonal system, which allows them to tend rice paddy,

DOI: 10.4324/9781003392033-7

other grain and vegetable crops, and combine this agriculture with limited pastoralism.

Despite these two groups' different cultural and livelihood practices and divergent socio-ecological systems, their relations with and experiences of their respective rivers are strikingly similar. As this chapter elaborates, the most notable similarities pertain to rituals involving *lu* and their perception of flowing water as exemplary of the Buddhist concept of *drelwa* (connection). For many Bhutanese and Khampa, *drelwa* is their everyday understanding and practice of *tendrel*, the Buddhist principle of interdependent origination. A general interpretation of *tendrel* understands all phenomena as arising in dependence on each other rather than existing independently (Samuel 2014, 33). This applies not only to the causes and conditions of existence but also to the parts and attributions of existence. Everything that exists depends on other things, and if it exists at all, it is because it arises due to causes and conditions external to itself. For many of the people in Kham

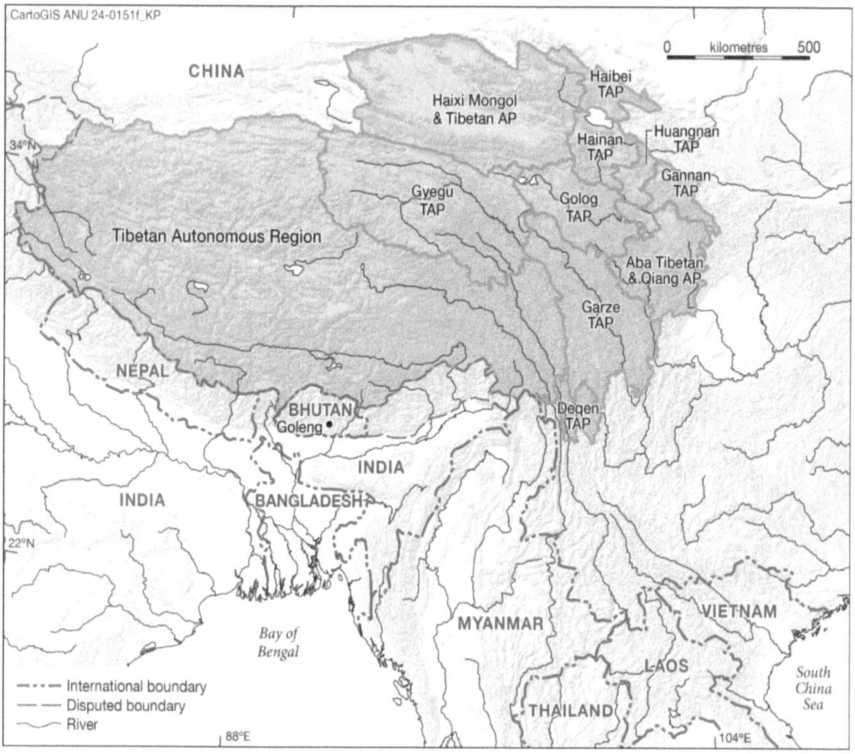

FIGURE 7.1 Map showing Goleng village and the Kham region of Sichuan and Qinghai provinces, and the TAR.

and Bhutan (indeed for much of the Buddhist Highlands), *drelwa* and *tendrel* mean that things and actions experienced in the present moment come about as results of previous instantiations of things and actions; and that subsequent things and actions are influenced by how one thinks and acts in the present moment. Both terms—*drelwa* and *tendrel*—are crucially inflected by Buddhist concepts of *tongpanyi* (emptiness) and *legyudre* (*karma*).

These connections are specific for distinct communities and river reaches. In Kham, for example, many people associate certain headwaters with *lu* (see Box 5b), and their actions around and with these headwaters will be conducted with an awareness of these deities' presence. As the rivers gather volume and flow downstream, people's interactions with the watercourses shift. There are no *lu* to be mindful of along these reaches, but people are more emphatic that these waters are connected to the rivers' downstream reaches. The specificity of relations toward distinct river reaches highlights how categories, such as "river," are not always regarded as a homogenous entity; rather, and based on the principle of interdependent origination, the Khampa and Bhutanese communities that form the ethnographic bases of this chapter point the way to a more complex and relational expression of human–river interactions. More generally, the ethnographic, participant-observation foundation of this chapter illustrates the contribution of anthropology to the transdisciplinary discussion in this book. The question "What is a human being" is necessarily both a "scientific"—in the sense of "empirically based inquiry" that Lévi-Strauss refers to in *La Pensée Sauvage* (2021)—and comparative endeavor that engages arguments framed both by similarities and differences across human communities.

7.1 The Rivers

The two rivers with which these groups have generational relationships, the Dri Chu and the Mangde Chu, are very different. The river that Khampas call the Dri Chu is known as the Chang Jiang, "Long River," on the Chinese plains and the Yangzi internationally. This 6,500-kilometer river is the Earth's third longest and flows from the Tibetan Plateau to the East China Sea. When its wetlands coalesce below the Dangla (Tanggula) Mountains they form three small rivers. From longest to shortest, they are the Dam Chu (Dangqu), Chu Mar (Red River, also known by its Mongolian name Ulan Moron), and the stream locals call the Dri Chu (Tuotuo He in Chinese).

Like other Vajrayana Buddhists, many Khampas pay particular attention to river sources, emphasizing their links to well-known mountains, lakes, or other auspicious places and give precedence to these connections over river length. They rely on authoritative (usually religious) sources like pilgrimage guidebooks to ascertain where a river begins. Based on these guides and teachings, they insist that the Dri Chu is the Yangzi's main channel and true source rather than the Dam Chu. These guidebooks and many locals also insist that the river flows from the mouth of a wild female yak living in a rock mountain among the Dangla Mountains. The association of the river's beginning point with an animal's mouth evokes the sources of the Khangtise's *Chuzhi* (Four Rivers) (see Box 7a at the end of this chapter) and the source of the Ganga (Ganges) River in India at Gaumukh, the Cow's Mouth (Sen 2019).

The Dam Chu and Chu Mar join the Dri Chu just north of Gyegudo (Yushu), a Khampa city in the Yushu Tibetan Autonomous Region, southern Qinghai Province, and the combined flow is also called the Dri Chu. In this region, it flows close to one of its tributaries, the Dza Chu (Upper Yalong), to its east. As the Dri Chu flows through a precipitous valley and the Dza Chu is surrounded by pastures, the pastoralists tend to favor the pastured river over the Dri Chu. The Dri Chu and Dza Chu offer a particularly stark contrast. The Dri Chu displays its famous golden color—this upper Yangzi is called the Jinsha Jiang in Chinese, the "River of Golden Sands"—and flows rapidly through narrow valleys. The Dza Chu is blue-green and braided, meandering across vast grassland plains dotted by the black tents of pastoralists. Much of this section of the Eastern Plateau consists of the river valleys and mountains between them. This is Chuzhi Gangtruk (Four Rivers, Six Ranges), the Khampas' homelands (see Box 7a at the end of this chapter and Figure 7a.1).

The course of the Dri Chu may be long, but it is far from equilibria (see Chapter 2). It descends almost 80% of its gradient in its first 1,000 kilometers (see Figure 1.2, Long Profile). In the past two decades, state-backed hydropower companies have begun exploiting this kinetic energy, installing the globe's second largest hydropower production capacity on the river's upper reaches. The globe's largest hydropower project is further downstream on the same river, in the Three Gorges east of the Sichuan Basin (see Chapter 9 and Xu and Pittock 2020a).

The other major anthropogenic impact on this river system is climate change. The catchment is beginning to experience the effects of global

FIGURE 7.2 *Left*, The Dza Chu at Dzachukha. *Right*, The Dri Chu at its confluence with one of its tributaries, the Tse Chu. Photographs, Gillian G. Tan.

warming, and these changes will increase. Melting glaciers and changes in climate patterns have already augmented the river's streamflow (see Chapters 3 and 4), flooding and droughts have increased (Yang et al. 2022b), and while glacial lake outburst floods (GLOFs, see Box 3a) are not as prevalent in this region as they are in Bhutan, the threat will worsen with warmer temperatures (Zheng et al. 2021).

Historically and culturally, Bhutan's Mangde Chu is a very different watercourse from the Dri Chu. It is a tributary of the Drangme Chu (Manas) rather than a mainstream, and as one of many rivers that provide water to Bhutan, it has supported local communities rather than the large civilizations that the lower Yangzi supports.

Bhutan sits within the catchments of four Brahmaputra tributaries in the Eastern Himalaya. From east to west, they are the Khangbu or Amo Chu (Torsa), a transnational river that descends from the Dromo (Chumbi) Valley in Tibet, the south-Himalaya draining Wang Chu and Puna Tsangchu (Sunkosh), and the transnational Drangme Chu. All of Bhutan's major river systems rise in the high mountains of the Eastern Himalaya, are glacier- and snow-fed, and run north to south, emptying into the Brahmaputra River in Assam, India, before it enters the Bay of Bengal, which is approximately 600 kilometers from Bhutan. The sea's proximity intensifies monsoonal rains in Bhutan's river catchments (see Chapter 3) and helps create intense biodiversity within its precipitous river valleys (Alam et al. 2017, Chapter 5).

The Drangme Chu catchment has tributaries that rise on both sides of the Himalayan watershed. The mainstream runs for 400 kilometers from Tsona County in Tibet, flows down through Tawang in Arunachal Pradesh to eastern Bhutan, and then onto the plains in Assam, where it is known as the Manas River. During this relatively short journey, it crosses three international borders: China–India, India–Bhutan, and Bhutan–India. Its central tributary, the Kuru Chu, rises on the Tibetan Plateau in Lhodrak County and descends directly into eastern Bhutan, where it joins the mainstream. The Mangde Chu is the Drangme Chu's most westerly tributary. It rises in Bhutan's Himalaya from the glaciers around the country's highest peak, Khangkar Punsum (White Snow of Three Brothers, 7,546 meters) and drains central Bhutan, traversing Trongsa and Zhemgang districts through deep gorges. In Zhemgang, it collects the Chamkhar Chu, which drains the northern Bumthang district, before joining the larger Drangme Chu near the Indian border.

This region is much moister than the Kham grasslands and, therefore, much more amenable to agriculture. Although the two rivers and their environments are very different in many ways, they are both being transformed by anthropogenic activities. Climate-change-induced GLOFs are more threatening in Bhutan than in most of the Asian Highlands (Williams 2016). According to the Bhutan Glacial Lake Inventory (BGLI) (2021), the rivers' catchments are studded with 567 glacial lakes with varying potential to trigger GLOFs. Bhutan has already experienced two GLOF disasters in the past three decades, in 1994 and 2015, both along the Puna Tsangchu in western Bhutan (Orlove 2016). There are 130 glaciers spanning an area of 11.86 kilometers within the Mangde Chu catchment (BGLI 2021). As they melt, any of these ice catchments could create a GLOF. Climate change is also affecting the non-glacial or naturally formed lakes further down the Mangde Chu catchment below 4,000 meters above sea level. These lakes have not yet been inventoried, but a few have already receded, while several others have dried up entirely in recent years. Bhutanese villagers attribute this phenomenon to the construction of roads near the lakes and the warming of their land by recent electricity connectivity in the villages.

Along with the effects of climate change, Bhutan's rivers have been transformed by the government's 2005 pledge to develop its hydropower resources. Ninety-nine percent of Bhutan's electricity comes from hydropower, and it exports more than it consumes (IRENA 2019). The dams have caused some displacement of local villages but have been generally well managed, which creates goodwill toward them among the public. However, they, like the dams on the Dri Chu, face an uncertain future as the catchment's climate changes; this has adversely impacted the rivers' aquatic ecologies (Wangchuk et al. 2017), and changed their socio-ecology.

7.2 Naming Rivers

Tibetic-language speakers—which includes speakers of the Khamke (the Khampas' language) and Dzongkha, the national language of Bhutan—have a wide range of terms that map onto the English word "river" (see Chapter 1). In the Khampa nomadic pastoralists' vernacular, one word, *chu*, is used to describe all kinds of water, including river channels, and it alludes to the primary role of water in all. Local informants often describe water's general character by referring to a quote from the well-known Tibetan writer and activist Tsering Woeser, who said of its waters,

> Taste a mouthful of the water of the Land of Snows, it is ice-cold and tasty, fresh and pure, clear and fragrant; when one drinks it, it will not hurt one's spleen or stomach, but it will moisten one's heart. This is Tibetan water with its eight virtues.
>
> *(Woeser 2009)*

In this quote, Woeser evoked the Buddhist master Atisha Dipankara Srijnana (982–1054 CE), who praised the pureness of Tibet's waters and declared that they possess eight virtues: purity, coolness, sweetness, softness, soothingness, peace and harmony, thirst-quenching, and nourishment. Khampas and many other Tibetic-speakers particularly associate these qualities with the rivers' *chugo* (water head), which are thought to be the abode of *lha-lu* (see Chapter 1 and Box 5b). When rivers are close to their sources or headwaters, they embody the qualities of Tibetan water. As rivers proceed downstream and, importantly, pass along and through human settlements, they become increasingly dirty (in a physical sense) and polluted. The Tibetic term *drib*, which describes these changes, also means "shadow," casting a shadow and darkening.

In central Bhutan, terms for rivers echo those used in Kham. The Tibetic word *chu* is employed widely across the Plateau to describe water, including rivers, along with the word *fay*, the word for water in the local Tibeto-Burman Khengkha language. The most common words used to describe rivers combine *chu* with other descriptors. Streams are called *rongchu* (ravine water), larger rivers are known as *chuchen* (big water), and some people also use the term *khang* (snow) to refer to rivers and rivulets. The association between these moving bodies of water and snow suggests ideas of purity, but only major rivers, such as the Mangde Chu, are called *tsangchu* (pure river). This echoes the Central Tibetan term for large rivers, *tsangpo*, and in both cases, probably indicates that their waters are pristine because they originate from mountain glaciers (see Chapter 1). Bhutanese villagers do not, however, extract water from the large rivers for drinking, as people view them as contaminated. Instead, they primarily drink from streams, lakes, and waterholes. Drinking from lakes and waterholes is especially prevalent among cow herders who move their animals from one pastureland to another.

For the Khampas and Bhutanese in this chapter, all rivers are water, but not all water comes from rivers. Most notably, water also comes from *tso* (lakes) and *gyatso* ("a great lake," a sea or an ocean). Furthermore, these waters are classified into *churig dun* (seven kinds of water): rainwater, snowmelt, river water, spring water, well water, water from roots, and saltwater. Another indication of the importance of water sources to these communities is that the source determines six of these seven categories. Only saltwater is defined by its ingredients.

Along with these mundane forms of water, Khampas, Bhutanese, and other Vajrayana Buddhists also speak of *drubchu* (miraculous or attainment water) that local people drink and place on their heads as a blessing. Khampas believe that *drubchu* is produced by *terton* (treasure revealers, see Chapter 1). The late nineteenth-century tantric masters Drime Ozer and Sera Khandro highlighted the self-arising and spontaneous aspect of miraculous water when they wrote:

> At that time, the aroma of a good smell permeated the valley, miraculous water sprang forth from the treasure site, the earth quaked, a sound came forth, and so on ... since I have extracted miraculous water in the manner

> of a treasure, everyone tasted it—it is the accomplishment of this place … I extracted *a self-emergent spring from a dry rock like this*!
>
> (*Jacoby 2015, 110–111*)

Miraculous water promotes well-being. Drinking it and pouring it over one's head brings fortune and helps prevent illness. In this way, it is part of an imaginary of actions associated with "cleansing" and "purity" in Vajrayana Buddhist lifeworlds.

7.3 Pragmatic Activities with Rivers

Among Tibetan nomadic pastoralists, water is vital not only for people but also for plants and animals, primarily the yak (*Bos grunniens*), without which life on the high-altitude grasslands would not be possible. Yaks provide pastoralists with milk that is transformed into various food products such as butter, cheese, and yogurt. The pastoralists also use their animals' hair, fur, hide, skin, and meat. These support a nomadic existence based on movement and black yak-hair tents (Tan 2020).

Highland pastoralists tend to be very specific about where they extract water for drinking, cooking, and cleaning. They only take water from streams, as close to the stream's source as possible, but not from the source

FIGURE 7.3 Inside a black tent of Kham nomadic pastoralists. Photograph, Gillian G. Tan.

itself. The source and the headwaters around them are the abode of *lha-lu* (and especially *lu*; see Chapter 1 and Box 5b). Kham's pastoralists have a strict code of orientation whenever black tents are pitched in a new location, and their choices encode an awareness of water flow when they collect drinking water. No single dwelling should monopolize a stream; therefore, all tents should be placed at a reasonable distance from waterways and not inhibit access by people living in other black tents. Yaks should be herded as far away from drinking streams as possible and not in line with the flow of their water. So, too, any ablutions must be undertaken at a significant distance from any drinking water source or flow.

Khampa pastoralists do not wash near lakes and rivers or bathe in them. Any bathing is done rarely, usually before the new year, at certain hot springs chosen from the many that dot the seismically active parts of the Plateau. All these activities are consciously directed toward separating drinking water from contaminating sources, whether human or animal. Water cleanliness is of utmost importance. Like the Bhutanese farmers, they do not extract drinking water from large rivers as they view such places as unclean. Some Khampas even regard the flowing movement of large rivers as a means to remove dirt and debris. In addition to underpinning the pastoralists' way of life, larger rivers may also impede it, blocking their passage for herding or trade, especially when a river is in flood. Historically, pastoralists used rope bridges and yak-skin boats to cross parts of waterways that horses could not manage (see Chapter 6). Those that plied the Dri Chu usually sat about ten people but were only rarely used for down-river transport as the Dri Chu's flow was so rapid. Such boats are no longer used regularly because they have been replaced by large concrete and steel bridges, often located at traditional crossing points.

In the Mangde Chu catchment, villagers use rivers for several pragmatic purposes. Their interactions with the river are modulated by the strength of its flows and currents. These are affected by the stream's hydrology and the season. The villages near the river (such as Goleng, Tagma, and Shobleng and those on the ridges above them) are all gifted with streams that can swell significantly in the summer due to heavy rain. It is common to have ephemeral, flowing bodies of water in the villages following heavy precipitation delivered by the Indian Monsoon. Villagers channel weak flowing streams to irrigate their paddy fields and turn the water-powered mills that grind the *druna gu* (nine types of cereals) that they grow and eat: rice, maize, wheat, barley, buckwheat, millet, pulses, mustard, and amaranths.

By contrast, strong flowing streams and rivers impede villagers' connections with other people and places. Before 2009, when a road connecting Goleng with the town of Tingtibi was constructed, the only way for those on the Mangde Chu's eastern bank to reach Tingtibi was by using a metal suspension bridge. As this bridge had a low load-carrying capacity, the people of Goleng, used the river to transport domestic animals such as cattle

FIGURE 7.4 *From left*, Bridge over the Mangde Chu River in Goleng. Paddy fields in Goleng. Mangde Chu in Goleng. Photographs, Kelzang T. Tashi.

and horses. The river's strong current (particularly during the long monsoon season) made this problematic. Hence, the Golengpas recall only migrating cattle to new pastures during the dry winter and fording the river in one of its shallowest sections. Even then, only adult cattle could generally manage the crossing; yearlings and heifers had to be transported via a bamboo raft. The Golengpas also used bamboo rafts to cross the river when the suspension bridge was damaged or otherwise unavailable. Those who wished to avoid the river during these periods preferred a rope-crossing method, still used across the Highlands, but arguably more precarious even than bamboo rafts.

Although the Golengpas and many other Vajrayana Buddhists believe that fishing produces negative karma (much like hunting, see Box 6a), some resort to fishing to supplement their livelihood. In recent years, however, it has become even rarer to fish rivers such as the Mangde Chu because legislated restrictions on fishing have reinforced social restrictions. The Department of Forests and Park Services now controls fishing in most of Bhutan's rivers, and potential fisherpeople must acquire a permit or license before catching a limited number of fish. Those caught fishing without a permit are penalized. This combination of karmic sanction and government restriction has encouraged most villagers who wish to eat fish to purchase them from the nearby markets, where they are sourced from India.

7.4 Relationships with Riverine *Lha-Lu*

In central Bhutan, Kham, and across much of the Highlands, ritual activities associated with rivers are not restricted to pragmatic concerns of the present life. Some ceremonies are designed to promote positive *tendrel* or interdependent origination—that is, thoughts and actions that influence the complex of subsequent thoughts and actions in positive ways. Thus local Bhutanese and Khampas imagine the achievement of higher rebirth and the unfolding of positive (or negative) karma. Therefore, along with sanctions on certain interactions with water described in the previous subsection, Golengpas and other Bhutanese villagers also use smaller streams to turn *chukor mani* (water-driven prayer wheels) to create positive karma. Using the water to turn the wheels connects this-worldly activities in the fields with the creation of karma for their future lives. The cylindrical water-driven prayer wheels are filled with thousands of Buddhist mantras and turned by streamflow. According to the Golengpas, the water turning them multiples the prayers' benefits by a millionfold or more and conveys blessings to people and places nearby. The merit generated by the water-driven prayer wheels can create a harmonious relationship between the villagers—who see themselves as *dronpo* (or *jonpo* meaning guests) on the land—and the local deities, whom they conceive as the land's *nepo* (supernatural inhabitants) (Tashi 2023; known elsewhere as *sadak* (earth lords) (see Chapter 1). In this sense, the flow of water generates blessings that mediate *zhide* (peace) between humans and wrathful *lha-lu*.

Many Highlanders have a fraught relationship with the *lha-lu*, who are their homelands' original inhabitants and, therefore, controllers (see Chapter 1 and Box 1a). Humans need to develop multiple relationships with various classes of *lha-lu* within their environment and not offend or insult them, as this could have disastrous consequences for themselves, their families, and their villages. The repercussions of insensitive actions operate in various ways. The strong karma of offending a buddha or bodhisattva—also known as *jiktenledepe lha* (supramundane deities)—incurs immediate repercussions. Buddhas are often associated with temples but also live in particularly sacred *ne* (*lha-lu* abodes), usually mountains and lakes. Many more *jikten lha* (worldly deities) live throughout the environment. These beings play a vital role in the region's Indigenous and geomantic lineages but are often understood to work for Buddhist adepts such as Padmasambhava (see Chapter 1). Offending them incurs less negative karma, because these deities not only feel the effects of human activities—mainly through actions that pollute or purify—but also respond with anger, jealousy, or beneficence as a result of these actions. Worldly deities may inflict great harm or bestow great fortune. The key to Highlanders' relationships with deities lies in the *drelwa* (connection) between them (Tan 2020).

The same is true of the *nepo* (supernatural inhabitants), *sadak* (earth lords), and *lu*, who are primarily associated with the subterranean worlds. *Nepo* and *sadak* can be used as a general category to refer to a region's *lha-lu* or specifically to those *lha-lu* that live underground (see Chapter 1). As we learned in Box 5b, the *lu* are primarily associated with aquatic spaces. They are often associated with the Indic *naga* (Mumford 1989) and are one of many water serpent beings that are a motif found in many other cultures, including Indigenous Australia, but there are significant differences in how people interpret and relate to these beings. In Kham, Bhutan, and many other communities across the Asian Highlands, *lu* are particularly resident in headwaters and springs, which are generally associated with "cleanliness." *Lu* are territorial about their abodes, and disturbing them (see also Ramble 2008) or visiting them when you are defiled by various pollutions, such as sex, birth, and death (Tashi 2021) will disrupt the harmonious relationship between humans and *lu*. When subjected to human pollution, *lu* may inflict varying degrees of illness, called *lu-ne* (*lu* disease). *Lu* disease usually manifests as skin afflictions such as *dze-ne* (leprosy and boils), as well as arthritis and aching joints. When biomedical treatments cannot remedy them, such illnesses may require a ritual intervention. Purifying rites may be performed by a Buddhist *lama*, a geomancer, or a Bon priest (Tashi 2023), depending on the custom in a particular area. Everyday activities that are believed to anger *lu* include polluting lakes, disturbing the serenity of their abodes, and ritually offending them, primarily by neglecting to make regular offerings.

Beings from the various categories of *lha-lu* can be found across the Highlands, but some cluster in certain areas. For example, the glacial lakes among the headwaters of Bhutan's major river systems are the home of many *tsomen* (lake spirits) who resemble mermaids. The lake spirits are hybrid creatures with a woman's upper body and a lower body that resembles a serpent or a fish. Four lake spirits are said to be the guardians of Tso Ngonpo (Qing Hai or Kokonor) in the Plateau's northeast (Buffetrille 1994), but their stories are distinct from Bhutan's lake spirits lore. The Bhutanese lake spirits are said to be transformed purebred *lu* who are related to the *lu* that live in central Bhutan, primarily in the non-glacial lakes scattered across the country (Tashi 2023).

The lake spirits and all *lu* types are alike, however, in their need for constant but appropriate propitiation. This will placate them and keep them well-disposed toward humans, conveying good fortune and prosperity. Conversely, if humans anger them, the *lu* will trigger natural disasters in the village, and people may suffer from incurable illnesses.

Like many places elsewhere in the Highlands, the central Bhutanese believe that *lu* dwell underwater and underground (see Chapter 1). When they live underground, they operate in much the same way as *nepo* and *sadak*. Reiterating the categorization of *lu* in the *Lu Compendium* (see Box 5b), a Bhutanese Bon priest from Bumtang explained that the various colors of *lu* correspond to certain psychological dispositions; white and yellow *lu* are well-disposed toward humans, while multicolored and black *lu* are inherently malevolent (Tashi 2023). He also localized this description by associating the four color-coded classes of *lu* with local snakes, which are more prevalent in central Bhutan than on the Plateau, which suggests that white and yellow snakes are more disposed toward humans than multicolored and black snakes.

7.5 Ritual Activities with Rivers

When Highlanders offend *lu* and other non-human entities that share their environments, they often call on expert help to placate them. These experts can be Buddhist, Bon, or geomantic specialists. An example of this occurred in a Khampa pastoralist community situated between the Dri Chu and the Dza Chu. After several community members contracted leprosy and others developed boils, its members attributed these afflictions to a local *lu* and invited a local Buddhist teacher to perform a ritual cleansing of their stream.

The pastoralists said the *lu* had been angered and could only be appeased by offerings of *sang* (fragrant smoke). The Buddhist teacher arrived and ascertained that the most appropriate ritual for this situation was not a *sang* offering. Instead, the locals should build and consecrate a *latse* (earth and stone cairn used for *lha-lu* propitiation) in the stream's headwaters, which were the *lu*'s abode. For several hours the pastoralists erected the stone cairn, and the lama consecrated it. During the ritual's final section, the lama placed a stone on top of the cairn. A few days later, the stone reportedly fell into the water and killed a snake. After that, the community was free from leprosy and boils; the *lu* had been appeased, and their anger abated.

Another example of rectification occurred when two floods in Bhutan (Orlove 2016), and the subsequent destruction of the paddy rice and other crops across the country by unseasonal rain, were attributed to the wrath of angry local deities, particularly the *tsomen* (lake spirits) and *lu* of upriver lakes whose habitat had been settled by humans.

To pacify the *lu* and lake spirits and to restore a harmonious coexistence between *lha-lu* and humans, communities across the Highlands engage in a practice called *dam* (*sealing*). Toni Huber (2004) described its application at Tsari, a *ne* (abode) on the border between southern Tibet and Arunachal Pradesh, which is particularly sacred because the buddha Demchok (Chakrasamvara) lives there. At Tsari, several classes of people were forbidden to access certain parts of the mountain.

This spatiotemporal restriction is embedded in Bhutan's cultural and religious landscape, within which the *lha-lu*'s volatility significantly shapes everyday concerns. For example, the residents of Buli, 25 kilometers north of Goleng, practiced *tsodam* (lake sealing) for nearly four months of the year. The ban lastes from immediately after the transplantation of paddy rice in spring through to the harvest season. During this time, not only visitors but also the residents themselves are banned from visiting the lake.

Sealing regions further north also prevents the *tsomens*' wrath from unleashing heavy and untimely rains and other natural disasters that would lead to poor harvests. In these regions, the locals practice *ridam* (mountain sealing), which restricts entry to entire mountain ranges that are the homes of local deities, and thus embargoes its populace from collecting forest products and allowing their domestic animals to roam in the mountains. These ritual practices are also common in parts of eastern Bhutan (see also Kuyakanon and Gyeltshen 2017).

7.6 Central Bhutan's River Mortuary Practices, an Example of *drelwa*

While the corporeality and physicality of connections are evident in almost all the Highlanders' pragmatic activities involving rivers, some rituals prioritize *drelwa* (connection), Buddhist funerary rites are an example of this kind of ritual. The four most widely known methods for disposing of corpses in much of the Highlands are burial, immersion, cremation, and sky burial (Gouin 2012). Sky burial was and is the common practice among Khampa pastoralists, and many—though by no means all—highly ranked monastics and others deemed deserving have been cremated (Gouin 2012, 52). In central Bhutan, however, inhumation was limited to those who died due to certain diseases, particularly leprosy and other epidemics. Sky burial was uncommon, even among Buddhist retreat masters living in the mountains. Instead, cremation was and remains the most popular method for disposing of adult corpses, and children's bodies were and are immersed. Hence, unlike in Tibet, cremation in central Bhutan is not reserved for religious adepts or high-status people; nor is it associated with individuals of lesser social status.

Like in other parts of Bhutan, in Gomeng and the villages around it, bodies of adult persons are cremated at the village crematorium after an astrologer identifies an auspicious day. The extent and elaborateness of the funerary rites differ between categories of people (see Tashi 2021 on these categories). The bodies of a few realized Buddhist adepts might be preserved inside a *kudung choten* (reliquary stupa). Some of these corpses are left intact before being interred in the reliquary stupa, and others are first cremated on private land or in a crematorium. These adepts' mortal remains (particularly bones) are often used to make *tsatsa* (small religious images made from clay pressed into molds) that will be placed inside their reliquary stupa. The community believes that preserving the adepts' remains in the reliquary stupa ensures the continuity of their blessings and teachings through a human rebirth. If they are not anchored to the world through the reliquary stupa, they may take a higher rebirth in buddha land that is inaccessible to all but the most advanced practitioners.

Cremated adult human remains are more commonly cast into a "pure river." Different families and communities adopt distinctive approaches to this disposal. Some families may deliver their family members' remains directly to the Ganga in India, which is considered particularly holy. In the villages in Trong and Nagkor counties, ashes are collected immediately after Buddhist ritualists perform funerary rites and then are carried to the Mangde

Chu and disposed of there. One of this chapter's authors, Kelzang Tashi, has also witnessed other families making *tsatsa* from their family members' remains (a process that brings them and the deceased karmic merit) and then disposing of them either on a mountaintop or, more commonly, in a river. The disposal of *tsatsa* in rivers calls to mind both the processes associated with adepts' funerals and the local practices of submersing ashes in the river.

Buddhist clergy and laypeople have strict prohibitions against placing human remains in smaller streams and rivers. Local tradition holds that leaving mortal remains in streams and lakes will bring demerit to the disposer and cause the deceased to be reborn in lower realms, either as an animal, spirit, or hell-being (see Chapter 1). Casting human remains in streams will result in the sedimentation of bones or ashes on its floor, thereby leading the deceased person to reincarnate as an earth spirit, particularly as a demon that dwells in ponds and slow-flowing streams. This subclass of local deities is wrathful and unawakened, and locals do not wish their relatives to take birth in this form. Instead, placing them in "pure water" (i.e., larger, purer rivers) will make the deceased more likely to enjoy rebirth in higher realms.

By placing their relatives' ashes in the Mangde Chu, the residents of Zhemgang ensure they do not take lower rebirths as local spirits that live in small streams and are instead carried by the rivers' strong flow to the Great Indian Ocean, from where they can attain rebirths in higher realms. The rivers' flow connects their deceased with the Indian plains' rivers, such as the Ganga and Brahmaputra, which are viewed as purer and holier than those of the Highlands. Indeed, the connections, or *drelwa*, with rivers are transcendental and connect the present with the future, local deities with higher gods, and the present life with the afterlives of people who inhabit the valleys drained by the big rivers.

7.7 Conclusion

Both the Khampa pastoralists and the Bhutanese agriculturalists, whose riverine practices this chapter describes, understand themselves to be connected to the river rather than extracting from it. Even though their engagement with the water that passes through their homelands is pragmatic, it occurs within the framework of *drelwa*, connection, and *tendrel*, interdependent origination, rather than seeing the water as an exploitable resource. The Khampas, for example, always draw their water from headwaters and springs but never use this water in situ because they respect the headwaters and do not want to sully them. The orientation to the well-being of others is clear: clean sources avoid making streams and rivers dirty for other persons; they also avoid "polluting," and thus angering, the *lu*. This practice holds different valences, some that are pragmatic and others that reflect religious imaginaries.

These communities consider headwaters, associated springs, and large rivers pure, but in different ways. Neither the Khampas nor the Bhutanese would drink from large rivers. Still, their flows and movement have a cleansing effect, and they foster *drelwa* that transcend the pragmatic and physical concerns of life. River mortuary practices in central Bhutan highlight the connection between these large rivers and the religious and soteriological aspects of life by acting as a conduit that takes the deceased to better rebirths.

Although we have distinguished these communities' complex and multifaceted relations with their rivers, these distinctions are not always reflected in the communities' everyday interactions with water. In these situations, the differences between the "physical" and the "metaphysical," "natural," and "spiritual" are never fixed and impermeable. Instead, through significant and intentional actions, both communities demonstrate the importance of connections between and among people, animals, *lha-lu*, and water. While it connects, *drelwa* also reveals the difference—or distance—between purity and pollution, upstream and downstream. The Khampa pastoralists are especially clear about this. They understand that actions upstream impact what happens downstream. They know flowing water is "clean" before it enters dense human settlements and becomes "dirty" by the time it leaves them. Flows of water are one way that such connections are generated and maintained. For both the Khampas and the Bhutanese, rivers provide an analogous backdrop for a broader cultural imaginary of connections.

One similarity is apparent: on the roof of the world, close to the pristine source of rivers that flow from the Tibetan Plateau and down through Bhutanese valleys, there is an integral sense that the movements of rivers—their ability to connect people who live near them to larger (albeit unknown) worlds—is important and, in some ways, sacred. This sense of sacrality is not the same as the accretions of holiness deposited on the Highlanders' *ne* (abodes), particularly their mountains and lakes, and South Asian Lowlanders' rivers like the Ganga. Still, given its analogy to *drelwa* and its position within broader sacred geographies, their flow continues to be significant. Importantly, this sense of *drelwa* and its inherent mutuality of causes and effects place Highlanders' interactions with rivers at odds with the logic of management and control. For them, rivers exist in webs of meaning that are intimately connected with their daily lives. By contrast, the logic of river management approaches rivers as resources to be used solely for utilitarian purposes.

BOX 7a THE *CHUZHI* (FOUR RIVERS) (CO-WRITTEN BY LHAMO KYAB)

Both the Yarlung Tsangpo-Brahmaputra and Dri Chu (Yangzi) rise near the headwaters of other large river systems. The Yarlung Tsangpo's headwaters mingle with those of the Indus and Ganga (Ganges) at the Khangtise (Gangdese) Mountains' base in the Tibetan Plateau's southwestern corner. The Dri Chu rises at the base of the Dangla (Tanggula) Mountains in the Plateau's northeastern corner, along with the Da Chu (Mekong) and western Gyelmo Ngul Chu (Salween). The Ma Chu (Huang He or Yellow River) rises immediately to its north.

Highlander cosmographies associate these two sets of complex interconnected river headwaters with the *chuzhi* (Four Rivers) that flow from Sumeru, the mountain at our world's center. Although Buddhists imported the idea of Sumeru and its Four Rivers along with much other Indian cosmography to Tibet, it was Bon practitioners who first associated this schema with the Highlands' rivers. They applied it first to the rivers that rise below the Khangtise Mountains in the Plateau's southwest and then to the rivers that rise below the Dangla Mountains in the Plateau's northeast.

7a.1 Chuzhi (Four Rivers) in Western Tibet

Khangtise (Kailash) is one of the Asian Highlands oldest (see Chapter 2) and most sacred mountains. It is the highest peak within the Khangtise Mountains, which stretch from the southwestern to the center of the Tibetan Plateau. Four religious traditions view this mountain as sacred: Bon, Buddhism, Hinduism, and Jainism. For the Bonpos, the mountain is a link to their *axis mundi*, or world center, Yungdrung Tseg, which sits at the center of the Bon pure land, Olmolungring (Karmay 1998, 107). For the Buddhists, it is the *mandala* of the tantric deity Demchok (Skt. Cakrasamvara), and analogous to Buddhism's Indic-origin *axis mundi*, Sumeru. For the Hindus, it is Kailash, the home of the powerful god Shiva. For the Jains, it is the site where their first religious teacher attained liberation (McKay 2016).

Khangtise is perhaps best known as a Hindu sacred site, and it is known internationally as Kailash because of this association, but the Bon and Buddhist cosmographies of this mountain are much older; Hindus and Jains only began visiting the area in the nineteenth century, after calling other mountains on the southern side of the Himalaya "Kailash" (McKay 2016). When they arrived, they encountered well-established Bon and Buddhist cosmographical lore that included the mountain and its complex and shifting hydrology. This model incorporated the two large lakes Mapam Yutso (Manasarovar) and Lagngar

(*Continued*)

Tso (Rakshas Tal) at its base and parts of the headwaters of three large river systems—the Indus, Ganga, and Brahmaputra. In these cosmographies, Mapam Yutso is an abode of deities, Lagnar Tso is the home of demons, and the region's complicated and changing hydrology is split into four discrete river systems flowing in four directions. This is the same number of rivers that descend from the Buddhist Sumeru (the massive mountain at the center of the earth in Indian cosmologies). Like Sumeru's rivers, Khangtise's waterways are said to flow from single sources in cardinal directions that resemble the mouths of Indian animals. The Indus, known locally as the Sengge Khabap (Lion Mouth Falls), flows from a spring resembling a lion in the north. The Sutlej, an Indus tributary, flows from the westerly Langchen Khabap (Elephant Mouth Falls), which resembles an elephant. The Karnali, a trans-Himalaya Ganga tributary, flows from the southerly Macha Khabap (Peacock Mouth Falls) and resembles a peacock. The Yarlung Tsangpo-Brahmaputra's marshy headwaters are said to flow from a single easterly spring that resembles a great horse, so this river is known locally as Tachok Khabap (Great Horse Mouth Falls).

This schema incorporates two common cosmographic images in Indic traditions: the enumeration of rivers and the association of river sources with animal mouths. This "source as mouth" image inverts the European idea of a river's terminus as its "mouth." The conceptual history of this Indic tradition in the Highlands is, however, surprising, as the first Tibetans to apply the Indian Buddhist four river, animal-mouth schema to Khangtise's rivers were the non-Buddhist Bonpos, whose religion developed in Tibet, not India. It was the Bonpos who named the headwaters around Khangtise the Khabap Zhi (Four Rivers Falling from Mouths).

7a.2 Chuzhi (Four Rivers) and Gangtruk (Six Mountains) in Eastern Tibet

After establishing Khangtise's cultural geography, centuries later, during the seventeenth to early twentieth centuries, Bonpos and Buddhists reused the Four Rivers trope in their representations of the eastern Tibetan Plateau (Blo bzang lhun sgrub rdo rje 2018; Kyab 2023).

The locals call this region Kham, and it has strong religious, linguistic, and cultural links to Central Tibet (see this chapter's main text, Chapters 6 and 8). The Bonpos redeployment of the Four Rivers motif in this region was part of a larger appropriation of Indic and Central Tibetan cosmographic paradigms in Kham as it became a center of religious activity. This re-configuration also included the creation of the 25 great sites of Kham by the *terton* (treasure revealer) Chogyur Lingpa (1829–1870) and the renowned scholar Jamgon Kongtrul (1813–1899) (Gardner 2006). The rivers' enumeration was much

(*Continued*)

more dependent on the region's riverscapes than cosmographic transplants from elsewhere, but the practice of enumeration became more diffusely and fluidly applied than it had been around Khangtise. Eventually, the idea of the four rivers became entangled with another descriptive enumeration, the name Chuzhi Gangtruk—"Four Rivers, Six Ranges"—a synonym for the entire Kham region.

These riverine schemas combine human perception and hydrology. Hydrologically, the headwaters around Khangtise are the complex, interconnected, shifting beginnings of three major river systems, and Chuzhi Gangtruk—Four Rivers, Six Ranges—covers the headwaters of seven rivers. Still, cosmographies and many modernist geographers divide them into discrete watersheds with four singular sources. This conception of four rivers with distinct sources, courses, and outputs informed human relations with both regions' hydrologies. Combining multiple river systems in a single geographical or cosmographical schema also connects them; their integrated basins are often described as a single country. But neither the rivers' flow nor the conceptions of their country are as fixed as their repeated evocation suggests. The idea of the Four Rivers persists, for example, but over the centuries, its attribution and river naming conventions have been remarkably fluid and reusable. Indeed, the practice of enumeration proved so adaptable that its repeated uptake and alteration makes it impossible to summarize the various four river schema across the Highlands here.

TABLE 7A.1 Chushi Gangtruk's Rivers and Ranges

Mountain range	*Rivers it separates*	*Provence*	*Main mountain*
Zalmo Gang	Ma Chu, Dri Chu	Qinghai	Nyenpo Yutse
Tsapa Gang	Gyelmo Ngul Chu, Da Chu	Yunnan	Rongtsen Kawa Karpo
Markham Gang	Da Chu, Dri Chu	Sichuan	Drala Chungchen
Pobor Gang	Dri Chu and Dza Chu-Nyag Chu (Upper and Lower Yalong)	Sichuan	Trozil Tromchi
Marja Gang	Dza Chu (Upper Yalong), She Chu (Xian Shui)	Sichuan	Trisho Kawalung
Minyag Rabgang	Nyag Chu (Lower Yalong), east Gyelmo Ngul Chu (Dadu)	Sichuan	Minyag Gongga and Zhara Lhatse

(Continued)

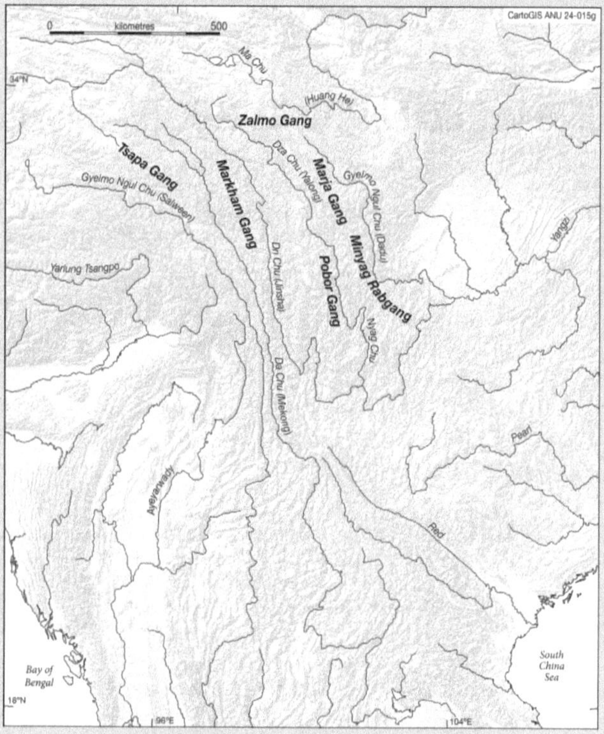

FIGURE 7A.1 Map of Chuzhi Gangtruk, Four Rivers, Six Ranges.

8

TERRITORIALIZING RIVERS

(Co-written by Dechen Palmo and Alexander E. Davis)

1800 CE—Now

Construction sites have mushroomed across Asian Highland river catchments in recent decades. On them, workers labor in the hot sun, monsoonal rains, and freezing winters to build dams, irrigation systems, embankments, riverside roads, and trains, barrages, and tunnels for state and state-affiliated corporations. Engineers in hard hats stand near them, scrutinizing blueprints produced by whole systems of laws, policies, finances, education, and expertise. The hard infrastructure the laborers build depends on the soft infrastructure (government and corporate services) that produced the blueprints. In contemporary political ecologies, these hard and soft forms of infrastructure work together to regulate the Highland river systems' geological, aqueous, ecological, and social flows.

The pamphlets and reports describing these projects invariably highlight their cutting-edge technology and call them "future-focused." But in many ways, these constructions are not so much future-focused as legacy projects. Their ultimate blueprint was developed a long way from the Highlands in lowland Europe two centuries ago when there was a pronounced shift in human–river relations. Humans had regulated rivers' flows before this, but not as effectively. Working within new social and knowledge paradigms such as private ownership, nation-state territorialization, and capitalism, the Europeans came to see rivers as resources that could be enclosed, extracted, securitized, and commodified. As a commodity, rivers' flows must

DOI: 10.4324/9781003392033-8

be regulated to provide consistent, non-seasonal flows and dependable water, power, and monetary surpluses. European governments came to see their rivers as a national asset, imagined political control over them could be absolute, and set their engineers the task of materializing this control (Cioc 2002).

After developing this new approach to European rivers, they then exported the idea of the modern, thoroughly regulated river to the world via their empires and colonies. Within a few centuries, the forces of territorialization, border demarcation, resource extraction, industrialization, and development had begun to entangle Highland river systems. The mid-twentieth-century decolonizing project presented a chance to undo these paradigms, but its leaders did not take it, choosing instead to maintain many colonial-era political and economic paradigms.

The following three chapters will explore this centuries-long global reshuffling and its consequences. As we will learn in Chapters 9 and 10, the colonial approach to rivers created dire socio-ecological results. It has only been "future-focused" in that it has locked the catchments into novel ecosystems (Hobbs et al. 2009) in which governance regimes are attempting to regulate the Highlands' dynamism through technology. This technological regulation is ultimately unsuccessful because it operates asynchronously from Highland Earth systems, exaggerating hazard risks and undermining vibrant societies and cultures. Unlike the Highlands' long-lived, light-touch socio-cultural systems that viewed environments as endowments, the Highlands' current governance systems tend toward increased control of resources, and even their conservation efforts operate from within this resource and extraction paradigm.

While the next two chapters look at the rivers' contemporary regulation, this chapter provides its historical background. It traces how histories of the Dri Chu and upper Brahmaputra catchments converged and diverged through the nineteenth and twentieth centuries. Initially, in the 1800s, they both accommodated multiple overlapping states and societies that operated through political, religious, trade, and family allegiances. These included smaller kingdoms like Nangchen, Chakla, Derge, Bathang, and Lithang in the Dri Chu Catchment and those in the upper Brahmaputra, Bhutan, Tawang, and Po. The Ganden Podrang government (1642–1959), based in Lhasa and led by the Dalai Lamas, was the most powerful Highland polity, politically and religiously (see Chapter 6), so the smaller states had tribute-like relations with it. All these states, including the Tibetans, also maintained relationships with two formidable Lowland empires, the Qing to the east and British India to the south, both of whom had designs on the Highlands.

The British were initially interested in "discovering" and mapping the strategically and scientifically important Highland rivers (Simpson 2023; Gardner 2021). However, the eastern Highlands' "frictions of terrain" (Scott

2009) posed a challenge to them, so rather than administering the Highlands directly, they schemed to establish it as a buffer zone between themselves and the Qing Empire (Gohain 2020b). The Qing Empire, which had previously established a *choyon* (patron–priest) relationship with the Dalai Lamas, was not impressed with British meddling in its sphere of influence and began developing its own imperial strategies for the Highland river catchments.

An important part of both the British and Qing governments' plans for the Highlands was trade networks. By the late nineteenth century, Tibetans, Chinese, and British traders transported, bought, and sold high-demand goods such as musk and wool across the region. This trade only increased in the twentieth century (Giersch 2010, 2019; Harris 2013), and the subsequent enclosure of land and water for state and commercial gain intensified extraction cycles. The Lowlanders who sought influence and wealth in the Highlands during this period were not constrained by subsistence agropastoral practices or religious beliefs. They operated at a distance and possessed more capacity to extract and accumulate resources, which they could then redeploy for further imperial expansion and extraction (Saito 2021).

The two catchments' histories diverged after the Qing Empire fell in 1911. Britain assumed regional dominance, but at the Shimla Conference (1913–1914), it failed to convince the new Chinese Republican government to accept an "Outer Tibet" ruled by the Dalai Lama's government and a border along the Eastern Himalayan peaks. When the People's Republic of China (PRC) (1949) and the Republic of India (1950) were founded, both inherited this border dispute and fought a war over it in 1962. Between the 1950s and 1970s, China and India's competitive territorialization consumed the catchments' earlier polities, leaving Bhutan the eastern Highlands' only remaining independent state.

Within a century, catchment governance had been transformed from a network of intersected polities to imperial borderlands and, eventually, three sovereign states. Once an international river connecting diverse societies and cultures, the Dri Chu has been reimagined as one segment of China's national Yangzi River. In contrast, the rivers of the Brahmaputra catchment, once trade and migration corridors, had been divided between four sovereign states: China in the north, India and Bhutan in the center, and Bangladesh in the south.

Colonization, trade, and territorialization introduced modern river regulation to the Highlands. The colonists introduced this technology to Lowland rivers in the nineteenth century, but it took over a century before they could use the Highlands' Gross Potential Energy (see Box 3a). These developments in hydropower extraction not only changed the rivers' character but also intensified their geopolitics—or more precisely, "hydropolitics" (Allouche 2020)—as the region's states made competing claims on the rivers' hydropower.

China and India began building large hydropower projects in the lower Highlands in the second half of the twentieth century. These large hydropower projects provided practical benefits for them and symbolized state power and development. India encouraged Bhutan to embark on its own large dam program simultaneously. Although the final report of the United Nations-sponsored World Commission on Dams (2000) discouraged large dam building, citing the dams' impacts on displaced communities and ecosystems, China and India continued to build large dams throughout the 2000s and gradually built them further into the Highlands. Tensions grew between provincial-level governments within China and India and between these two large states, who competed to enclose and exploit the transboundary Brahmaputra catchment (Davis 2023).

In the last fifteen years, China, India, and Bhutan have changed track, investing in run-of-the-river hydropower constructions that combine smaller reservoirs with tunnels rather than twentieth-century, multi-purpose large dams. As this chapter explains, these dams are supposed to decrease the tensions of hydropolitics and displace fewer people during construction, but this dam type can also cause long-term social and ecological damage in this tectonically active and minoritized region (Huber 2019; Li et al. 2022).

While this gradual enclosure has been the general flow of river history over the past two centuries, this chapter also recognizes its Highland countercurrents, particularly the Indigenous communities that have preserved their relational knowledge lineages in the face of colonization and marginalization, and the force of the rivers themselves. Contemporary states may claim to be naturalized, monolithic sovereign entities with defined boundaries, but this chapter will show this is a cultivated illusion. States are historically contingent and impermanent, and their borders are only maintained by labor and force, often at the expense of minority groups who occupy bordered lands. Many of these communities continue to resist territorialization in multiple ways, and the rivers are their allies. Cycling between the earth and the sky (see Chapter 3), they can never be fully enclosed, commodified, or owned. Even as states and corporations seek to regulate the rivers, social and physical forces loosen their constraints.

8.1 Shifting Polities (1850–1913)

As we learned in Chapter 6, the fifth Dalai Lama, Ngawang Lozang Gyatso (1617–1682), founded the Tibetan state in Lhasa in 1642. He did this with the support of his sponsor, the Khoshut Mongol ruler, Gushri Khan (1582–1655), who had defeated the Tibetan King of Tsang in battle. This victory established Gushri Khan and the Dalai Lama as Central Tibet's most powerful rulers and their school of Vajrayana Buddhism, the Geluk, as its dominant creed. The Dalai Lamas' *choyon* (patron–priest) relationship with

Gushri Khan was superseded by an ongoing relationship with the Qing Emperors (1644–1911).

The Ganden Podrang did not exercise sovereign control over a defined territory. Instead, it operated through a network of allegiances. Monasteries, aristocrats, and the estates of reincarnating teachers managed farmland and rangeland, which they tenanted out in exchange for tax obligations. The tenants' obligations to powerful institutions sometimes overlapped, and some aristocratic families and monasteries located in Ganden Podrang territory also held estates outside its sphere of influence (Samuel 1993; Schwieger 2023).

These networks stretched across the Highlands, often blended with religious affiliations, into the lands of the river systems' smaller polities. These smaller polities usually maintained autonomous relationships with the Qing and British empires as well as the Ganden Podrang. In the Dri Chu Catchment, the Qing court had given the special title of *tusi* to the rulers of four kingdoms: Chakla (whose capital was at Dartsedo), Derge, Bathang, and Lithang. There were also smaller polities at Nangchen in the catchment's far north and Gyelthang in its south (see Figure 8.1). These kingdoms benefited greatly from the trade routes established by Qing emperor Kangxi (r. 1661–1722) through their territories. By the middle of the eighteenth century, traders were transporting musk, deer antlers, pelts, yak tails, wool, and medicinal herbs along these routes and beyond through exchange networks that stretched around the globe (Giersch 2010).

The Tibetic-speaking states that refugees from Central Tibet developed in the upper Brahmaputra on the south side of the Yarlung Tsangpo-Brahmaputra watershed during the seventeenth century had a different character and more complicated relationships with Lhasa. The Drukpa Kagyu reincarnate Zhaptrung Ngawang Namgyel (1594–1651) founded Bhutan after he defeated an invasion by the Kings of Tsang in 1627. He subsequently resisted invasions by the Ganden Podrang in the seventeenth century. Competition between Bhutan and the Ganden Podrang led to the latter's annexation of the Drangme Chu (Manas) headwaters around Tawang, where it established a Geluk monastery. In the same century, the Namgyel Dynasty, including Nyingma refugees from Central Tibet, founded the Kingdom of Sikkim in 1642, annexing Rong (Lepcha) and Limbu territory (Mullard 2011). A lesser-known kingdom named Po or Poyul controlled the terrain around the Yarlung Tsangpo Gorge. Between the seventeenth and twentieth centuries, its kings supported the inward migration of Tsangla-speaking migrants from eastern Bhutan and religious and political refugees from the Dri Chu Catchment. Migrant settlements in the middle and lower gorge extended Po rule south through the sizeable sacred site of Pemako into the lands of the Adi people (Grothmann 2012; Gamble 2022). Like the Dri Chu Catchment states, these polities maintained trade routes across their territories

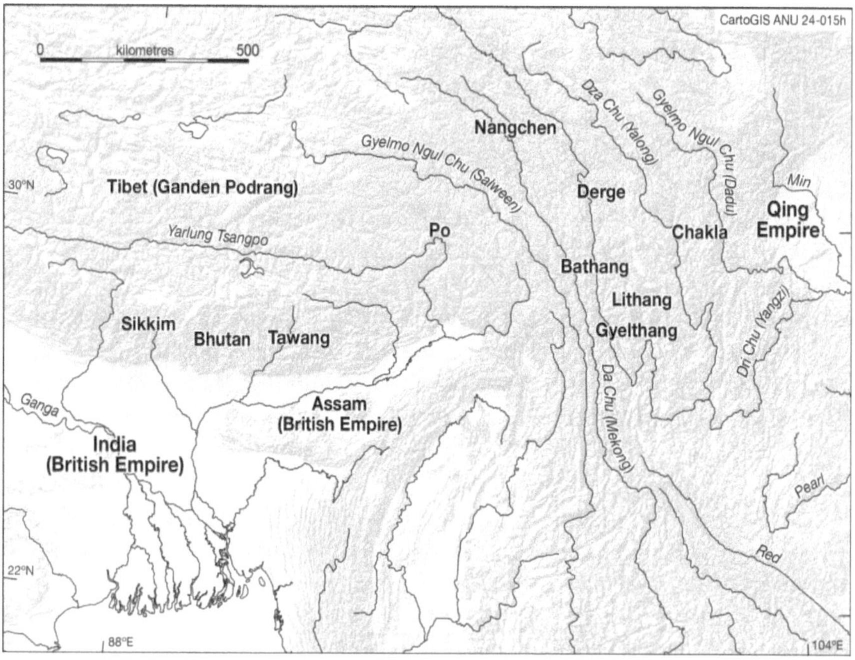

FIGURE 8.1 The major and minor polities of the late nineteenth century in the eastern Asian Highlands.

that increased their wealth, particularly as the British Empire extended its control into Assam in the nineteenth century.

Most of both catchments' polities operated similarly to the Ganden Podrang, with tenants offering tax obligations in exchange for land and protection within complex allegiance networks. They all had either treaty or tribute relationships with the Ganden Podrang, with whom they shared religious and cultural affiliations. Most also maintained treaty, trade, or tribute relationships with the Qing court. The states south of the Himalayan watershed had to balance their treaty relations with the Qing and Tibet with their unequal treaties with the British.

The relationship between the Qing Empire and the Ganden Podrang was particularly complicated. The Qing, a colonial empire (Perdue 1998), expanded its territories and established Han settlements far beyond Chinese territory to its north, west, and south, but it lacked the technology to settle and control the Tibetan Plateau, so its influence in the Highlands fluctuated. After defeating the Zunghar Mongol invasion of Central Tibet in 1720, the Qing established a small, permanent military presence in Lhasa. An *amban*, usually a Manchu or Mongolian representative of the Qing, headed this force and ostensibly controlled the Plateau for the Qing. On the ground, however, Qing influence peaked in the eighteenth century and was virtually nonexistent by the mid-nineteenth century. While both governments maintained the artifice of their *choyon* arrangement, they were in direct imperial competition in Kham (Giersch 2019).

The British were the region's other major political entity. By the mid-nineteenth century, they either directly or indirectly ruled most of South Asia and large sections of the Yangzi Catchment. The British East India Company had arrived in South and East Asia to trade with the Mughal (1658–1757) and Qing empires in the seventeenth century but slowly increased their influence, eventually operating as a quasi-state. The Company annexed Assam in 1828 before the British government disbanded it in the 1850s and established direct crown rule of India. During the British reign (or Raj), Indian government influence crept further into the Highlands. Britain fought the Duar War (1864–1865) with Bhutan and then forced a protectorate treaty on it, helping to install Bhutan's first King, Ugyen Wangchuk (1862–1926) in 1910. In 1861, it forced the Kingdom of Sikkim into an unequal treaty that forfeited Darjeeling and Kalimpong to the British, giving them greater access to the mountains.

British colonialism had a significant impact on India's river systems. Their colonial governments prioritized private property and extractive capitalism. They set up irrigated tea plantations in Assam, Darjeeling, and Kalimpong, and at the beginning of the twentieth century, hydropower plants.

The Qing curtailed British influence in China for centuries by restricting the foreigners' operations to several treaty ports, principally Guangzhou, at the mouth of the Zhu Jiang (Pearl River) in Southern China. In the mid-nineteenth century, however, the British won two Opium Wars (1839–1842, 1856–1860), which allowed them to increase sales of India-grown (and particularly Brahmaputra Catchment-grown) opium to China and stretch their trade and political influence into the interior. By the late nineteenth century, Britain controlled the Yangzi's waters up to Chongqing (Reinhardt 2018).

By the end of the nineteenth century, the Qing and British Empires were competitors in both catchments. The catchments were "multistate margin" (Giersch 2010) for the empires, in which human, trade, and water flow connected the Highlands and Lowlands. They also represented a frontier, a space where land and water could be enclosed and trade imposed (Baruah 2020). The competition between the two states intensified when the British used Sikkim as a base to invade Tibet in 1904, killed between two and three thousand Tibetans, and forced an open trading policy on the Dalai Lama's government that included the establishment of permanent trade offices on the Plateau (McKay 1997; Harris 2013).

When the British invaded, the thirteenth Dalai Lama, Tupten Gyatso (1878–1933), fled to Mongolia and China seeking support. None was forthcoming. Instead, Qing officials responded by intensifying their attempts to assimilate Kham into the Qing Empire and the Chinese cultural sphere. Inspired at least partly by British settler colonialism, Sichuan Governor General Lu Chuanlin (1836–1910) prescribed a forward policy that took an interventionist stance in Kham (Relyea 2019; Gros 2023). The Qing built roads and bridges that enabled more access to the region and Han businesses to increase their operations (Giersch 2019). The primary focus of the forward policy, however, was agriculture. Qing governors like Lu determined that Chinese civilization was based on agriculture, and the intensification of rice growing through Han settlements in Kham would sinicize—or civilize—the Khampas.

This project pushed cultivation beyond its environmental limits and strained the region's ecological and hydrological capacity, bringing hardship to Tibetan nomads and Han settlers (Frank 2019). The Khampas rose repeatedly against this dispossession, but following every armed resistance, the Qing sent in more troops. In 1905, Khampa rebels in Bathang assassinated Fengquan, who the Qing court had appointed assistant *amban* to Tibet. To avenge his death, the Qing sent an army to Kham. It was co-led by a man who became the most influential leader in the Highlands for the next six years, Zhao Erfeng (1845–1911). Zhao held several senior positions within the Qing administration, including Sichuan-Yunnan Frontier Commissioner. He was committed to transforming Kham into Chinese provinces and argued for European-style settler colonialism to bring this about (Relyea 2019; Frank 2019).

While Zhao focused on Kham, Tibet's last *amban*, Lianyu (in office 1906–1911), tried to conduct similar reforms in Central Tibet. After meeting with resistance, he asked the Qing to send an army to support him. In 1909, a 2000-strong Qing army marched to Lhasa from Sichuan, fighting Tibetan resistance as it advanced (Relyea 2019). The army eventually took over Lhasa and caused the Dalai Lama and his Council of Ministers to flee to British India.

This Chinese-led rule of central Tibet is remembered for its brutality. It was also short-lived. The Qing Empire fell within a year of the troops arriving in Lhasa. Revolutionaries across China declared a new Republic of China (1912–1949), and in Sichuan, a group of them executed Zhao. When the Dalai Lama returned to Tibet, he expelled the remaining Qing troops and proclaimed Tibet an independent state. In Kham, ruling families, warlords, merchants, and republican Chinese competed for power (Giersch 2019). Soon after, the Dalai Lama's reformulated government took advantage of Kham's situation and invaded, extending Lhasa's control past the Dri Chu (Relyea 2015, 1006–1009).

The region's only remaining imperial power, British India, was concerned about this instability in its buffer zone and called a trilateral conference—Tibetans, Republican Chinese, and themselves—to settle Highland politics.

They proposed the Highlands be divided into three entities: British India south of the Himalayan watershed, "Outer Tibet," the Dalai Lama's realm, west of the Da Chu-Dri Chu watershed, and "Inner Tibet" within the Republic of China to its east. The Tibetans begrudgingly agreed, but Ivan Chen, the Republican representative, refused to sign the treaty, presenting a counterclaim of Chinese sovereignty over the Tibetan regions Zhao Erfeng had bureaucratized (Relyea 2022, 95). His stand reflected the Republican government's claim that China was made up of "five races under one union"—Han, Manchu, Mongols, Tibetans, and Hui Muslims—and anywhere these races lived was China (Zhao 2004). The repercussions of this failed treaty have echoed down the generations and underpin the region's contemporary geopolitical tensions. One of the many things the British failed to understand in their negotiations at Shimla was the variant understandings of land among the conference's participants.

8.2 Drawing Rivers: Cartography and Enclosure (1850–1913)

Highland polities had agreed upon boundaries before the British arrival, but they were mountain pass checkpoints that were agreed locally rather than by centralized governments. The military movement was monitored and trade across them was taxed, but they were not cartographically defined as lineal borders (Gamble and Davis 2024). In 1727, for example, a border-marking stele was erected on the Bum La pass at the Dri Chu–Da Chu watershed on the road between Kham and Central Tibet to indicate the border between lands controlled by the Ganden Podrang and the Qing (Gros 2019, 48). Another stele had been erected on the pass between Sikkim and Tibet (Harris 2017, 154). The spaces on either side of these passes were not monitored.

The fact that the Shimla Conference's main priority was to set lineal borders between three sovereign states looks unremarkable in hindsight. At the time, however, it was transformative. Its introduction of territorial geopolitics—the European conception of mapped, territorialized nation-states—marks a shift in the Highlands' geographical imaginary. European empires demanded mapping, and "indigenous conceptions of space [were] overwhelmed by the need to have a location that [could] be recognized by [this] political power" (Fox 2002). This process did not stop at delineating international borders; it also territorialized land within borders.

Rivers played a fundamental role in these intertwined bordering, cartographic, and territorializing projects. Borders rely on maps, and delimiting river systems is a fundamental element of mapmaking. The British needed to track river courses through territory to fill out their maps. Furthermore, as they deemed watersheds and thalweg (the line following the deepest part of a stream along its length) the best sites for borders, knowledge of river courses and their watersheds was fundamental to their "science" of border-making (Gardner 2021).

The borders the British hoped to establish at Shimla (see Figure 8.2) would have been linear and to their advantage. By establishing Tibet as a buffer state between China and British India, they expected to limit Chinese influence in the Highlands and increase their own (Baruah 2020). They also ensured that the eastern section of the border—named the McMahon Line after British India's foreign minister, Henry McMahon (1862–1949)—passed below the strategically important Tawang territory. Tawang had been linked to Tibet for centuries, but its passes gave access to Assam, so the British determined it should be part of their territory (Gohain 2020b).

Their official argument for keeping Tawang was at least partly based on its position south of the Yarlung Tsangpo-Brahmaputra watershed, whose contours the British had been painstakingly mapping for decades (Edney 1997). Determining the flow of rivers was a core part of this colonial project, and finding the sources of rivers that flowed into their territory was considered strategically important and a heroic achievement for imperial adventurers. British explorers attempted to find the Yangzi's source during expeditions from a base they established in Chongqing after their victory in the Second Opium War. But a French group led by Captain Ernest Doudard de Lagrée

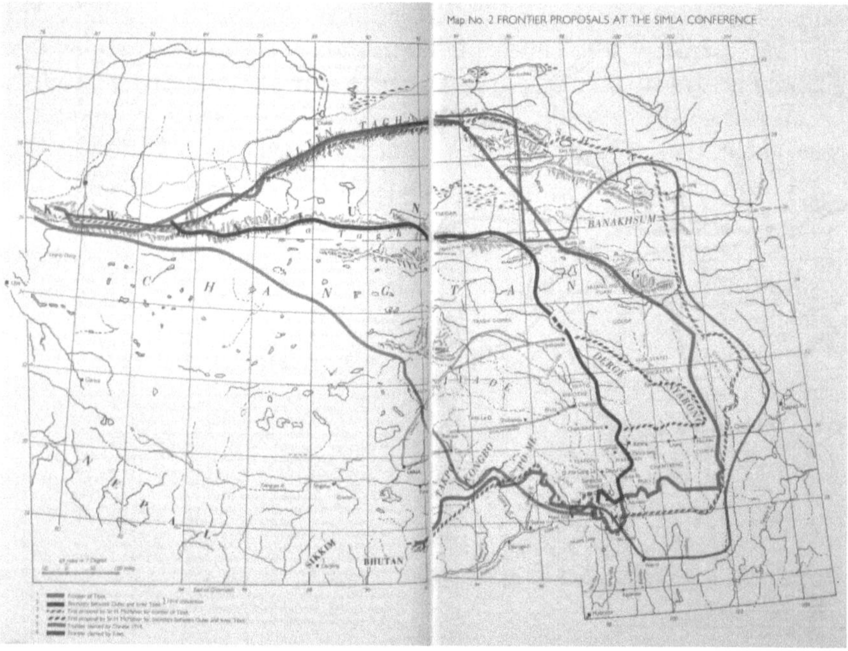

FIGURE 8.2 The various borders suggested at the 1914 Shimla Conference. Map by Hugh Richardson.

beat them to it, "discovering" the Mekong and Yangzi's parallel upper reaches in 1870 (Wong 2010, 26). Later, in 1905, the Swedish explorer Sven Hedin traveled to western Tibet at the behest of the British Indian government to find the sources of the Indus, Brahmaputra, and Karnali rivers. He found all these sources at the base of Khangtise (Kailash) (Hedin 1909; see Box 7a).

The longest-standing riverine cartography puzzle was the course of the Yarlung Tsangpo, which, given the security and flooding threat it posed to British assets in Assam, was considered a matter of imperial importance. Multiple Highlander sources, textual and human, provided evidence for the river's course, but the Europeans dismissed or missed them. At least one text, a record of the upper Brahmaputra catchment composed for the "treasure revealer" Jigme Lingpa (1730–1798) by his disciple Jangchup Gyeltsen describes the river's course from Kongpo to Assam in detail (cited in Ardussi 1977a). Several Himalayan "pundits," whom the British government sent into Tibet surreptitiously to map the region after the Tibetans denied it entry, also described the river's course.

Nain Singh (1830–1882) consistently referred to the Yarlung Tsangpo as "the Brahmaputra" in the reports he compiled in the 1860s (Lange 2020). The unnamed monk who composed the Wise collection maps in the same decade described the Yarlung Tsangpo as the "Brahmaputra … [which] rises from the Kylās [Kailash] and runs near L'hasa, and then called Sungehin or Tsang Chin. Also Yarou tsanpo" (Lange 2020, 279; see Figure 8.3). The report of another pundit named Kinthup, who hailed from western Sikkim and endured kidnapping and enslavement to map the entire river's course, was also deemed unreliable. Instead, the British and French continued to argue about the Yarlung Tsangpo's course through the seventeenth and

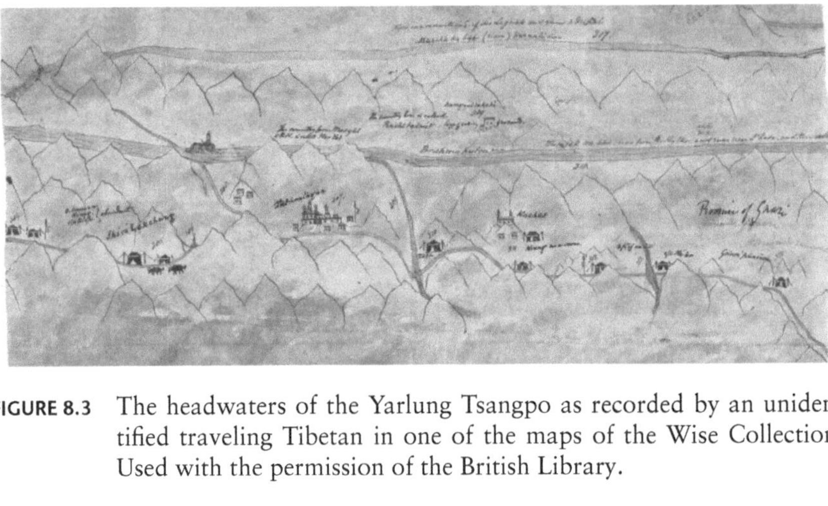

FIGURE 8.3 The headwaters of the Yarlung Tsangpo as recorded by an unidentified traveling Tibetan in one of the maps of the Wise Collection. Used with the permission of the British Library.

eighteenth centuries. Some said it flowed into the Ayeyarwady; others said it was the Brahmaputra (Simpson 2023).

The imperative to find the river's course intensified after a 1900 flash flood carried "a number of bodies then supposed to be of Tibetans and a quantity of coniferous trees" down to British stations in Assam (Bentinck 1913). The 1904 British invasion of Tibet gave them more "authoritative" information on the river's course. But it was not until two British veterans of the 1904 invasion, Frederick Bailey (1882–1967) and Henry Morshead (1882–1931), used their experiences and Tibetan contacts to travel through much of the Yarlung Tsangpo Gorge that the puzzle was deemed solved. Bailey and Moreshead were successful because of their respect for local knowledge and belief, which extended to their post-discovery efforts to have Kinthup's efforts officially recognized (Simpson 2023).

Along with their riverine cartographical fixations, the British were instrumental in changing the approach to rivers in Lowland India and China, and their hydrological cultures would later be carried into the mountains. Their riverine impact was most evident in India. They were perplexed by monsoonal climate cycles (see Chapter 3) and worked to insulate the agrarian sector from their associated floods and droughts (Amrith 2018). They used new technologies to transform seasonally inundated canals into perennial irrigation systems, with barrages and weirs, including permanent headworks across riverbeds. Their large-scale hydrological works included the world's most extensive irrigation program in the Indus River catchment. After engineering the Indus, they set their sights on the Brahmaputra River in Assam, where they had recently developed tea plantations. They built a network of embankments along the rivers' lower reaches that changed its flood regimes, disconnecting the river from its flood plains (Saikia 2019), and, in 1897, established India's first small-scale hydroelectric power project in Darjeeling (Chowdhury and Rajhans 2021). This "great hydraulic transition" (D'Souza 2019) overrode vernacular water governance models (Boelens et al. 2022; Chowdhury and Rajhans 2021) and disconnected agriculturalists from floodplains and seasons. As flood-dependent agricultural communities became flood-prone, they entered disaster regimes.

Along with this hard infrastructure, British authorities introduced the soft infrastructure of "river basin management." As David Gilmartin explains, this introduced science "intersect[ed] with new forms of colonial administration and electoral politics to influence new visions of provincial and 'national' identity" (2020). This soft infrastructure proved even more durable in many ways than the large hydrological projects of the British. Their centralized Public Works Department established a precedent for top-down, engineer-centered decision-making that continues today. As it was positioned within Imperial Civil Services, this department also reproduced racial, caste, and class hierarchies (Chowdhury and Rajhans 2021).

British colonialism operated differently in China and India, so its influence on riverine transformation also differed. The British (and, to a lesser degree, French) colonialists occupied treaty ports and conducted enforced "free trade" on its rivers. Chinese states had a long history of riverine enclosure (see Chapter 1), and the British did not construct most of the dams, barrages, and other infrastructure along the Yangzi. But, as Corey Byrnes (2018) argues, the British commodification of the Yangzi and their presentation of it as a threat (see also Courtney 2018; Gao 2022) changed perceptions of the river and encouraged later river regulations. "For China to 'see like a [modern] state,'" Byrnes (2018) argues, "it first had to see like a modern imperial power."

In both regions, it took decades for the technology to improve sufficiently for states and corporations to begin accessing the Highland's globe-leading Gross Potential Energy, but these colonial interventions established the social and political structures that enabled its eventual exploitation. By the time this exploitation grew exponentially in the new millennium, the two rivers' histories had diverged dramatically. The Dri Chu had been incorporated into China's nationalist project, and the Yarlung Tsangpo had been divided between multiple states and territories.

8.3 Constructing a National River: The Dri Chu (1913–1959)

After the Shimla Conference failed to establish a Tibet–China border, the two "self-consciously modernizing" states (Giersch 2019) intensified their competition in the Dri Chu Catchment. The Chinese Nationalist Government, based in Beijing, claimed the entire region but did not control it directly. Instead, it operated through warlords who shared influence with local religious figures, aristocrats, and traders. Between 1918 and 1933, the thirteenth Dalai Lama's armies attempted to assert control in the eastern Highlands. In 1918, Tibetan troops crossed the Dri Chu and marched toward Nyarong, Bathang, and Lithang. Before this skirmish developed into a full-scale war, Eric Teichman (1884–1944), the British Consul to Western China, negotiated a settlement that set the Dri Chu as their border between the Tibetans and the warlords. This did not stop Tibetan bandits from crossing the Dri Chu again in 1923 and fighting across several Qing-era bridges to come within 30 kilometers of Lijiang. The warlords fought back, and in 1932, the Dri Chu at the Kamtok river crossing was re-established as the border between Tibetan and Nationalist territory.

Fighting between these two polities and their various supporters continued into the 1930s as the catchment's provincial borders took shape. In 1928, the Nationalists created a new province on the Plateau named Qinghai, ruled by the warlord Ma Bufang (1903–1975) and his family, which split the river's headwaters and the ex-kingdom of Nangchen from the rest of Kham. In 1930, another Tibetan militia crossed the Dri Chu, this time to intervene in a dispute between two monasteries in the upper Nyak Chu (Yalong) catchment, Dargye and Beru. Sichuan's governor, Liu Wenhui (1895–1976), marched an army to meet them and worked with Ma Bufu to push the Tibetans back to the western side of the Dri Chu again.

With the Dalai Lama's death in 1933, Tibetan incursions into Kham ceased, but unrest continued in Kham as Khampas started the first of several "Khampa rule for Kham" protests (Tsomu 2018). Liu Wenhui put down the first of these rebellions and, in 1935, received Nationalist government support to establish a new province named Xikang in the Dri Chu's central and southern catchment. Xikang's capital was Dartsedo (population around 1.5 million), and although some of its claimed territory was on the western side of the Dri Chu (see Figure 8.4), its governance was concentrated on its east side. Xikang continued to exist through the Sino-Japanese War (1937–1945) and both periods of the Chinese Civil War (1927–1937, 1945–1949) as Liu expanded regional Qing-era sinicization and settler colonial policies, encouraging Han settlers to establish farms within its boundaries. Liu's approach differed from Qing-era policies, however, in that it focused on implementing modern and experimental agriculture that would

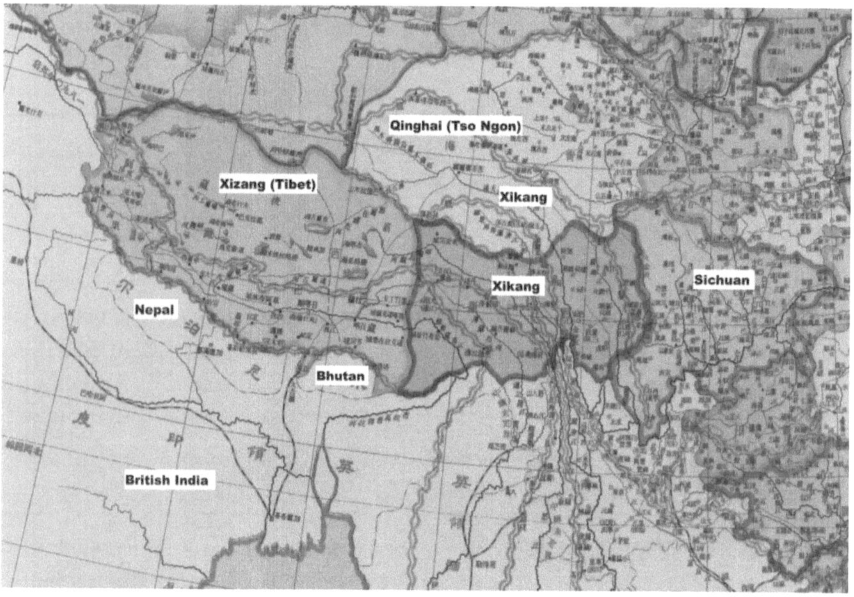

FIGURE 8.4 Adapted extract from the approximately 1933 map *Zhonghua min guo fen sheng xin tu* (*New provinces of the Republic of China*), published by Wuchang Asia Information Society, Wuchang, and available from the Library of Congress open access collection. Note that it includes Sikkim as part of British India and much of what is now Arunachal Pradesh as part of Xikang.

expand the region's high-altitude agricultural boundaries (Frank 2016). He also installed a state-of-the-art hydroelectric plant in Dartsedo (Yang 2017).

When the Chinese Communist Party emerged victorious at the end of the Civil War in 1949, it established the PRC and claimed all lands its predecessors had purported to govern. The government of Tenzin Gyatso (1935–), the young fourteenth Dalai Lama, agreed to join the new entity in exchange for internal autonomy. For many Khampas, caught between Tibet and Chinese governments again, the only recourse was armed resistance. The most well-known of these resistance groups, Chuzhi Gangtruk (Four Rivers, Six Ranges) (see Box 7a), named for the region's topography, formed in 1958 and continued to fight well into the 1970s, aided by the CIA (McGranahan 2010, 163–184).

During the 1950s, despite this resistance, the People's Liberation Army gradually took control of Kham and rearranged its internal borders. After designating Xikang as the PRC's first "Tibetan Autonomous Region" in 1950, military officials dissolved it in 1955 and split the erstwhile province along the Dri Chu. Land east of the Dri Chu was incorporated into Sichuan Province, areas south of it into Yunnan Province, and regions west of it into the Tibetan Autonomous Region (created in 1965). The Dri Chu's new position within a sovereign state's territory but outside any province was a product of the catchment's relatively recent turbulent history. This intra-state position and its intense Gross Potential Energy would make it a prime candidate for future "nation-building" hydropower projects.

8.4 Constructing an International River: The Yarlung Tsangpo (1913–1962)

While the Dri Chu was being reshaped as a national river, the upper Brahmaputra catchment was being divided among several emerging states, not all of whom would survive through the next turbulent decades. For Central Tibet, the era began hopefully as the thirteenth Dalai Lama, bolstered by his engagement with the British Empire in India, started a process of modernization and nation-building. In 1913, he declared Tibet an independent nation and established a mutually recognizing treaty with Mongolia (Gupta 2021). He also made efforts to modernize Tibet's soft and hard infrastructure. One of these efforts was to dispatch four aristocratic students to study in England in 1913. For the most part, this was an unfruitful exercise, but one of the students, Rigzin Dorje Ringang (1900–1948), found his calling as an engineer. While in England, he trained in ammunition making and electrical engineering before returning to Tibet in 1924, where he helped to establish the Lokdon Khang (Office of Power and Electricity) and install hydroelectrical machinery he had brought from Kolkata on the Dogdé Phuchu north of Lhasa. This construction project became Tibet's first hydropower station, which delivered electricity to both Lhasa and the Dalai Lama's summer residence at Norbulingka (Shakya 1986) until it was destroyed by a flood in 1946.

Elsewhere in Central Tibet, several local leaders copied the Dalai Lama's example and began pursuing their own modernizing hydraulic missions. The Sakya hierarch Trichen Dzamling Chegu Wangdu (r. 1895–1915) built an imposing irrigation system. It operated through a series of pools and new aqueducts that transported glacial and river water around Sakya's relatively dry valley (Ekvall and Downs 1963). In 2021, the International Commission on Irrigation and Drainage listed Sakya's irrigation system as a World Heritage Irrigation Structure (*Xinhua* 2021).

Much of this modernizing impetus stalled in 1933 when the thirteenth Dalai Lama died. After his death, the Tibetan government fought internal battles and failed to establish itself as an independent actor in the international arena. One of the things it forgot or could not comprehend was the British territorialization project, which it was committed to through the Shimla agreement. Tibetan tax collectors continued to operate below the McMahon line in Tawang (Gohain 2020b) and Pemako (Gamble 2023). Distracted by two world wars, the British did not push their claim in these areas and left the lines deliberately fuzzy for decades.

This situation changed, however, when the newly independent and partitioned India determined to establish its borders formally. This required complicated negotiations with the Tibetans, and then, after Tibet was incorporated into the PRC in 1950 by signing the "Seventeen-Point Agreement" with the Chinese Communist Party.

At first, both the PRC and Indian governments worked to win the hearts and minds of borderland people (Guyot-Réchard 2016). But border disputes slowly eroded their relationship over the 1950s, especially after the Lhasa uprising of 1959 forced the Dalai Lama into exile in India, and in 1962 the two countries fought the Sino-Indian War. This was a conflict over territory rather than the defense of borderland people (Gohain 2020b), and it resolved little. China won the war but then retreated to its previous position on the McMahon Line. The most tangible outcome of the war was a harder border.

Within a generation, the region's geopolitics had been thoroughly transformed. Tibet was no longer an independent state, and two new entities, the PRC and the Republic of India, had positioned their armies along a disputed Line of Actual Control (LAC) on the Yarlung Tsangpo-Brahmaputra watershed. At this stage, Sikkim and Bhutan remained independent states, but India annexed Sikkim in 1975. The North East Frontier Tracts were reconstituted as the North East Frontier Agency (NEFA) in 1951, the NEFA Union Territory after the 1962 war, and the state of Arunachal Pradesh in 1972 (Gohain 2020a). By the mid-1970s, three sovereign states within unresolved borders and three provincial-level governments operated in the upper

Brahmaputra catchment, and another sovereign state, Bangladesh, governed in the river's lower reaches (Davis 2023). As the Cold War unfolded and states increasingly focused on development, the geopolitical conditions for intense river regulation began to unfold.

8.5 The Multiple Technologies of Transformation (1960s–)

Despite or because of their unresolved high-altitude borders, India and China followed the 1962 war with decades of Highland state-making. By this stage, the rules for and technologies of state-making had changed. The post-World War II reconfiguration of the international order meant that states now gained legitimacy through mutual recognition, and all space must be contained within borders (Ludden 2003). Just occupying an area was not deemed sufficient to cement sovereignty; states combined occupation with emotive national narratives and imagery. Many of these narratives suggested the states were not only "incontestable and timeless" (Billé 2016) but also anthropomorphized, organic entities (Billé 2014) or "geobodies" (Winichakul 1997). This is particularly true of multi-national states like China and India, whose majoritarian territorialization minoritized multiple other nations and whose claim to their borderlands is recent and contested. Their "cartographic anxiety" (Krishna 1994) and postcolonial insecurities (Chatterjee Miller 2020) have encouraged state operatives to present eternal, indivisible, embodied, but incommensurate and non-negotiable narratives.

This process devastated the communities and nations whose territories were subsumed into the new nation-states. The authoritarian PRC set about organizing its minority frontiers through an all-encompassing ethnicity calculus. Most of its citizens belonged to one group, the Han, and all non-Han, about 7.5% of the population, were divided between 55 ethnic minorities (Leibold 2006). These groups became the "target of sweeping biopolitical interventions" that disempowered and marginalized them (Gros 2023) and ensured that their homelands would be sites of extraction rather than accumulation.

Democratic India also categorized and marginalized its minorities. India's constitution and legal system recognize multiple caste and ethnic groups, who are given special rights and protections. The Indigenous peoples of its central hill tracts and its Highlanders are called "tribes" within India's legal system, and the fifth and sixth schedules of India's constitution give these groups special protections and rights (Gergan 2020). The government's approach to these groups has vacillated between paternalistic apartheid and assimilation (Baruah 2020).

After fighting the 1962 war, both countries spent decades extracting resources from their borderlands to fuel their state-building projects. Moreover, these initiatives coincided with the Cold War (1947–1991) emphasis on "development." Like enclosure and territorialization, "development" was a Western idea—an offshoot of the Post-World-War-II Marshall Plan to revitalize Europe—that was exported globally. Its primary tenet is that human flourishing comes from economic growth—which, in turn, comes from industrialization. This industrialization can be achieved by injecting capital into markets or public sector interventions (Kothari 2019, 5–12). During the Cold War, it took communist and capitalist forms. China followed the communist development model. India was politically unaligned and nominally socialist, receiving aid and development from communist and capitalist nations. Neither side factored ecological sustainability into its development models, and the impacts on the environment were exaggerated by the onslaught of the region's longest and severest drought in centuries (see Chapter 3).

Instead, in the first decades after the establishment of these two states, the world passed through a period of "high modernism," during which many polities, both communist and capitalist, operated with an "uncritical, unskeptical, and thus unscientifically optimistic" faith in the "possibilities for the comprehensive planning of human settlement and production" (Scott 1998). Concurrent technological advances allowed for this modernism to climb further into the mountains. Much of the PRC's Highlands was transformed through state farms, the penetration of intensive agriculture, greenhouses, and upgraded transport infrastructure (Yeh 2013). The Highland development models of India, Bhutan, and Sikkim were patchier. Paternal apartheid policies left sections of the Eastern Himalaya undeveloped for decades (Baruah 2020), while other mountainous regions were the site of large-scale modernist projects.

Large dams were regularly presented as the quintessential modernist undertaking. According to the logics of modernity and development, large dams fulfilled a state's "hydraulic mission" to provide water to all its citizens (Moore 2018); they also generate hydropower and prevent flooding. In addition, such projects also represented a highly visible example of state building. The PRC government began constructing large dams in the 1960s, but problems with its USSR-sponsored Sanmenxia Dam set large dam projects back decades (Pietz 2015, 183–184), and it only began building them again in the 1990s. India inherited the British-built Indus River irrigation scheme and extended it into the Himalaya's foothills; the large Bhakra-Nangal project on the Sutlej River was completed by 1963. While inaugurating this dam's construction, Jawaharlal Nehru (1889–1964), India's first prime minister, famously called Bhakra-Nangal "the new temple of resurgent India, the symbol of India's progress" (*The Tribune*, October 22, 1963).

During the 1970s, Bhakra-Nangal's water played an essential role in India's Green Revolution, during which agricultural outputs increased through intensified irrigation, fertilizer use, and groundwater harvesting. The Indian state also installed the Farakka and Teesta Barrages in eastern India, which impeded water flow to newly independent Bangladesh (Afroz and Rahman 2013) and kickstarted the region's intense hydropolitics. While the Green Revolution increased India's food supply, it also damaged its environment. These dams have produced salinity and waterlogging (Dharmadhikary 2005), intensified agriculture has led to eutrophication and groundwater depletion (Cullather 2010), and increased black carbon production through crop burning. The soot drifts onto the Highlands, pollutes glaciers, and contributes to rising temperatures (see Chapter 4).

For some communities, this development transformed their lived experience. On the Lhasa Plain, for example, the PRC began its modernization program as early as the 1950s; it mobilized the People's Liberation Army to repair embankments and dig more than 110 irrigation canals (Yang and Hu 2013). It almost succeeded in a plan to drain the Lhalu Wetlands (Yeh 2009; Khetsun 2007, 225) and made those accused of participating in the 1959 uprising engage in "reform through labor" by building a new hydropower plant at Ngachen (Najing) on the Plain's eastern edge. These laborers began by extracting sand and boulders from a quarry, but the quarry was abandoned when an outburst flood killed 11 workers in the early 1960s, and they began mining the riverbed (Khetsun 2007, 66–70). In 1977, they constructed the Yangpachen Geothermal Power Station on a site north of the city renowned for its hot springs. This remained the primary source of Lhasa's power until the Yamdrok Yutso Hydropower station was built in 1988. It relies on water running from the lake down to the Yarlung Tsangpo.

At the same time, modernist development reached its global peak in the 1960s and 1970s, and international environmental conservation movements began to reform in response to it. Many of this movement's participants were Global North experts committed to development, who merely added layers of environmental governance to development projects. But the movement also included community groups across the Global South that protested local environmental destruction, like the Narmada Bachao Andolan (Save the Narmada) movement that opposed the construction of the Narmada River dam (Baviskar 1997). Narmada Bachao Andolan and other global anti-dam movements encouraged the World Bank and the World Conservation Union to establish an independent World Commission on Dams (1997–2001). The Commission's 2000 report found that, on balance, large dams produced no benefit for communities or countries (WCD 2001, 98–99).

The report led the World Bank to withdraw funding from large dams for a decade, but it did not stop Chinese and Indian state-backed hydropower companies from building more dams. Even after the Commission handed down its report, several south Indian states began large dam projects, and the Arunachal Pradesh government installed the state's first large dam. China's rejection of the Commission's findings was even more blatant. The state-owned China Three Gorges Corporation continued to build the world's largest dam on the Yangzi's middle reaches as the commissioners finished their research. The Three Gorges Dam had displaced millions of people by the time it began operating in 2003 (Wilmsen, Webber, and Duan 2011), and decades later, it is still a cause for multiple environmental concerns.

In the new millennium, both countries increased the number and size of their hydropower constructions, making the Highland's edges, with their intense stores of Gravitational Potential Energy (GPE), the center of the worldwide hydropower industry. Several factors underpinned this intensification. First, in 1999, the Chinese government launched its Western Development program, which included the construction of a dam cascades down the Dri Chu, Dza Chu (Yalong), eastern Gyelmo Ngul Chu (Dadu), and Da Chu (Mekong) rivers (Harrell 2023, 327–373). Those built on the Dri Chu were particularly large and included the world's second, fourth, seventh, and tenth-largest dams (see Chapter 9). Second, the United Nations and the World Bank responded to the climate crisis by establishing the Clean Development Mechanism (CDM). This program included hydropower within its remit and encouraged the World Bank to reverse its earlier decision to refuse large dam funding. Three-quarters of all CDM dams have since been built in China and India (Erlewein 2014). Several more are being constructed in Bhutan. And third, in the 2010s, the semi-privatization of hydropower through the creation of state-backed companies led to the construction of more, smaller, hydropower-only dams. Most of these dams are run-of-the-river and use tunnels to drop water from one river reach to the next. These kinds of dams have been built in Sikkim, where 9 such dams have been constructed, 15 are being constructed, and 28 more are planned (*Hindustan Times* 2023). The PRC has also constructed such dams in the Gyatsa Gorge between Central Tibet and Kongpo, and there are plans to exploit the intense drop between the Yarlung Tsangpo's pre- and post-Gorge reaches by drilling a tunnel between them (Shan 2020, see also Chapter 9).

Governments have created laws preferencing this style of dam construction because they displace fewer people during construction. However, their opponents contend that they cause landslides and earthquakes and intensify outburst floods (Huber 2019). They point to the October 2023 glacial lake outburst flood's interaction with the Teesta III dam, which was washed away in a catastrophic dam collapse, as an example of post-construction risk (*Hindustan Times* 2023). The community in Sikkim, as elsewhere in the Indian Himalaya, remains split on the benefits of hydropower (see Box 9b).

8.6 Conclusions

Within 170 years, the political ecology of the Dri Chu and Brahmaputra catchments had been utterly transformed. While at the beginning of this period, they housed a network of interconnected, allegiance-based polities connected by religion, trade, and community connections, they ended it divided by the boundaries of three sovereign states and multiple province-level governments. The Dri Chu, which once connected the many cultures and polities of the Highlands, was reimagined as one section of the Chinese Yangzi River. The upper Brahmaputra, whose river valleys connected uplands and plains, was bifurcated by unresolved Sino-Indian and Sino-Bhutanese borders. The urge to regulate and control these rivers, driven by hydro-political rivalries, increased with the rise of new sovereign states, which regulated the rivers' flows and transformed human relationships with the rivers. There has been some pushback against this ever-increasing trend of regulation, however, as Indigenous peoples, other local communities, and the rivers themselves continued to defy the attempts to control them. These countercurrents are important, not because they point a way backward to a pre-declension environmental idyll—which is, after all, an impossible retreat (D'Souza 2023)—but because they demonstrate the many alternative currents of history, its ever-changing nature and therefore, the many other possibilities that exist in the present.

9

MANAGING RIVERS

1970s–Now

Over the past two centuries, the environmental, social, cultural, economic, geopolitical, and scientific transformations within the upper Brahmaputra and Yangzi catchments have transformed people's understanding and use of the eastern Asian Highland river systems. The territorialization process outlined in the previous chapter has established a zero-sum-game competitive model between the catchments' nation-states, and particularly between China and India, whose relationship is complicated by an unresolved border dispute. Local, often Indigenous, reciprocal relationships with the river systems operate alongside intersecting state and commercial imperatives. States need the rivers' water and Gravitational Potential Energy (GPE, see Box 3a) to fulfill their hydraulic missions to bring water to all citizens, and commercial entities—often working with governments—seek profits from these same resources. At the same time, these hydraulic missions and extractive practices need to be curtailed by national and international commitments to maintain the region's diverse ecological and hydrological systems. The river management regimes that have developed to balance these competing interests differ between the states that claim sovereignty over various parts of the river catchments. They are social and political artifacts dependent on the individual states' histories (see Chapter 8). They also shape the rivers' governance and, therefore, their environmental health.

DOI: 10.4324/9781003392033-9

Since 1975, when Sikkim became an Indian state, four countries have shared governance of these two river catchments: China, India, Bhutan, and Bangladesh. Within China and India, powerful provincial or state governments also administer the rivers. The Chinese provinces of Qinghai, Sichuan, Yunnan, and the Tibetan Autonomous Region (TAR) have authority over part of the catchments. In India, six states—Arunachal Pradesh, Assam, Meghalaya, Nagaland, Sikkim, and West Bengal—manage parts of the Brahmaputra catchment. These governments' attempts to manage the rivers are compromised by the competing demands on river resources and competition between national and subnational governments, along with the rivers' complex hydrological characteristics and intense biodiversity.

As we will show in this chapter, these river management regimes are affected by the histories of the countries through which they flow and broader river management trends, which are constantly evolving as hydraulic engineers, conservation biologists, and other water resources specialists revise existing models. In recent decades, a significant trend in river management has emerged, pivoting toward more holistic approaches, notably integrated water resources management (IWRM). IWRM acknowledges the interconnected nature of river systems and aims to balance the various socioeconomic demands with environmental and ecological sustainability. It is a multi-stakeholder approach that involves governmental agencies, industries, and civil society, empowering communities to participate in decision-making processes. This collaborative approach fosters better understanding and coordination between different user groups and aids in mitigating potential conflicts over water use. However, implementing these new approaches to river management often happens slowly and depends on government amenability and ability. This, in turn, depends on their historically constructed systems and priorities.

This chapter examines how history and circumstances affect river management by comparing, contrasting, and questioning the river management regimes of the two catchments' four sovereign states. All four countries look to these rivers to fulfill their hydraulic mission to provide safe and reliable water services throughout their territory (Allouche 2020). Their primary focus has been providing water for drinking, irrigation, industry, and hydropower, but they have approached this task differently. Our analytic approach to their river management systems is to outline river management generalizations, namely that (1) China's technocratic approach to river management overreaches; (2) India's overly bureaucratic approach mismanages its rivers; (3) Bhutan approaches but does not reach a sustainable standard of river management; and (4) Bangladesh must adapt to the management outcomes of these three upstream riparian countries.

We ask how these governments' river management institutions operate in diverse societies and how they cooperate in transboundary governance.

We also examine how all these river management schemes diverge from current international river management standards, which favor the conservation of mountain catchments over the preservation of river valleys. Such management schemes aim to conserve "intact" forests and scenic landscapes, supply water, and minimize hazards such as landslides, erosion, and floods rather than conserving a representative range of river biodiversity (Worboys et al. 2015). Conserving certain river sections may seem to underscore the importance of conserving a representative range of river biodiversity. However, in practice, this implementation often merely meets the minimum environmental impact assessment requirements. Such a piecemeal focus on protecting specific, often iconic or charismatic, species overlooks the broader ecological importance of preserving an intact riverine environment. Proper conservation should prioritize the integrity of the entire river environment, from its landscapes to its biodiversity. While it might seem economically advantageous in the short term to focus on specific species, neglecting the comprehensive health and interconnectedness of the river systems could lead to unforeseen long-term consequences ecologically, economically, socially, and culturally. As we discuss at the end of this chapter, these long-term consequences will also encourage further government intervention, which in turn may lead to further state control of vulnerable groups. These inevitabilities lead to questions about the effectiveness and sustainability of current governance models.

9.1 China: Technocratic River Management

China shares 19 international river catchments with neighboring countries. In most of these catchments, it occupies the upstream position, and its neighbors are dependent on water that descends from these areas (Biba 2014). In the main, since the People's Republic of China (PRC) was founded in the late 1940s (see Chapter 8), China has maintained an ideological, high-modernist approach to river management based on the phrase *rending shengtian* (humans can control nature) (Jiang et al. 2020a). Within this ideology, scientific management is the most significant tool for managing rivers and water resources, and river management schemes emphasize high-tech engineering approaches (Olson 2015). This infrastructure-based, engineering-heavy paradigm enables China to allocate water accurately and efficiently, and this, in turn, is going some way toward solving its dire domestic water crisis, which threatens 600 million people and two-thirds of Chinese cities with water scarcity (Crow-Miller et al. 2017). Taking this technocratic approach to managing rivers also embodies China's national identity as a high-tech state that enjoys substantial economic progress.

China's technocratic approach to rivers manifests as intense interventionism, particularly through the construction of dams. By 2020, China had built over 6,500 large dams and more than 95,000 small dams (Xu and Pittock 2020a). It also constructed one of the largest inter-catchment water transfer projects in the world, the South-North Water Transfer Project, which moves water from the Yangzi River catchment to water-poor regions of North China, most of which are in the Huang He (Yellow River) catchment (Lin 2017). Across China, widespread and *ad hoc* development has resulted in extensive, often damaging, environmental and social change (Xu and Pittock 2020a; Xu et al. 2022). This change has been particularly devastating in the Asian Highlands, where it has wreaked havoc with the region's globally significant integrated geological, climatological, and ecological systems and its socio-cultural systems.

Working through state-owned enterprises, the Chinese state began building hydropower stations and damming the mainstream of the Yarlung Tsangpo in the early 2010s. The Chinese government contracted the Gezhouba Group to build Zangmu Dam, the first of a series of dams within the Gyatsa Gorge, which sits between the historic Tibetan regions of U and Kongpo and now divides Lhokha (Shannan) and Nyingtri (Linzhi) districts. It was constructed in the early 2010s and began operating in 2015 (Wang et al. 2020d). It was the first of eight dams the government planned to build within the Gyatsa Gorge in cooperation with various state-owned enterprises. By the end of 2023, two further dams were built in the Gyatsa Gorge, Jeixu and Jiacha, another, Dagu, was under construction, and there were plans for another four in this gorge. Across the Yarlung Tsangpo, they built 27 dams and planned another six on the mainstream and the Yarlung Tsangpo's major tributaries (see Figure 9.1). Zhang et al. (2019) found that the maximum impact from five recently constructed dams on the Yarlung Tsangpo contributed to about 8–11% reduction of mean annual discharge from the Brahmaputra River into the Bay of Bengal.

The TAR Government has planned to construct more dams by 2035 to accelerate decarbonization and extend irrigation zones to ensure food security (People's Government of the Xizang Autonomous Region 2021). This includes one of the world's highest-energy hydropower projects, planned for the lower Yarlung Tsangpo, near the border with India. The Mutuo Dam was first envisaged in the 1980s and is one of China's most ambitious but unrealized projects. It would sit between two 7,000-meter peaks, Namche Barwa and Gyela Pelri, and exploit the river's significant Gross Potential Energy (see Chapter 1 for a graph of its long profile through this section). It could produce 40,000 MW of energy, almost double the Three Gorge Dam's 22,500 MW. Preliminary financial assessments indicate that the project's initial phase might cost more than 500 billion yuan, with total investments possibly surpassing 1 trillion yuan. Given its location at the junction of tectonic plates and a history of significant geological activity, including earthquakes, the project's engineering feasibility remains a challenge, and there are serious

FIGURE 9.1 The hydroelectric and irrigation dams of the Yarlung Tsangpo Catchment.

concerns regarding its environmental impact and engineers' ability to design it in a way that addresses international concerns.

The cascade of dams state-owned hydropower companies have built in the Dri Chu Catchment also rivals the Three Gorges's hydropower-producing capacity. As part of the Western Development program launched at the turn of the millennium, the central government planned a series of cascades down the Dri Chu, Dza Chu (Yalong), and eastern Gyelmo Ngul Chu (Dadu) rivers in the Dri Chu Catchment, and on the Da Chu (Mekong) and western Gyelmo Ngul Chu (Salween) rivers that run beside it. Only the dams on the Gyelmo Ngul Chu were shelved, and the rest have been built. The cascade of dams on the Dri Chu is particularly sizable. It is the world's largest hydroelectric generating system, including the Wudongde (10,200 MW), Baihetan (16,000 MW), Xiloudu (13,860 MW), and Xiangjiaba (6,840 MW) Dams, which together produce twice as much electricity as the Three Gorges Dam. There are plans for 21 smaller dams on the upper stream with a total generating capacity of 8,980 MW (see Figure 9.2).

Along with these large constructions that negatively impact riverine environments, the Chinese government has implemented some technocratic hard (or constructed) and soft (or governance) infrastructure to alleviate the effects of the dams and associated buildings on freshwater biodiversity and river health. Its engineers designed and installed fish-friendly turbines that reduce fish injuries and mortality across all affected river systems and constructed fish

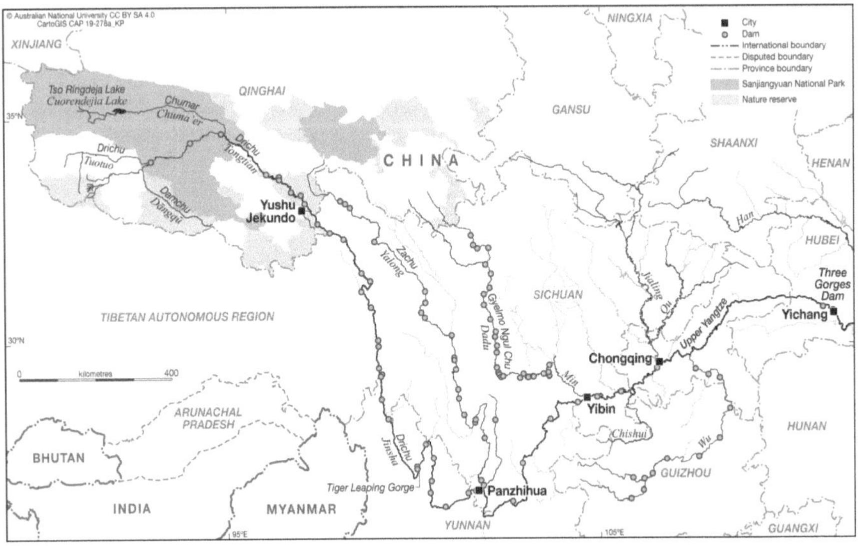

FIGURE 9.2 The hydroelectric and irrigation dams of the Dri Chu Catchment.

lifts to transport fish around three dams on the Da Chu (Mekong) (Xu and Pittock 2019), and their latest inventions seek to alleviate the shock to native fish species when they encounter cold water from dam outlets. This sudden drop in temperature can disrupt native fish species' metabolic and reproductive processes, many of which are adapted to specific temperature ranges. To lessen these effects, local and national governments have installed stratified water outlet towers on some dams that mix cold bottom water and warm top water before it is released (Xu and Pittock 2019). They have placed artificial fish breeding nests and created wetlands downstream of dams. This provides extra habitats for riverine species and gives them a space in which they can survive the intense, cold flows of peak releases from hydropower stations (Wang 2022c).

All these measures go some way to alleviate the effects of hydropower and other dams on fish, but there is no definitive evidence that these technocratic mitigation measures are effective (Xu and Pittock 2018, 2019). The environmental changes dams cause are difficult to offset. For instance, even if fish can migrate over dams using fish passes, the dams still fragment their habitats, leading to more vulnerable, isolated populations. There are also long-term concerns for such species' genetic diversity. Additionally, the complexity of river ecosystems means that assisting one species might not necessarily benefit the entire community. Furthermore, the overall effectiveness of these solutions remains uncertain.

The Chinese government's technocratic approach has also influenced the legislation and bureaucracy they use to manage the river. River management is conducted through national and subnational departments, including provinces, municipalities, district or county administrations, and township and village committees. The Ministry of Water Resources sets water management policies at the national level, while the Ministry of Ecology and Environment focuses on pollution control and conservation. Subnational counterparts then adapt their national guidelines to fit local needs and conditions. Regular communication between the two levels (national-to-province, province-to-municipality, municipality-to-district/country, district/county-to-township, township-to-village committee) is intended to ensure that nationwide conservation goals are met, but with enough flexibility to address the unique challenges faced by each region. While departments at each level often collaborate to manage water and environmental goals, tensions can arise due to differing priorities and local challenges.

Local governments are laboratories for policy innovations in China, so some of the most innovative governance responses to riverine management have occurred at the local level. One of these is the creation of a "River Chief" system. It was initiated in 2003 and applied nationally in December 2016 to enforce rules, improve accountability, and manage shared rivers across administrative regions and institutional mandates. River chiefs are leaders within the Communist Party at provincial, municipal, country, and township levels (see Figure 9.3 maps their responsibility). They promote water governance by incorporating all stakeholders into the water management framework under the unified leadership of county or municipal governments; they also seek to foster cooperation and mobilize resources to achieve water management goals.

One example of how this operates in practice is the way these departments worked together to conduct extensive fish surveys that concluded the rivers' tributaries provided habitats for most fish species. They used these findings to argue that the rivers' mainstems should be dammed for hydropower and their tributaries should be reserved as fish habitats (Xu and Pittock 2019, 2020a). This program has impressive results on the upper Mekong River's largest tributary, the Buyuan River, in southern Yunnan. This river provides a habitat for around 100 fish species, equivalent to 72.8% of the lower Mekong's total species (Xu and Pittock 2019). If this strategy is implemented rigorously for all rivers, most aquatic fauna could be conserved. Still, strict compliance with conservation regulations is required for these fish reserves to work as biodiversity sources. The Chinese government's response to the need for compliance is, as in many other areas of governance, rigorous technology-dependent monitoring, and strict enforcement. In the Dri Chu Catchment, for example, the Ya'an County Government has installed cameras, speakers, and microphones along the river to monitor its use in real time (see Figure 9.3).

To further ensure compliance and restrict illegal fishing in designated fish reserves, some local governments have extended riverine restrictions to

FIGURE 9.3 *Left*, A map outlining the parts of the upper Mekong in Yunnan for which the River Chief is responsible. *Middle*, Monitoring system that observes the rivers and riverbanks 24 hours a day. *Right*, Floodplain overpass on the 54-kilometer-long recreational bike path in the Chishui National Fish Reserve. Photographs, Hongzhang Xu, 2019.

conserve and restore entire mountain catchments. In some instances, this has restored localized ecosystems, but these programs have also accelerated livelihood changes for people who live there, and come at the cost of local genealogical knowledge (Xu et al. 2021a). By restricting local access to the river's ecological system services, these environmental programs have worked with other hard and soft infrastructure development—particularly transport and communication networks—to accelerate broader national trends toward industrialized agriculture and urbanization. These factors dismantle riverine lifeways by incorporating local communities into the broader national economy (Xu et al. 2021b).

While some tributaries and large sections of the rivers' mainstreams are dammed out and their catchments repurposed for intensive agricultural production, hydropower generation, and other industries, those conserved as reserves are repackaged as eco-destinations. In these restored tributary catchments, local communities are transformed into a commodified experience and become the labor force for a growing tourism industry (Xu and Pittock 2020b). Tourism development has also led to the construction of new infrastructure, like bridges and sealed roads, in some protected areas. One example of this kind of infrastructure is the 54-kilometer-long bike path overpass that now runs through the Chishui National Fish Reserve in the upper Yangzi Catchment in Guizhou Provence (see Figure 9.3).

Although it will ultimately help stabilize the Highlands' climate, China's ambitious climate policy will further fragment river systems. To peak emissions by 2030 and achieve net zero by 2060, the government must dramatically increase hydropower capacity and output across the country. It is focused on three primary renewable energy sources: solar photovoltaics, wind power, and hydropower. Around 60% of renewable capacity additions since 2000 have come from solar photovoltaics and wind power, with a record installation of 127 GW in 2020 and 92 GW in 2022. Wind and solar combined provide 11.7% of power production in 2021 (Zhou et al. 2022). Hydropower has provided 35% of total renewable capacity additions since 2000. This is set to increase as the government looks to exploit as much of the country's GPE as possible.

However, the degree of this fragmentation will depend on the type of hydropower the government chooses. If it decides to build fewer large-scale, hydropower-producing dams and smaller, pumped-storage hydropower facilities, this would have less impact on China's rivers. Pumped-storage hydropower facilities are more flexible than large-scale facilities and could complement the high temporal variability of wind and solar power production and act as "batteries" for these sources. When there is an excess of power generated from wind or solar, rather than letting that energy go to waste, it is used to pump water from a lower reservoir to an upper one. This water can then flow downstream and create hydropower when no other energy sources are available. With 30 GW of installed pumped-storage hydropower capacity, China already leads the world in this technology (Gilfillan and Pittock 2022a). The latest published Pumped-Storage Hydropower Development Plan (2021–2035) commits the government to extend this capacity to 120 GW by 2030 (NEA 2021). The increasing uptake of hydropower and pumped-storage stations will help China to achieve deep decarbonization of its electricity sector with increasing uptake of solar and wind (Wang et al. 2022c).

Placing pumped hydro projects off-river would be more environmentally friendly than those that have already been installed on-river. Moreover, there are more options for siting off-river developments, which would enable those with lower environmental and social impacts to be prioritized. Pumped-storage hydro projects are smaller than conventional hydropower dams and impact fewer habitats and species. Moreover, they can be integrated with many existing reservoirs (Pittock 2019). In some cases, the development and efficiency of pumped-storage hydro might even lead governments to reconsider the utility of the already-installed large-scale dams. If pumped-storage hydro can effectively store and provide necessary power, some large dams—especially older ones with significant safety, environmental, or social concerns—could be decommissioned. Any increase in dam construction, whether for pumped-storage hydro or conventional hydropower, will have some environmental impacts, but these should be limited as much as possible.

9.2 India: River Mismanagement

India's population size is similar to China's—around 1.4 billion—but it has a very different governance structure and geography. India's government is a democratic constitutional federation within which a central government negotiates with 28 states and 8 union territories. The balance of power between the states and the center has changed depending on the state and federal governments' relative wealth and changing politics. After independence from British colonial rule in 1947, India's central government adopted a planned, protectionist approach to its economy, colloquially known as the "License Raj." The government then liberalized its economy in the 1990s, which led to exponential economic growth (see Chapter 8). Since 2014, when the right-wing, nationalist Bharatiya Janata Party (BJP) government was re-elected, the power pendulum has swung toward the center. It may swing back to the states again.

India inherited a large hydraulic bureaucracy from the British colonists and maintained and expanded it after independence. For the first four decades of its existence, the Indian government single-mindedly pursued development through government planning. After it began to liberalize and open its economy following a balance of payments crisis in 1991, it pursued growth through government-led and free-market initiatives. This combination of historical factors led the state's water-focused agencies to systematically promote water infrastructure development at the expense of the ecological and cultural values provided by free-flowing rivers (Molle et al. 2009).

India's post-independence governments have pursued a "project approach" to development, which has aided their cause in multiple ways. First, having projects to announce has enabled the spectacle of development (if not actual development) and has produced multiple headlines and, at least in theory, stirred nationalist pride in India's postcolonial achievements. Second, the project approach has made it easier for the central and state governments to mobilize international capital from organizations like the World Bank and national capital from the country's most prominent corporate entities. Unfortunately, this highly centralized corporate control of river development has enabled powerful interests to enclose common pool resources, gain benefits, and perpetuate undemocratic institutions and social processes (D'Souza 2002). The project-based paradigm has also perpetuated itself as new initiatives are proposed to deal with previously created problems.

An early pre-liberalization example of this project-based approach was the Bhakra Dam, which sits on the Sutlej River (Langchen Tsangpo in Tibet) in the Indus River catchment. As noted in Chapter 8, Jawaharlal Nehru (1889–1964), then Prime Minister of India, likened this multi-purpose, hydropower, irrigation, and flood-prevention dam to a modern temple when he inaugurated it in 1963. "This dam has been built with the unrelenting toil of man for the benefit of mankind and, therefore, is worthy of worship," he said. "May you call it a Temple or a Gurdwara or a Mosque, it inspires our admiration and reverence" (Asthana et al. 2014, 64).

Following the large anti-dam movement against the giant, multi-purpose Sardar Sarovar Dam on the Narmada River in the 1980s (Baviskar 1997) and economic liberalization in the 1990s, two trends have shaped Indian dam building. First, the dams being built have tended to be smaller, single-use hydropower dams, and second, they have tended to be built in the Himalaya (Baruah 2020). These smaller projects are created within an approach known as "competitive federalism"; economic centralization and political decentralization encourage Indian states to compete to facilitate business expansion by expediting project approvals (Khalid 2020; Richards and Singh 2002). Within this political economy, individual states treat the social consequences of all large developments—the displacement of people, environmental degradation, and social polarization—in teleological terms as minor sacrifices for the greater good (D'Souza 2004). Those who uphold this model view the construction of dams in a positive light, particularly for those states in the Himalaya, within whose borders lie intense levels of GPE.

According to the National Register of Large Dams, there are 5,264 completed large dams in India, and 437 more are under construction (Central Water Commission 2022). Their construction often attracts local protests (see Box 9b). Still, they do not usually receive the same international attention that the anti-Narmada River dam movement garnered in the 1980s and 1990s. Many policymakers and developmental theorists assume that achieving adequate mitigation for these projects is possible by reducing the size of dams, paying compensation, and taking other steps to deal with the ecological damage and culture fragmentation (Ministry of Power 2022). Consequently, they have not systematically studied the socio-cultural impacts of these dams or required systematic freshwater biodiversity conservation in and around them.

Along with interstate development competition, the push to extract more hydropower from the Himalaya and elsewhere in India has also been driven by the central government's ambitious decarbonization targets. Although its per-capita emissions are relatively small, India's large population makes it the third largest carbon emitter after China and the United States. Prime Minister Narendra Modi (1950–), leader of the BJP-led governing coalition, has set a target of net zero greenhouse gas emissions by 2070. By 2030, Modi says India will reduce the greenhouse emissions intensity of its economy by 45% (Menon 2022). Modi has also pledged to increase non-fossil fuel energy capacity to 500 GW by 2030, which will account for 40% of all electric power. Hydropower will play a starring role in this transition. India has created the world's largest pipeline of new hydropower projects to fulfill the country's target of 500 GW of renewables by 2030. In the next decade, it plans to construct 91 GW of hydropower stations, nearly double the number China intends to build (Ferris 2022). Even though the likelihood of all these dams being constructed is relatively small as they have been inflated by bureaucratic and political corruption, rent-seeking, anti-dam campaigns, and legal appeals (Alley 2017), even this smaller number of dams will transform the Himalaya's river systems.

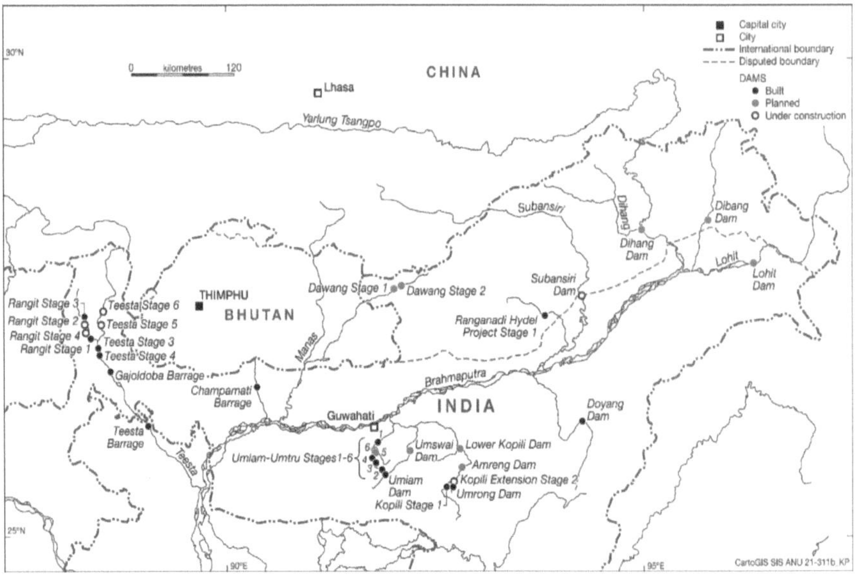

FIGURE 9.4 Hydropower projects in the upper Brahmaputra Catchment, Northeast India.

The pressure to build hydropower projects in the Indian Himalaya does not stem from the Indian state alone. International organizations such as the World Bank have also encouraged it. After the condemnatory World Commission on Dams report, the World Bank pulled out of dam financing during the 1990s and 2000s but began re-encouraging their construction over the past decade for carbon reduction purposes (see Chapter 8). International funders such as the Clean Development Mechanism (CDM) have begun to fund hydropower construction in India, particularly in the Himalaya. The CDM funded the Rangit II (2015), Jorethang Loop, and the recently destroyed Teesta III hydroelectric projects in Sikkim (see Box 9b), as well as the projects that were planned near Tawang (2023) before being abandoned after a long-running protest movement (Gohain 2020a).

A dramatic increase in the use of pumped-storage hydropower may aid India's ambitious plan for new and low-cost domestic energy. As wind and solar provide 41.72% and 42.04% of renewable power, respectively, the energy storage that pumped-storage hydropower provides is desperately needed (Haldar and Modak 2022). As of 2023, about 2.5 GW of pumped-storage hydropower is operating, and projects that will store another 3.1 GW are under construction (Buckley and Shah 2019). Despite the low levels of infrastructure in this field, India is still on the path to becoming a world leader in pumped-storage hydropower.

While India's states—particularly those in the Himalaya—compete to develop hydropower, its central government is trying to manage India's potentially catastrophic water scarcity. India is one of the world's most water-stressed countries; about 1 billion people live with water scarcity. Growing competition over finite water resources, compounded by climate change, has profound implications for India's food security, farmers' livelihoods, and economic development (Singh and Kumar 2021). Half of India's annual precipitation falls in 15 days, which means that floods and droughts are a fact of life (Sarkar 2022). In 2016, around 78% of water in India was used in agriculture (Gulati and Banerjee 2016). Rapid industrialization and urbanization further increase competition for water (Chopra and Ramachandran 2021). Government policy and local use have relied on unsustainable extraction from rivers and groundwater to feed their communities' insatiable and ever-increasing water needs. The practice of extracting water from rivers and through tube wells (also called bore wells) has existed for millennia but was intensified by mechanization and engineering projects during the British colonial period and shifts in agricultural practice during the Green Revolution of the 1970s and 1980s (see Chapter 8).

India's "project-led" development approach and severe water scarcity underpin the drive to exploit its rivers' water for domestic, agricultural, and industrial uses. Along with Mexico, it was one of the first two Global South countries to develop and adopt a national water policy in March 1987. Since 1987, various iterations of the center's water ministries have revised this policy twice, in 2002 and 2012 (Kumar 2018). These three plans have focused on the catchment as the primary hydrological unit in planning, developing, and managing water resources (Pandit and Biswas 2019). Despite this approach, no catchment plans have been developed in India. When borders dissect catchments, each state is separately responsible for planning and managing water resources within its jurisdiction. For catchments within a state, the planning and management of water resources is led by individual projects. This is especially true for large water infrastructure (Pandit and Biswas 2019). As the country's demand for water has increased, river water has often been overallocated and is regularly extracted outside the replenishing monsoon seasons (Parween and Singh 2021).

Groundwater exploitation has led to even more drastic reductions in available water. Since the 1970s, India's farmers have increased their use of tube wells to produce a second seasonal crop and convert large areas of previously rain-fed fields into groundwater-irrigated agriculture (Dangar et al. 2021). This has helped produce many more food crops for India, which has not experienced a widespread famine since the Green Revolution, but these practices have also depleted groundwater supplies by about 2.8 centimeters per year (MacAllister et al. 2022). India is now the largest user of groundwater worldwide; it pumps 25% of all the groundwater extracted globally.

Over half of India's districts are threatened by groundwater depletion or contamination. If current trends persist, 60% of India's communities will likely see groundwater tables fall to critical levels within two decades, which will place at least 25% of the country's agriculture at risk of failure (Dangar et al. 2021).

A manifestation of the trend within India to produce project-led solutions has been the development of the National River Linking Project, an extensive infrastructure plan to "solve" India's uneven water supply by directing water from water-rich areas like the Brahmaputra catchment to more arid regions in the country's northwest and center. Engineers within what was then called the central government's Ministry of Irrigation were the first to conceive this idea in the 1980s. It was not implemented until the BJP government took office in 1999 and began championing it (Gupta and Deshpande 2004). The plan was abandoned after the Congress Party came back into power between 2004 and 2014, but the BJP reinvigorated it when they were re-elected in 2014. As of 2023, this large infrastructure project is being led by the National Water Development Agency, which sits within the powerful Ministry of Jal Shakti (Water Resources), a mega-department created in 2019 through the amalgamation of three water bureaus, including the erstwhile Department of Irrigation. The National River Linking Project has three components: in the Himalaya, the southern peninsular, and between large states in the country's center. In 2022, the central government identified 30 river-linking sites, 14 of which are in the Himalaya (*The Hindu*, December 8, 2022).

Many civil society representatives in India have been and remain critical of India's river management (Pandit and Biswas 2019). They claim government management practices do not consider the complexities of IWRM, which has led to a lack of river catchment management organizations and catchment-level planning, unregulated groundwater extraction, and water pollution (Pandit and Biswas 2019). The combination of the large-project-based approach from the central government and economy-centric dam development strategies within states has disempowered the local communities and generated multiple geological, environmental, and social hazards (Huber 2019).

Many civil society organizations have consistently opposed engineering-based water management, such as large dams, river linking, and canal irrigation, and the related multiple threats posed by climate change, particularly glacial lake outburst floods (GLOFs) in the Himalaya (see Box 4a), have intensified these protests. This threat is evident from the Dhauliganga River floods in Uttarakhand in February 2021 (Li et al. 2022) and the 2023 GLOF in Sikkim (see Box 9b), which were deadly and destroyed hydropower constructions. This energized anti-dam social movements from Indigenous activists along the Tawang, Siang, Teesta, and Subansiri rivers, and their efforts have already slowed down and even stopped the construction of dams (Bej 2020; Bentley 2021; Gergan 2020).

Internationally, India operates as a middle riparian nation, upstream from Pakistan and Bangladesh and downstream from China, Nepal, and Bhutan. It shares three large river systems with China and other neighbors: the Indus catchment with China, Pakistan, and Afghanistan, the Ganga catchment with China and Nepal, and the Brahmaputra with China, Bhutan, and Bangladesh. While China has often been portrayed negatively in the Indian press regarding its role as an upper riparian nation, the reality is more nuanced. Beijing has made limited efforts to engage in transboundary water cooperation and—except for three years following the 2017 Doklam standoff—has shared limited hydrological data on the Yarlung Tsangpo's flow with India. However, concerns persist in India over China's dam constructions and potential water diversions, which has led to calls for greater transparency and collaboration between the two countries.

Although not an upper riparian polity like China, India is nevertheless a hydro-hegemon regarding resource utilization and in how it employs control strategies. Like China, India has consistently opposed measures to agree on multi-lateral obligations for managing shared rivers sustainably and equitably, such as the UN Watercourses Convention (Barua et al. 2018). Instead, it has entered multiple bilateral water-sharing treaties and transborder hydropower projects with its riparian neighbors. The 1948 Inter-Dominion Accord shares the Indus catchment's waters between India and Pakistan. India has also signed four—arguably unequal—treaties with Nepal over the largest Ganga tributaries, and many regard India's plans for water diversion, river linkages, and dam-building as a "resource capture" strategy to take water from downstream Bangladesh (Ho 2016). The National River Linking Project, for example, includes plans to connect 37 rivers through 31 links with 9,000 kilometers of canals to divert flows from the Ganga and the Brahmaputra. Bangladesh has raised strenuous objections to this proposal and challenges any reduction in its share of Brahmaputra and Ganga waters. If India ignores its opposition and conducts the project as planned, Bangladesh could face an estimated 10–20% reduction in Ganga flow, which would desiccate large areas of its land (Ho 2016).

9.3 Bhutan: A Nature-based Approach to Managing Rivers

Bhutan's development policy is conservation-centered and prioritizes nature-based solutions (see Chapter 10). According to the Bhutanese Government's Bhutan Water Vision 2025, the overarching goal of its water management strategy is "Supporting life, prosperity and gross national happiness" (National Environment Commission 2015). By taking this approach, Bhutan's planning sets aside water for nature conservation; this includes the prevention of natural disasters, which serves as a guide to all water allocation decisions (National Environment Commission 2015). Along with sustainable use and management of water resources, this policy also insists on equitable water use rights that respect local traditions. It emphasizes that water administration should be responsive, transparent, and accountable (National Environment Commission 2015).

Five of the Brahmaputra's Himalayan tributaries descend through Bhutan: the Amo Chu (Torsa), Wang Chu, Puna Tsangchu (Sunkosh), and Drangme Chu (Manas) (see Chapter 7). These rivers combine with the South Asian Monsoon (see Chapter 3) to make Bhutan a water-rich country. Its per capita, per year water availability is more than 100,000 cubic meters (National Environment Commission 2015). Although Bhutan has only tapped 1.4% of its hydropower potential, this hydropower generation has already made it one of the world's first negative-emissions countries (Yangka et al. 2019). However, Bhutan still has water management issues. Some of its regions experience water scarcity; there is frequent flooding and seasonal water shortages for drinking and agricultural purposes. About 22% of Bhutanese people do not have safe potable water, and only 12.5% of arable land is irrigated (National Environment Commission 2015). Moreover, its rapidly increasing population and growing demands for water and energy create water resource management challenges (Yangka et al. 2019). Another issue the country's water managers must confront is its rivers' increasing sediment loads and, therefore, the decreasing productivity of its hydroelectricity projects and the shrinking of its dams' life spans (Vaidya et al. 2021).

It could be argued that Bhutan's positive conservation focus and abundant water supplies overshadow other issues with its water management approach. The Ministry of Agriculture oversees irrigation, land use, and hydropower development. The Ministry of Economic Affairs is responsible for hydro-meteorological and GLOF data collection (National Environment Commission 2015). This lack of river management integration forestalled the creation of a National Integrated Water Resource Management Plan until March 2016 Commission (National Environment Commission 2016). The plan is implemented by five river catchment committees that are responsible for the country's major rivers. The National Environmental Commission oversees and coordinates the committees' work Commission (National Environment Commission 2016).

Although Bhutan—unlike other states in the region—has adopted an IWRM approach to river management, its fragmented institutional arrangements and an absence of effective and practical coordination mechanisms have limited IWRM's effectiveness (Tariq et al. 2021). This effectiveness is substantially undermined by Bhutan's lack of a central water resource agency that could oversee water management (Lhamo and Chhetri 2022).

Changing extraction and climate patterns also threaten Bhutan's river management. The country's constitution mandates that forests cover 72% of its territory, and more than one-fourth is protected area (Dorji 2016).

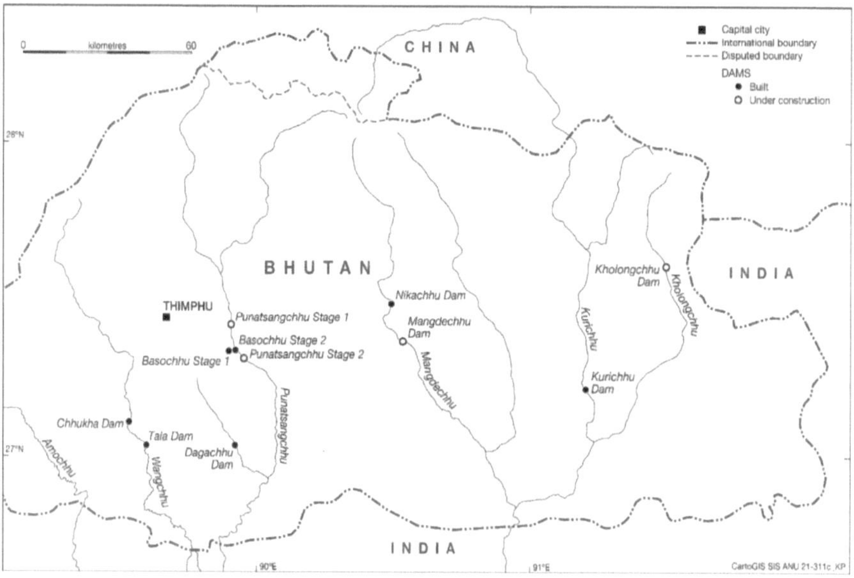

FIGURE 9.5 Operational and under-construction hydropower projects in Bhutan.

Although timber is harvested under sustainable management plans, increasing demand for wood, firewood, and non-timber forest products has extended the exploited area. Consequently, encroachment and bushfires have become new challenges for catchment health and freshwater biodiversity conservation (National Environment Commission 2015). Climate change has led to an increase in the occurrence of GLOFs (Hagg et al. 2021), and Bhutan must improve its river management systems to deal with both this threat and its increasingly fluctuating water resources.

Along with the need to manage internal threats and ineffective systems, Bhutan's efforts to maintain river health are also compromised by its commitment to hydropower, both to maintain its negative carbon emissions status and to balance its trade with India. At the fifteenth Council of the Parties on Climate Change (COP15) in 2009, Bhutan committed to remain carbon neutral (National Environment Commission 2020). To do this, it needs to mitigate its growing emissions from economic development and a booming population. The government laid out its plan to maintain carbon neutrality and adapt to a changing climate in the Climate Change Policy of the Kingdom of Bhutan (2020). Although the policy adopts an "ecologically balanced sustainable development" approach to pursue carbon-neutral development, it has also allowed dam developers to leverage its emission reduction credits scheme to expand hydropower projects and, by doing so, gain carbon credits they can sell to mining, transport, and other carbon-intensive sectors (National Environment Commission 2020).

This situation may be shifting. Growing domestic electricity demand and the need to import electricity from India in the dry season are impacting Bhutan's balance of trade. India's control of electricity export and import markets via the Indian Energy Exchange also disadvantages Bhutan. Consequently, the Government of Bhutan is looking to develop greater energy self-sufficiency through solar and pumped hydropower developments, which might reduce environmental and social impacts. Bhutan plans to incorporate a further 300–400 MW of solar capacity by 2026 (Khan and Fenton 2023). The government is also exploring green hydrogen and other alternative renewables (Lhaden 2023).

Bhutan's complicated relationship with its southern neighbor and largest trading partner, India, which sees Bhutan as a *de facto* protectorate and is another source of pressure to expand its hydropower industry. The Indian central government and state-affiliated companies have investments in most of Bhutan's hydropower dams and import the majority of the electricity they generate (Yasuda et al. 2017). This situation is unlikely to encourage Bhutanese administrators to adequately consider all environmental and sustainability impacts when assessing a hydropower project's viability. It may also be part of the reason that, to date, there has not been a comprehensive study of the ecological effects of hydropower projects in Bhutan (Yasuda et al. 2017).

9.4 Bangladesh: An Adapting Approach to Managing Rivers

Bangladesh's land mass is created by the sediments that flow down from the four large river systems that converge within it: the Brahmaputra-Jamuna, Ganga-Padma, Surma-Meghna, and Karnaphuli (Ministry of Water Resources 1999). All four of these rivers originate in other countries, and Bangladesh is the last riparian—and, therefore, the most vulnerable—country along their course. Its water management challenges depend more on what upstream countries do than its own actions. It is particularly impacted by the decisions made by its closest neighbor, India (Qadir 2008). Within these four large river systems, 57 tributaries and mainstreams cross borders to enter Bangladesh; 54 flow into Bangladesh from India (Chowdhury 2009). Decisions made upstream threaten Bangladesh's riverine health by intensifying riverbank erosion, creating floods, diminishing water flows, and diminishing dry-season groundwater availability (Chowdhury 2009).

Bangladesh has more issues with floods than droughts. Since pre-independence, avoiding flood-induced famines and attaining food security have been the defining motivations in Bangladesh's water management (Gain et al. 2017). However, like much-drier India, disease rather than desertification has forced its people to rely on tube wells. Bangladesh experienced catastrophic cholera epidemics and growing irrigation demand, both of which drove extensive investments in shallow tube wells. These wells are still a primary water source for many in Bangladesh. Because of Bangladesh's geography, the country will not face the same groundwater depletion rates as India, but it still faces the consequences of overextraction, particularly around Dhaka, its capital city, where groundwater extraction exceeds recharge.

Unlike India, Bangladesh's groundwater often contains unhealthy levels of heavy metals, including arsenic (Parvin 2021). Arsenic is produced throughout the Asian Highlands by dynamic geological processes (see Chapter 2). The Highland rivers then deliver it to Bangladesh, where it settles in groundwater. The intensified use of tube wells for drinking water and agriculture has led to mass arsenic poisoning in Bangladesh, where thousands die of arsenic poisoning every year (Pitt et al. 2021). Bangladesh's government must find a way to increase water usage without increasing or intensifying this poisoning.

The Brahmaputra-Jamuna system's sediment flows are not only problematic because of their arsenic content but also because of their sheer volume. The Brahmaputra system is one of the planet's most heavily sediment-laden large rivers (see Chapters 2 and 3). Accretion reduces navigability without adequate dredging and management as the water becomes too shallow for boats, particularly those with deep hulls. Reclaiming the navigability of Bangladesh's river networks is essential for improving regional integration and enhancing transboundary economic cooperation. However, dredging must be managed carefully for multiple social and ecological reasons. Without adequate safeguard measures and proper technologies, dredging activities can be captured by illegal sand miners seeking construction materials. Unmanaged sand mining and dredging can cause significant upstream and downstream river bank erosion, unearth toxic pollutants, and cause severe damage to biodiversity (The World Bank 2020).

Bangladesh's government must manage all these riverine issues as its economy transforms from primarily agricultural to industrial and service sectors. This shift not only throws water management into flux but also creates more demand for this scarce resource. Industry accounts for only 2% of water withdrawal at the beginning of the 2020s, but by 2050 is expected to increase by 440% (The World Bank 2020). Consequently, leveraging regional cooperation has become Bangladesh's focus for water management in the short to long term (The World Bank 2020).

Bangladesh's National Water Policy prioritizes river catchment planning and management (Ministry of Water Resources 1999). It is also the strongest advocate in the region for the catchment-wide management of its international rivers. As the Bangladeshi government sees it, international river management should include water-sharing agreements, data exchange, resource planning, and sustainable management (Ministry of Water Resources 1999). Bangladesh needs these river management regimes to help it adapt to other increasing vulnerabilities, including climate change. It cannot address water shortages and agricultural disruptions without a sustainable water management relationship with other riparian countries.

Although India and Bangladesh signed water treaties and have a Joint Rivers Commission to address transboundary water disputes, they also have disagreements on how to allocate water from the Ganga, Brahmaputra, Meghna, Teesta, and several other shared rivers (Baten and Titumir 2016). Bangladesh's water management experts are especially concerned about how India's River Linking Project will affect them. India's plans to transfer water from the Ganga-Brahmaputra to the Mahanadi catchment to mitigate water scarcity in western and southern India will result in the diversion of this much-needed fresh water away from Bangladesh (Baten and Titumir 2016). Bangladesh began expressing concerns about this project in 1982, soon after it was proposed. Since then, India has published more than 16 feasibility reports on the project without consulting Bangladesh (Baten and Titumir 2016). Compared to interactions with India, Bangladesh's much-less-fraught, limited cooperation with China and Bhutan has proceeded smoothly (Samaranayake et al. 2016).

Bangladesh must manage crucial water issues to adapt to changing circumstances, including groundwater depletion, contamination, sea-level river, flood, drought, transboundary water sharing, and climate change-induced water vulnerabilities (Gain et al. 2017). In the Bangladesh Delta Plan 2100, Bangladesh has initiated an ambitious integrated water resource management strategy in which water is central to socioeconomic development (Hadi 2019).

9.5 River Management as State Activity

State efforts to regulate the Asian Highland rivers demonstrate an objectives-oriented view toward managing and using them. This orientation has led Chinese, Indian, and Bhutanese state and business interests to construct large-scale, concrete-heavy projects that often significantly affect the hydrological cycle. In China, this objectives-oriented view is merged with its "techno-political regime" (Crow-Miller et al. 2017). It was this technocratic approach that facilitated the establishment of the corporate-government partnership that constructed the world's largest dam on the Yangzi's Three Gorges and the South-North Water Transfer Project. Crow-Miller et al. describe this regime as intent on going beyond mere infrastructure or ecology, manifesting doubly as the "control of nature" and a "technocratic vision of national development" that implicates water with political subjectivities (Crow-Miller et al. 2017, 234). One of the implications of this approach is its integration of human societies into the management and control of rivers. While this is particularly true of the Chinese state's technocratic approach to river management, it is also true of other states. Even Bhutan, whose more nature-orientated water management is less technocratic and more sustainable than either China or India, struggles to balance coexistence with available water resources as its population booms.

Moreover, most states' large-scale riverine activities have both intended and unintended consequences. The resettlement of communities for dam construction is a case in point. The Chinese government intentionally displaced an estimated 1.13 million people (according to official statistics) to allow for the 632-kilometer-long reservoir they created behind the Three Gorges Dam (Wilmsen et al. 2011). At the same time, it implemented a reportedly involuntary displacement program with a combination of compensation schemes, migration strategies, and development promises, which—despite disputes and disagreements—were carried out as intended (Wilmsen et al. 2011). The Indian government's involuntary resettlement program to facilitate the construction of the Sardar Sarovar Project on the Narmada River was much smaller but threatened entire Adivasi (Indigenous) communities (Garikipati 2002).

Communities often respond to these projects in ways that authorities do not anticipate. Against all expectations, Adivasi groups resisted the Narmada River's damming and their resettlement for two decades. In the upper Brahmaputra catchment, Rong (Lepcha) and Bhutia groups protested the hydropower dams that transformed the Teesta River (see Box 9b), and Monpa offered up fierce resistance to a dam cascade on the Nyamjang Chu in Tawang, Arunachal Pradesh. Two anti-dam campaigners were killed while protesting the Nyamjang Chu dams, but the movement stopped their construction by questioning the project's impact on the nesting grounds of the endangered and sacred black-necked crane habitat (Gohain 2020b).

The entanglement of environmental management and impacts on—and from—local human populations, with both expected and unexpected outcomes, "positive" and "negative" consequences, may be expected. What is less immediately apparent is that the state-led techniques of government—for example, policies of national development and regulations on environmental strategy and use—ultimately cannot be separated from the larger project of governmentality itself, namely the control of conduct and the legibility of human populations (Foucault 2003; Scott 1998). To understand how governments and governmentality work, we must re-assess our understanding of the state itself, which often goes un-examined in conservation and environmental analysis. This analysis leads us to question whether states are the inevitable outcome of the evolution of human society. As James C. Scott (2017) demonstrated in his study of early states' archeological records, there is much evidence, by contrast, for the consistent *in*stability of this form of social organization. For states to continue to exist, they must constantly assert and re-assert their dominance. Correspondingly, coercion and labor are required to maintain a state, and increasingly, one contemporary mode of assertion is environmental governance.

These forms of state coercion are particularly evident in the Chinese state's governance of the Tibetan Plateau's high-altitude pastures. Yeh (2005) has called this "green governmentality," and, more recently, Shen and Jiang (2021) have described it as "authoritarian environmentalism." These pastures include Sanjiangyuan (the Three Rivers Region), which lies east of the Dangla Mountains, where the Da Chu (Mekong), Dri Chu (Yangzi), and Ma Chu (Huang He, Yellow River) headwaters coalesce. Government discourse first reframed Sanjiangyuan and other territories as environmentally "degraded" and then used this reframing to justify environmental "improvements" that included the resettlement or management of the territories' communities.

These policies may have been developed away from the affected territories in separated organs of state, but they were enacted on them. We know from Foucault's work on governmentality as "the conduct of conduct" that human subjects are not removed from these processes and that policies are never only top-down. The work of a complex of subjects, from officials and citizens to non-government workers and local—usually Indigenous—people combine to enact the continuation of the state as the dominant form of human social organization. In his study of environmental governance in the Western Himalaya, Agrawal (2005) used the term "environmentality" to describe how individuals who had been previously ambivalent toward village forests later began to participate in forest regulation after involving themselves in institutions of environmental regulation. His description of this transformation provides insights into the organization of human actions and societies and their impact on environments and rivers, while underscoring the ways humans and environments co-form complex socio-ecological systems.

9.6 Conclusions

Focusing on the status quo, policy progress, and future challenges of country-level river management in China, India, Bhutan, and Bangladesh has emphasized how individual states approach river management as a 'zero-sum' game, placing nation-states, especially China and India, in competition with each other. Yet, as this book's findings have already highlighted, river systems are complex socio-ecological systems that are to be understood holistically rather than through the lens of national and territorial boundaries.

As all regional governments, particularly China, India, and Bhutan, try to decarbonize, they will likely increase the number and size of hydropower projects within areas of high GPE. If done well, with a focus on low-impact pumped-storage dams, these projects could reduce carbon emissions, complement solar and wind power, and manage disasters and increasing runoff from melting glaciers. However, the global warming-induced melting and thawing of the cryosphere is severely altering the region's hydrological regime. The complex set of interacting processes that this heating has unleashed is destabilizing landscapes throughout the region. It has caused glaciers to retreat and detach, permafrost to thaw, and increased the frequency and intensity of landslides, rock-ice avalanches, debris flows, and outburst floods from glacial lakes and landslide-dammed lakes (Li et al. 2022).

How to deal with the trade-off between climate change mitigation and adaptation will remain an issue for all states within the entire Brahmaputra and Yangzi catchments. Territorialization has established a competitive model and the nation-states should switch to a more cooperative model enabling more dialogues and fewer speculations. The most effective response would be for governments to work across borders and with local communities to balance social, economic, and environmental needs. But, given the region's complex social and geopolitical issues, this seems unlikely in the foreseeable future.

BOX 9a ECOSYSTEM SERVICES

Over the past half-century, biodiversity conservation has progressively embraced more integrative and holistic frameworks. One of the most influential adopted by international conventions is the "Ecosystem Approach," a comprehensive strategy advanced by the Convention on Biological Diversity (CBD) that has since been promoted through case studies and technical guidance (see CBD 2009).

The approach applies scientific methods to ensure ecological systems management takes account of the processes, functions, and interactions that exist among organisms and their environments. Importantly, this means it also recognizes that culturally and socially diverse peoples are an integral component of ecosystems. This adaptive management approach considers the dynamic nature of ecosystems and the surprises and uncertainty that arise from our absence of knowledge about them. Recognizing this lack of knowledge, the Ecosystem Approach acknowledges that, in some instances, it may be necessary to implement management measures even when cause-and-effect relationships are not clearly established.

The CBD's Ecosystem Approach operates through complementary principles (CBD 2009) and provides a five-point operational guidance for application (CBD 2009):

1 *Focus on the relationships and processes within the ecosystem.* This focus enables a better understanding of ecosystem resilience and the effects

(*Continued*)

on biodiversity loss and habitat fragmentation, the underlying causes of biodiversity loss, and the determinants of local biological diversity in management decisions.

2 *Enhance benefit-sharing*, particularly those that benefit the stakeholders responsible for the systems' production and management.
3 *Use adaptive management practices and include a learning process.* These help to adapt methods and practices to the ways in which systems are being managed and monitored.
4 *Carry out management actions at an appropriate scale.* This geographic, governmental, or community scale should suit the issue being addressed and be decentralized to the lowest appropriate level.
5 *Ensure intersectoral cooperation* in developing and reviewing national biodiversity strategies, action plans, and the integration of the ecosystem approach. This should be implemented in agriculture, fisheries, forestry, and other production systems that impact biodiversity.

In the same decade that the Convention was developing the idea of an "Ecosystem Approach," the Millennium Ecosystem Assessment (MEA 2005) defined the related concept of "ecosystem services" as the benefits people obtain from ecosystems. The Assessment identified four broad categories of sometimes overlapping services:

1 *Provisioning:* products obtained from ecosystems, including food, fresh water, and fuel.
2 *Regulating:* benefits derived from the regulation of ecosystem processes, such as erosion control, storm protection, climate and water regulation, water purification, and waste treatment.
3 *Cultural:* non-material benefits obtained from ecosystems, including cultural diversity and heritage, aesthetic values, and recreation and tourism.
4 *Supporting:* services that are necessary to produce all the other services, such as soil formation and retention, primary production, nutrient and water cycling.

The underlying principle of this system uses economic values to define the worth of different ecosystems. One goal of monetization of the environment is to make its worth more obvious to economy-focused government systems. This approach is often cited in arguments against the system. However, while it is often cited in arguments about the conversion or degradation of wetlands for purported economic purposes, proponents have not developed a global application of the argument in favor of conservation (or restoration) (MEA 2005).

Wetland conservationists had previously developed a suite of less well-known systems to value and evaluate wetlands' ecosystem services (e.g., de Groot

(Continued)

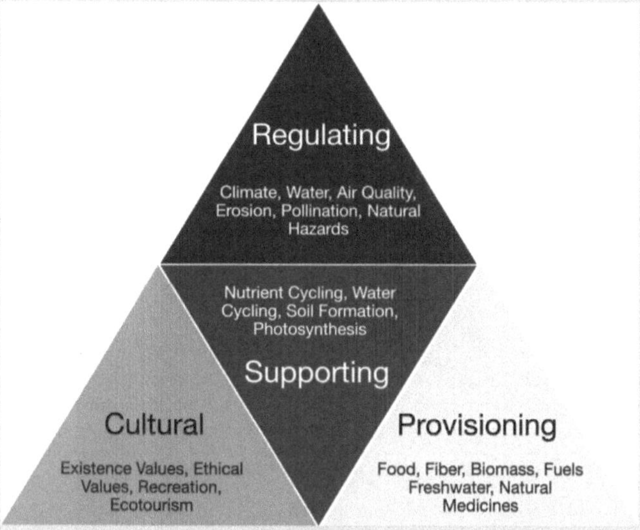

FIGURE 9A.1 General relationship between ecosystem services and the constituents of human well-being. Based on Achaempong 2020.

et al. 2006). They include quantitative and qualitative approaches and deal with valuation, evaluation, and communication. While a number of these valuation techniques are debated, the increasing use of approaches that focus on the marginal cost of change and more valid responses of people in stated preference surveys make estimating the value of wetland ecosystem services more practical and useful (Hatton et al. 2010).

In the Asian Highlands, Chinese institutions have explored and applied aspects of this ecosystem services approach as part of centralized and technocratic policy efforts to enhance sustainability through applying scientific knowledge (Chen et al. 2014b). China's ecosystem service regime has been expanded and applied to other countries in the Himalayan region, such as Nepal (Kandel et al. 2021). Recent studies, particularly from the Tibetan Plateau, demonstrate that from 1990 to 2015 the overall ecosystem service values increased; this was primarily driven by the growth of aquatic systems and a decline in barren lands and glaciated regions (Jiang et al. 2020b). Jiang et al. (2020b) argue that the elasticity analysis of these values, though not highly sensitive, can be instrumental in pinpointing regional anomalies and emphasizing areas that urgently require ecosystem-focused interventions. Bhutan and India have not applied these concepts through national policy in this way.

(Continued)

TABLE 9A.1 Ecosystem Services from Highland Ecosystems

Ecosystem services and the cryosphere	
Cultural services Spiritually significant sites Places of cultural practice Recreation and Tourism	
Regulating services Maintains river flow Natural water storage	**Provisioning services** Water Habitat for specialized species
Supporting services Maintains regional climate	
Ecosystem services and free-flowing rivers	
Cultural services Spiritually significant sites Places of cultural practice Recreation and Tourism	
Regulating services Maintains river flow and sediment transport Water purification through natural processes	**Provisioning services** Water Fish and other aquatic life for food
Supporting services Maintains regional climate Supports riverine biodiversity	
Ecosystem services and dammed rivers	
Cultural services Spiritually significant sites Places of cultural practice Recreation and Tourism	
Regulating services Maintains river flow and flood prevention Potential hydroelectric power generation Controlled sediment retention	**Provisioning services** Water Power Aquaculture
Supporting services Maintains regional climate Sediment capture and altered nutrient cycling	

BOX 9b ANTI-HYDROPOWER PROTESTS IN SIKKIM

The Teesta River flows for 414 kilometers from the glaciers at the base of Kangchenjunga (8,586 meters) down through the gorges of Sikkim to the plains of West Bengal before emptying into the Brahmaputra River in Bangladesh. As it travels through Sikkim, it drops precipitously from 5,000 to 330 meters above sea level. Through this steep descent, the Teesta possesses some of Earth's most intense GPE (see Box 3a) and can produce 3,735 megawatts of electricity (Chauhun 2023).

Since the mid-1990s, various governments and hydropower companies have proposed at least 45 projects to capture this energy. Seven have been built in Sikkim, 15 are being constructed, and 28 are planned. Two more large dams have been built on the Teesta just south of the Sikkimese border in West Bengal (Chauhun 2023). Given Sikkim is only 7,000 square kilometers (approximately the same size as Greater Chicago), these projects are transforming the small state's society and environment. The dams' proponents, including the Sikkimese, West Bengal, and central Indian governments, insist that they produce clean energy, create employment, and generate revenue. But the construction of the dams has also provoked large protest movements inside Sikkim.

Indigenous people whose homelands are threatened by dam construction and experts concerned by environmental and safety hazards have led these protests. The protests began in the 1990s and continued for decades, but their intensity has fluctuated with the state's changing political ecologies. They gain momentum whenever a new dam is proposed, or a disaster occurs. In October 2023, the state experienced its worst flooding disaster in decades when a GLOF burst from South Lonak near Kangchenjunga, decimated Sikkim's largest dam, the Teesta III at Chungthang in North Sikkim (see Figure 9b.2), and then traversed the state, killing at least 50 people, washing away bridges, and destroying property. Despite the ongoing protest movements against dams, this tragedy, and the multiple ongoing threats dam building poses to Sikkim and its people, neither the local nor national government has deemed it necessary to halt dam construction.

9b.1 Background

Sikkim is an Indian state in the Eastern Himalaya, bordered by Nepal to the west, the Tibetan Autonomous Region (China) to the north, and Bhutan to the east. The state's borders are the Teesta's watershed. Its famous mountain, Kangchenjunga, crowns an area of intense biodiversity, which UNESCO recognizes as a World Heritage Site. Around a third of the state's area in the north is situated within the Kangchenjunga National Park and the adjacent biosphere reserve.

(*Continued*)

Sikkim is a multiethnic state. The Lhachenpa and Lhachungpa people are Indigenous nomads whose homelands are Sikkim's northernmost and highest-altitude valleys. The Rong (also known as Lepcha) homelands are British-ruled, in Sikkim's middle hills, which are confusingly still called "North Sikkim." The Limbu people's lands stretch from southwestern Sikkim into neighboring Nepal. Another community, now known as the Bhutia, moved into Sikkim from neighboring Tibet or, more specifically, according to their oral histories, Minyak in the Dri Chu Catchment around the sixteenth century (see Chapter 8). The Bhutia understood Sikkim to be a special kind of *ne* (*lha-lu* abode) called a *beyul* (hidden land), which had been blessed by Guru Rinpoche (Bhutia 2021). They settled in West Sikkim, establishing the Namgyel Kingdom there in the seventeenth century, and gradually took control of the entire Teesta catchment uplands, from Kangchenjunga in the north to the Bengali plains in the south.

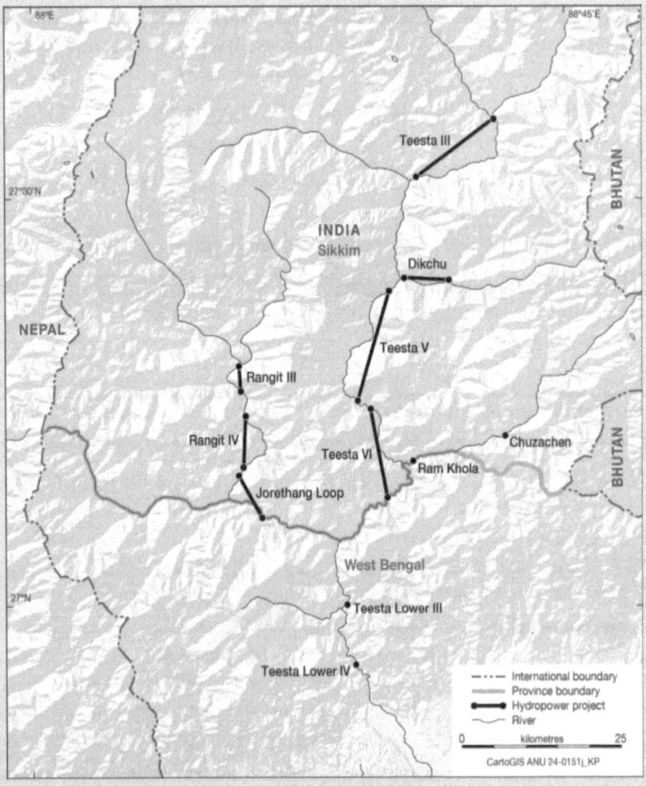

FIGURE 9B.1 Major hydropower projects operational (or under-repair after 2023 GLOF) in Sikkim.

(Continued)

The Namgyel Dynasty lost its southernmost districts, Kalimpong and Darjeeling, to invading Gorkha armies in the 1780s and again to the British a few decades later. Sikkim became a British protectorate from the 1880s to 1917 and was the launchpad for their Tibetan invasion (see Chapter 8). However, the kingdom regained its independence in 1917 and chose not to join the Indian Union in 1947 (Chawla 2023). Kalimpong and Darjeeling had been incorporated into Bengal, and became part of the Indian state of West Bengal (McKay 2021). During the British period, many Nepali traders and laborers migrated to Sikkim, Kalimpong, and Darjeeling, eventually becoming the region's majority ethnic group. As the Kingdom's ethnic composition changed, the Royal Government established Dzongu in North Sikkim as a reserve for the Rong (Lepcha) people.

The change in Sikkim's ethnic population eventually led to an unsustainable situation in which India's democratic government supported an unrepresentative Bhutia monarchy. The monarchy lost power in 1975, and Sikkim joined the Indian Union, becoming India's twenty-second state. At this time, the state's residents were promised that its culture, heritage, and environment would be preserved. Article 371 F of the Indian Constitution emphasized the socioeconomic advancement of the population and recognized the Lachungpa, Lachenpa, Rong (Lepcha), Limbu, and Bhutia as "scheduled tribes" with special rights to own land and manage national resources in their homelands. Dzongu remained a reservation for the Rong.

9b.2 Sikkim's Fluvial Political Ecology

The London-based Burn and Company Limited attempted to build Sikkim's first hydropower plant in 1913 but could not get approval from the Royal Sikkim Government (Lepcha 2020, 2). After the government achieved independence from the British, in 1927, it built its own plant, a micro-hydro dam on the Rani Khola that produced 50 kilowatts of electricity, which the royal family and select other Gangtok residents used. They started a second project on the Sang Khola, 30 kilometers from Gangtok, in 1965. It provided two additional megawatts of electricity to eight of Sikkim's towns. By 1975, when India annexed Sikkim, the new state was producing three megawatts of its own electricity and began sourcing more from hydropower projects in neighboring states (Wangchuk 2018).

The outlook for Sikkim's hydroelectric future changed profoundly in 1995 when the Indian government released its Mega Power Policy, which set out plans to develop Himalayan hydropower. The central government's Water Commission estimated the state's hydropower potential to be around 8,000 MW (Wangchuk 2018) and produced several reports labeling Sikkim "India's future powerhouse" (Gergan 2020). Three years later, in 1998, the state government established the Sikkim Power Development Corporation to scope

(*Continued*)

potential dam sites. India's public sector hydropower company, the National Hydroelectric Power Corporation (NHPC), began building Rangit Dam, completed in 1999, and Teesta Stage V, which was completed in 2008. The NHPC also began building two large dams on the Teesta just south of the state border in West Bengal, Teesta Low Dam III and IV, although they did not become operational until the early 2010s. The commission proposed yet another NHPC dam at Rathong Chu in West Sikkim, but it was abandoned after Sikkim's first anti-dam protests.

In the 2000s, the central government amended its Mega Power Policy to allow for the privatized construction of hydropower projects (Wangchuk 2018). This policy started a race between Sikkim and Arunachal Pradesh to see who could construct dams more quickly and corner the market (Wangchuk 2018). Both states were the homes of multiple Indigenous communities, "scheduled tribes" in Indian law. They were also primarily rural and needed revenue for infrastructure, employment, and regional development (Gergan 2020; Huber 2017). The push for hydropower turned them into capitalist frontiers, where investors with limited or no hydroelectrical experience began operating in the industry (Huber 2017).

To Sikkim's detriment, it won the competition with Arunachal Pradesh. Although Arunachal distributed more Memorandum of Understandings to potential investors, rent-seeking, logistical difficulties, and intensive protests in Tawang forestalled the construction of all but one of its dams (Mishra 2023). By contrast, private hydropower companies built five new projects in Sikkim, including the largest, Teesta Stage III at Chungthang in Northern Sikkim.

Most of the new dams were run-of-the-river. These dams impound smaller reservoirs on high-altitude river reaches and then drop water through tunnels to turbines on lower reaches. Government agencies believed that such dams cause less human displacement and negative environmental impacts than earlier larger dams (see Chapter 8), so they were often rushed through environmental approvals. However, this approval process only focused on a dam's construction and did not consider the post-construction risks of tunneling and water impoundment (Huber 2017). These risks are intensified by Sikkim's geologically and hydrologically dynamic features; earthquakes, landslides, and GLOF from its 644 glacial lakes are common (Sattar et al. 2021).

Central and state governments approached Sikkim's Earth Systems in a paradoxical way. They cast the region as a "hazardscape" (Chakraborty et al. 2021) that required intensive government oversight for local community activities, but allowed multiple hydropower projects to be built without long-term socio-environmental scrutiny and accountability. The state and federal governments contradictory governance decisions only make sense to colonial and techno-optimistic logics. They were not putting the Sikkimese people or

(Continued)

their environment first. As Mabel Gergan (2020) has noted, India's "territorial notions of 'remote' mountainous spaces coalesced with ideologies of racial difference to produce an imaginary of the Eastern Himalaya as peripheral to and racially distinct from 'mainland' India."

This marginalizing imaginary underpinned some of the arguments for dam construction in Sikkim—including its small population and large forest areas. It ignored the fact that displacing these small Indigenous communities would lead to the loss of entire cultures (Lepcha 2020). It also ignored these communities' special connection to the forests, which they viewed not only as a shared resource but also as sacred land. This special connection with the land encouraged Rong (Lepcha) and Bhutia community members to protest the dams' construction.

9b.3 The Hydropower Protest Movement

The first protest movement formed in opposition to the Rathong Chu Hydropower Project, which was to be built on the Teesta tributary Rathong Chu in West Sikkim in the mid-1990s. Anti-dam protests began in meetings at Pemayantse Monastery in West Sikkim and had the backing of leading Buddhists, Bhutia, and Rong. They protested locally and in Sikkim's capital, Gangtok, wrote petitions and made statements in Sikkim's state legislature. In 1995, Sonam Paljor Dezongpa, a leading anti-dam advocate, began a hunger strike that lasted for 28 days. Protestors' actions encouraged further activism, and the project was abandoned on social and environmental grounds a few years later (Arora 2009; Wangchuk 2018).

The longest continuing anti-dam protest, which objected to dams on the Teesta mainstream, began in 2002 when the Join Action Committee of North Sikkim organized protests against the Teesta Stage V hydropower project near the town of Dikchu on the border of North and East Sikkim near the Rong's Dzongu Reserve. The unrest lasted for several years but did not stop the dam's construction. Several years later, in June 2007, the Affected Citizens of Teesta (ACT) launched their organization, this time focused on a series of cascading dams to be built on the Teesta. They stood against the entire cascade, particularly the Teesta IV dam, the construction of which was planned within the Dzongu reservation. On June 20, 2007, ACT's General Secretary, Dawa Tshering Lepcha, and his colleague, Tenzing Lepcha, started a relay hunger strike that continued for more than two years. During the hunger strike, the two activists were often joined by other participants, asked Buddhist monks and other religious figures to pray for them and evoked Mahatma Gandhi's peaceful resistance to British rule by sitting under his photograph and calling their protest a *satyagraha* (struggle for truth).

(*Continued*)

A few weeks into the hunger strike, the Sikkimese government agreed to review the dams but only allowed one anti-dam campaigner to join the review committee. Instead, they took advice from Sonam Gyatso, a Lepcha Member of the Sikkim Legislative Assembly, who insisted that despite the protests, the residents of Dzongu welcomed the dam (Wangchuk 2018). Eventually, however, the government only built two of the ten proposed dams on the mainstream. Teesta Stage IV in Dikchu on Dzongu's border and the smaller Pannan Hydroelectrical Project inside Dzongu went ahead.

The 2023 GLOF confirmed the dangers that the anti-dam activists and scientists had predicted. Traveling down the Teesta's entire length, this outburst flood destroyed one dam and severely damaged all other infrastructure it encountered. It also eroded local land and washed away bridges and roads. Experts say the dams intensified the GLOF's force (Chauhun 2023). Neither the state and central governments nor the involved hydropower companies have commented on the Teesta dams' future.

FIGURE 9B.2 *Top*, Teesta V at Dikchu before and after 2023 GLOF. *Bottom*, Teesta III at Chungthang, before and after 2023 GLOF. Before photographs, Ruth Gamble, 2022. The post-flood photos were widely circulated on social media.

10

NGOS AND RIVERS

1970s CE –Now

As we have learned in the past two chapters, the Asian Highlands and their rivers face many environmental issues and a few crises. Enclosure and extraction, unsuitable and unsustainable development, territorialization, and competitive geopolitics have all taken their toll on the health of these complex river systems. The primary response to these crises has been to conserve critically affected river systems (see Chapters 5, 8, and 9). This chapter examines how non-state actors, specifically environmental conservation non-government organizations (NGOs), offer a model to manage the Asian Highlands' complex socio-ecological river systems that is cooperative across agencies and with local communities. Their approach is a distinct counterpoint to the competitive and zero-sum approach of the nation-state regulation of riparian environments. The chapter also approaches environmental conservation, the ideas underpinning it, and the actions inspired by it from the perspective of human discourse and social organization. From this perspective, it develops on the previous chapter's conclusions that conservation is neither naturalized nor inevitable but a human activity that shapes intertwined human societies and ecologies. Conservation efforts exist alongside state projects *as human activities*; they balance or counter national development plans, often in the context of environmental exploitation. Sometimes, conservation efforts are programmed in parallel with state plans, such as the fish conservation efforts of the Chinese state along the Yangzi River (see Chapter 5). In other

DOI: 10.4324/9781003392033-10

situations, they emerge via the activities of NGOs that intend to redress the effects of development on riverine ecosystems.

This analysis of non-state conservation discourses and practices highlights the distinction between local environmental knowledge (LEK)—a situated knowledge of relations and practices between humans and their environments—and the enumerated and objectives-based approach of bioconservation. Tensions between these approaches have existed since the inception of conservation practices over a hundred years ago. Recently, many NGOs and other conservationists have attempted to bridge this gap by including local perspectives through community collaborations. Evidence suggests, however, that the incommensurability of these two knowledge lineages means that terms such as "participation" and "inclusion" often act against their stated intentions, occluding the views and voices of local people. This tension acts as a limit to the NGOs' cooperative model. The chapter also analyses the conservation movement's attempts to attach value to "nature" through the ecosystem services model, elaborating on the tensions between this approach, LEK, and state imperatives. Instead of finding ways to synthesize these approaches, it offers an anthropological, comparative analysis that gauges their extent, efficacy, and possibility.

10.1 Conservation, the Environment, and Environmental Conservation

The environmental ideas in current global bioconservation practice are rooted in the same seventeenth-to-nineteenth-century shifts in European intellectual, social, and economic history that led to the territorialization outlined in Chapter 8. Enlightenment thinkers separated external "Nature" from human "Culture" (Plumwood 1993). The landed gentry enclosed and exploited land and water. The rise of capitalism during the Industrial Revolution encouraged laborers to move from farms to factories, creating a metabolic rift between humans and their environments.

Movements to protect the separated Nature first arose in the countries most affected by the Industrial Revolution, the United Kingdom and the United States. After experiencing decreased air and water quality and dwindling forests or "wilderness," citizens of these two states formed multiple societies to combat threats to Nature, including the Commons Preservation Society (1865), the Lake District Defence Society (1883) in England, and the Sierra Club (1890) in the United States. On both sides of the Atlantic Ocean, romantic authors such as William Wordsworth (Yoshikawa 2020) and Henry David Thoreau inspired conservation movements (Buell 2023, 70–100), and these initiatives often advocated for the establishment of nature reserves. John Muir (1838–1914) co-founded the Sierra Club to push for the creation of more national parks after his successful campaign to establish Yellowstone National Park. Many of these parks were created by enclosing First Nations' lands (Merchant 2003), and advocates for national parks and other enclosing conservationist practices often ignored the vital role local relationships with the land had played in conserving these wilderness areas.

While conservationist societies worked to enclose and set aside large tracts of land for Nature, another group of conservationists, often working within imperial governments, took a different approach. They advocated for the proper management of nature's resources through state control. In 1878, the German-British naturalist Dietrich Brandis (1824–1907) established the Imperial Forestry School at Dehradun in the Western Himalaya to manage the extraction of timber from Himalayan forests (Weil 2006; Gadgil and Guha 1993), and the US government established its own Forest Service a few decades later, in 1905. During the first half of the twentieth century, American and imperial British governments extended this "management" approach to an increasing area of forests and waterways—once again, often at the expense of Indigenous peoples (Merchant 2003; Gadgil and Guha 1993).

Following World War II, the United Nations (UN) adopted this nature conservation framework, establishing two international agencies to protect Nature: the United Nations Educational, Scientific and Cultural Organization (UNESCO 1945) and the International Union for Conservation of Nature (IUCN 1948). As their consultants began implementing their programs, their discourse expanded from a focus on the conservation of pristine parts of nature to encompass the more expansive "environment," which included all human surroundings. This more expansive vision led to new fields of study, within which experts accumulated data to describe the environment (Warde et al. 2018). This led to a shift in conservation theory and practice as scientists began using data to determine specific sites of ecological import within the broader environment. A growing number of threats intensified the data-driven, scientific impetus to conserve these environments.

Through the 1960s and 1970s, this scientific knowledge inspired popular, non-governmental, "ground-up" environmental movements and national and international government agencies. The formative ground-up initiative to ban the insecticide dichlorodiphenyltrichloroethane led to the establishment of the non-governmental Environmental Defense Fund and a new department in the US government, the Environmental Protection Agency. Environmentalists celebrated the first Earth Day on April 22, 1970 and established multiple international organizations such as Friends of the Earth, Greenpeace, and Conservation International as the decade progressed. Although these groups were founded in Global North countries, they were often inspired by and supportive of Global South movements, such as the Chipko Andolan (Hugging Movement), whose participants hugged trees in the Western Himalaya to stop them from being felled (Gadgil and Guha 1993).

Pressure from these conservation collectives and scientists led governments to sign several important treaties and establish associated international environmental authorities. Seven countries signed the Ramsar Convention to protect wetlands in 1971; 50 had joined by the end of the 1980s, and it now has 172 signatories (see Box 5a). The UN established the United Nations Environment Programme (UNEP) after the first UN Conference on the Human Environment in Stockholm, Sweden, in 1972. The same year, UNESCO's General Conference adopted the World Heritage Convention to recognize cultural or natural heritage sites. These places were often already national parks or national monuments, so the national and international conservation structures reinforced each other.

In Mace's (2014) historical overview of conservation biology, she identifies four distinct phases in its development. Before and throughout the 1960s, conservation biologists promoted a "nature for itself" phase associated with the nineteenth-century mindset of prioritizing wilderness and intact natural habitats. Their focus, informed by scientific foundations in wildlife ecology, natural history, and theoretical ecology, was on species conservation and protected area management. Then, in the 1970s and 1980s, the field entered a phase she terms "nature despite people." This phase emerged because humans were increasingly destroying habitats and causing species loss. It depended on the scientific analysis of environments to determine threats to species and strategies to reduce risk, such as establishing minimum population sizes and sustainable harvesting levels. Mace notes that "by the late 1990s, there was ample evidence that pressures on habitats were ubiquitous and persistent and that the best conservation endeavors were failing" (2014, 1558; see also Colby 1991).

The next phase in conservation's history focused on "nature for people," as the realization grew that "nature provides crucial goods and services that are irreplaceable yet had been consistently ignored" (2014, 1558). The scientific focus correspondingly shifted from species specificity to general ecosystem networks. The management of protected areas had to be integrated across the ecological system. In the last phase, Mace identified a shift from an overly utilitarian perspective on nature as a service provider for human beings to one that explicitly acknowledges the dynamic relationships between people and nature. It emphasizes "the importance of cultural structures and institutions for developing sustainable and resilient interaction between human societies and the natural environment" (2014, 1559). It also includes insights from resource economics and the social sciences. Since Mace wrote her article, there has been yet another shift as conservationists look for "nature-based solutions" (Qi and Dauvergne 2022). To non-conservationists, these seem strikingly like the phase in the 1960s identified as "nature for itself."

Mace's overview refers mainly to science-based literature and evidence in conservation biology. She acknowledges that the multiple phases reveal how scientific tools and techniques have not always kept pace with the objectives and practices of conservation activities, namely measuring conservation success, designing ecosystem management, and assigning economic value to nature. Nonetheless, another broad subset of literature on conservation practice in the social sciences has shown more fundamental differences in the unfolding processes of conservation efforts around the world. These have manifested mainly through stated polarities, including those between environmental conversation and economic development. These studies note the Western, Enlightenment-era, Nature-Culture binaries within which many conservationists and conservation programs operate and the lack of such binaries in the local and Indigenous communities where conservation activities occur. Moreover, a key early work in this field by Escobar (1998) proposed that "biodiversity," the bedrock of environmental conservation, might itself be understood not as a true object of science but rather as a historically produced discourse. As discourse, it provides a new way of understanding reality and, crucially, a different way of expressing "the relation between nature and society in global contexts of science, cultures, and economies" (1998, 55).

10.2 Conservation in the Asian Highland River Catchments

Conservation efforts in the upper Brahmaputra and Dri Chu river catchments have been and remain intimately connected to the political ecologies of its governing regimes. The earliest conservation regimes in the region were those established by the British government, and they followed the general script of other imperial conservation processes. That is, a group of concerned locals made petitions to the government about environmental degradation, and the government created reserves and national parks at the expense of local people who had been removed from their land. The largest enclosed space in Northeast India is Kaziranga National Park, which the British government transformed from a game reserve into a rhinoceros conservation area in the early twentieth century, and the government of independent India made a National Park in 1974 (Saikia 2009). While the conservation groups advocating for these parks tended to stay on the plains, another group of imperial environmentalists known as "naturalists" or "plant hunters" such as Joseph Dalton Hooker (1817–1911), Joseph Rock (1884–1962), and Frank Kingdon-Ward (1885–1958) began making expeditions into the hills. On these journeys, the Plant Hunters "discovered" a multitude of flora species, extracting their cuttings and seeds and sending them back to Europe and the United States, where they were cultivated and sold to wide markets (Mueggler 2011). The eastern Highlands' international reputation as a space of biodiversity can be traced back to this trade.

Although India established multiple National Parks on the site of British conservation areas after independence, there was generally a lull in conservation efforts in the eastern Highlands between the 1960s and 1990s. The Chinese state pursued high-modernist reforms that had a disastrous environmental effect but did not allow conservationist dissent (Yeh 2013). Although India became famous for its conservationist and anti-dam movements during this time, there was much less development in Bhutan and Northeast India than elsewhere in South Asia, so there was also a smaller conservation movement (Saikia 2019; Gamble 2023). The 1980s bioconservation panic over the theory of Himalayan Environmental Degradation led to intensive interventions in Nepali Highlanders' lifeways in the following decades (Chakraborty et al. 2021), but affected Eastern Himalaya highlanders' lives much less. However, conservation movements proliferated in the catchments following the push to develop the region that followed India and China's economic liberalization policies of the 1980s and 1990s (Gamble 2021, 2023).

Norman Myers's concept of "biodiversity hotspots" (Myers et al. 2000) was a powerful influence on the conservation movement's approach to the eastern Asian Highlands. It became oriented toward species and habitats and tried to mitigate the impacts of dam construction and reservoir operation on the natural landscape and species modification (Ghosh and Chakraborty 2021). Apart from fish conservation (see Chapters 5, 8 and Box 6a), other species of particular concern for Himalayan riverine and wetland systems include birds (Dendup et al. 2021), otters (Gupta et al. 2020), and amphibians such as bullfrogs (Tenzin and Dhendup 2017). A variety of factors affect these species' sustainability. Hydropower, other infrastructure and urbanization, farming and grazing in vulnerable habitats, the collection of plants and animals for medicinal purposes, and the indirect ecological impacts of habitat loss and ecosystem shift can all affect them adversely.

Large and small international NGOs have worked for decades on environmental protection in the Asian Highlands. International organizations such as the World Wildlife Fund, Conservation International, Flora and Fauna International, and the International Crane Foundation have implemented environment and conservation programs across the eastern Highlands. The number of smaller international and local NGOs operating in the southern, non-China Highlands has increased significantly in the past two decades. One estimate of registered NGOs in Nepal, for example, put their number at almost 40,000 (Karkee and Comfort 2016). International (or overseas) NGOs were allowed to operate freely in China from the 1990s to 2016. At the start of 2017, however, the national government passed an "overseas NGO law" that constrained their presence and activities. The government has allowed some local NGOs to continue functioning with fewer restrictions, but this depends on whether their focus aligns with government aims.

NGOs generally offer an approach to environmental conservation that is distinct from the competitive, zero-sum game approach of sovereign states discussed in the previous chapter. As smaller and more agile institutional entities, NGOs tend to be more responsive to on-the-ground needs or to emerge from local communities themselves. They also tend to work cooperatively amongst themselves and with local communities in ways that national governments do not. This cooperative model for the management and regulation of riparian environments has been the basis for some well-known successes in the conservation efforts of the Asian Highlands' river catchments. One of these is the work Bhutan's Royal Society for the Protection of Nature (RSPN) has done for black-necked cranes.

The black-necked crane (*Grus nigricollis*) is an endangered species, the only alpine species of crane, and only around 10,000 of these birds live outside captivity. In Tibetic languages, the birds are called Trung Trung or Cha Tsering (the long-life bird) and represents both fertility and the Dalai Lamas. The cranes spend their summers nesting in the Tibetan Plateau's high-altitude wetlands in Qinghai, Sichuan, northern Tibet, and Ladakh. These wetlands provide shelter for their nests and food sources such as sedge tubers, plant roots, earthworms, frogs, shrimp, and small fish. Wetlands are an essential part of Highland river systems (see Box 5a), and the crane's dependence on them highlights the support rivers provide to the habitats of various endangered species. As these wetlands are susceptible to impactful anthropogenic activities such as draining, damming, levee construction, and irrigated agriculture, the black-neck cranes' situation also highlights the fraught relationship between these birds and high-altitude development.

During winter, black-necked cranes seek refuge in lower valleys in Central Tibet, northern Yunnan, and Bhutan. In Bhutan, they have become a significant symbol of environmental protection. This symbolism persists despite the relatively lower number of birds that nest within Bhutan's wetlands compared to their other winter habitats (WWF 2023a). Phobjikha (Wangdiphodrang district), Bomdeling (Tashi Yangtse district), Tangmachu (Lhuntse district), and Gyetsa and Thangbi (Bumthang district) are all well-known black-neck crane habitats. Each contains or is located adjacent to monitoring and protection sites for cranes in national parks.

The Bhutanese government began officially conserving black-neck cranes in 1995 when it implemented the Forest and Nature Conservation Act (FNCA). They followed this initiative with the 2006 Forest and Nature Conservation Rules (FNCR), which listed black-necked cranes under Schedule I as a protected bird species (Phuntsho and Tshering 2014). Since 2002, the cranes have been protected under Bhutan's Forest and Nature Conservation Act. The government implemented these measures following reports in the 1990s of significant declines in the number of black-necked cranes and reports that some Bhutanese farmers were killing them to protect cash crops (Reinfeld 2003). However, Phuntsho and Tshering (2014, 13–14) found that while there were specific declines in some areas over periods from the 1970s to the early 1990s, the average growth rate of black-necked cranes increased by almost 2% over the past 28 years (from 1986/7 to 2013/4).

The Royal Society for the Protection of Nature (RSPN), Bhutan's only non-profit organization, has supported the conservation push, with initial backing from the International Crane Foundation. In 1998, it initiated a festival at Gangtey Monastery in the Phobjikha Valley, Wangdiphodrang District, to celebrate the black-neck crane's arrival. Since 2013, the one-day annual festival held on November 11 charges the approximately 250 tourists (2014 numbers) who attend a Nu 500 (about USD 8) entrance fee (Phuntsho 2016). The festival also provides indirect economic opportunities from these visitors' presence and an opportunity to solidify the black-necked crane in Bhutanese culture. The birds were traditionally regarded as "heavenly," and their image was inscribed into the motifs of Bhutanese handicrafts. However, the RSPN's efforts have made the cranes more present and vivid in peoples' lives.

Reinfeld (2003) considers the RSPN's approach to crane conservation an "evolution" in the often oppositional ecological stances taken in conservation practice. In many instances of Highland conservation, this oppositionality pits conservation practices—including restrictions on access, use, harvesting, and hunting—against other community benefits derived from land and resource use, such as economic gain and cultural practices. Defusing this oppositionality in black-neck crane conservation is complicated because the anthropogenic impacts on the cranes are not always direct or obvious. Humans are much more likely to degrade their habitats than to harm the birds directly. Black-necked cranes avoid overly planted or forested areas since these might harbor predators and are particularly impacted by riverbank and riverbed monoculture tree plantations (mainly willows), which cause habitat loss (WWF 2023a).

The RPSN's programs in Bhutan demonstrate the viability of locally created responses to modern conservation efforts and the implications of a successful cooperative model for watershed conservation. By drawing on the bird's cultural significance *and* using its symbolism to create alternative economic pathways for local people—particularly those directly impacted by conservation rules that restrict land and resource use—the NGO has succeeded in increasing black-necked crane numbers in their wintering habitats in Bhutan. It is equally important, however, to recall that the birds migrate and join the flow and movement of other species across the Asian Highlands, which means they rely on various parts of its river systems. A longer-term positive outcome for the cranes would require other successful state and NGO programs to operate across their habitat areas. If these were functioning, these programs' socio-ecological impacts could be compared to Bhutan's apparently successful conservation initiatives. But other Highland crane preservation efforts are limited and do not follow this cooperative model.

FIGURE 10.1 Images of black-neck cranes in Asian Highland environments. *Clockwise from top left*, black-neck cranes in a field next to the Yarlung Tsangpo in Central Tibet, photograph by Ruth Gamble, 2018; Phobjikha Valley, Bhutan, photograph by Ruth Gamble 2023; detail of a mural in Bhutan, photograph by Ruth Gamble 2023; and the annual crane festival at Gangtey Gonpa, Bhutan, photograph by the Bhutanese Tourism Bureau.

10.3 Traditional Ecological Knowledge

While conservation practices have relied on the knowledge of scientists and experts, there is also a significant overlap and complicated exchange in the Asian Highlands between NGO conservation projects like those that protect cranes in Bhutan and local communities' relationships with their environments. These interconnections are often called Traditional Ecological Knowledge (TEK). Specifically, TEK is a term used to describe how local, often Indigenous, peoples relate to their environments and communicate these relations across generations within their communities.

Anthropologists first framed TEK within their discipline several decades ago, but others have picked it up and used it in various fields and settings. Anthropologists began discussing TEK in the 1980s, primarily in response to increasing empirical data on folk biological classification systems. These data sets revealed that Indigenous people demonstrated detailed and extensive local knowledge of their environment (see Conklin 1957 for early work). Apart from knowledge of varied ethno-classificatory schemes that the anthropologists studied, a key result of this foundational research was an intensified disciplinary focus on the relationships—usually functional—between people and their environments (Dwyer 2005, 15). Efforts to codify such relationships as "knowledge" led to TEK, and, importantly, experts understood this knowledge to be extractable from the contexts in which it had been practiced. As an extractable resource, it could then be packaged as a commodity for use in discussions on improving conservation efforts. Conservationists and others viewed TEK's use as a way to include local knowledge and participation. Through this process, TEK became both an approach to conservation and a body of literature that evaluates the viability of TEK within a dominant framework of Western scientific knowledge. (This chapter uses the term in both ways.) After they realized how implicated TEK had become in conservation practices, those writing critically about conservation began using another term, local ecological knowledge (LEK). LEK refers to general descriptions of Indigenous environmental practices and approximates anthropologists' earlier expressions of human–environmental interactions.

Studies of TEK initially focused on systematically treating local knowledge of the environment as "a cumulative body of knowledge, practice, and belief that developed by adaptive processes, handed down through generations" (Berkes et al. 2000). This situated knowledge about the relationship of living beings (including humans) with one another and with their environment. Advocates of TEK perceived this relationality as the primary contribution TEK could make to the conservation of biodiversity, rare species, protected areas, ecological processes, and sustainable resource use more generally. Berkes et al. observed that such TEK practices were varied and fell within three broad categories:

1 They may align with ideas of conventional resource management, such as traditional restrictions on gathering from village commons in the Asian Highlands that regulate extraction or regenerate fragile species (Jodha 1998, 293–294).
2 They may continue in local areas even though conventional resource management has abandoned them. An example is the *jhum* system of shifting agriculture practiced in Arunachal Pradesh, Bhutan, Sikkim, and Yunnan. This system burns a site to clear it, cultivates it for several years, and then leaves it to regenerate.
3 They may exist only in local areas and not be found in conventional resource management. An example is pest management among the Warlis of India, who place certain kinds of tree branches in paddy fields to attract birds that eat insects. This planting creates buffers against pestilence outbreaks (Pereira 1992).

Berkes et al. relate such practices to an acquired practical understanding of the dynamics of complex socio-ecological systems. TEK's situated and adaptive qualities aligned it exceptionally well with ecological resilience, even though not all traditional practices are ecologically adaptive.

Nazarea (2006, 322) identifies different generations of literature relating to TEK. The first generation focused on "content, comparing and contrasting local knowledge with scientific knowledge and legitimizing it in terms of Western standards." This induced efforts to integrate TEK with bioconservation, using local knowledge to complement scientific models. The second generation was more focused on process and transformation. It was based on the observation that "local knowledge is not just intrinsically dynamic and situated but is often contingent on external opportunities and constraints" (2006, 322). This second-generation approach liberated TEK from comparisons with Western science and the burden of legitimization but added another complexity, namely authenticity. In this space, the politics of establishing priority—deciding whose voice matters most—became the focus of debates, often at the expense of understanding how local Indigenous people regard their environments and how their interactions with them were adaptive and practice-based.

While many TEK studies have attempted to go beyond positing a stark dichotomy between Indigenous and scientific knowledge, evidence also suggests the actual practices of integration are fraught and often complex. In a foundational article published in the late 1990s, Nadasdy reversed the question from the challenges of integration to the underlying problem with integration, namely that failures to integrate knowledge for the successful and sustainable implementation of conservation projects are complicated by influential yet unacknowledged power relations between Indigenous people and Western conservationists. Nadasdy emphasized two aspects: the dominance of centralized institutional practices (both of states and NGOs) and control of abstractions in discourses around TEK and its integration into resource management (rather than vice versa). For example, a study on fisher knowledge in Puerto Rico (García-Quijano 2007) found that despite the increased recognition of local knowledge for understanding environmental change and ecosystem management, the onus was always on translating local knowledge into scientific language and never the other way around. Reversing Walter Benjamin's original statement that a good translation betrays the destination language and not the source language, Viveiros de Castro (2004, 5) notes how bad translations betray the original [language]. This point reveals how such a "betrayal" of translation is one manifestation of the domination expressed by a hegemonic knowledge system over others (see also Stengers from Chapter 1).

The common thread through these episodes is that as knowledge of relations and practices between humans and their environments, TEK differs significantly from—and often conflicts with—modern scientific discourse of human–environment interactions. As Nazarea observed, however, a second generation of literature points to the politics around TEK, complicating the dichotomy between the knowledge systems. Agrawal (2002) traces these politics through database creations that collect, store, and preserve Indigenous knowledge. He argues that through this process of particularization—namely, the identification and separation of *useful* knowledge—or knowledge deemed important by development practitioners—what emerges is a specific kind of power relation. The politics of classification precisely decide what is valued as useful or provable knowledge, and to be useful or provable, such information must stand up to repeated validation. For example, the practice and use of herbal medicine such as *neem* might involve a complex of rituals, words, symbolic gestures, and movements that accompany the prescription of that medicine. However, through the processes identified by Agrawal (2002, 291), the medicine is encoded into a database through its strictly useful components. The contest relating to its use is not included in what is deemed valuable. This politics is hegemonic when it does not even consider the possibility that there are not only other fundamental ways of categorizing or knowing but also vastly different approaches to understanding how and what exists in our environments.

10.4 Assigning Values to Nature

Along with expert-driven conservation programs and the complicated use of TEK for conservation purposes, another way that NGOs have pursued conservation goals in the Asian Highlands is by linking them to international programs that ascribe a value to Nature. Adherence to or criticism of this approach has proved to be one of the subjects in which the multiple channels of our multidisciplinary studies of rivers have diverged (see Chapters 1 and 11). Like the rest of the chapter, what follows is written from a primarily anthropological perspective and, therefore, may contrast with the description of these programs elsewhere in the book (see Chapter 5 and Box 9a, for example).

The conceptual division between the human "cultural" world and the "natural" world, mentioned earlier in this chapter as a foundation of modern environmental conservation, is questioned most convincingly in the range of anthropological literature on human–non-human relationships. By describing and analyzing the multiple ways that humans regard non-humans such as animals and plants as kin relations, for example, in the Amazon (Descola 1994) or Aboriginal Australia (Glaskin 2012), or in the varied expressions of human–river and human–deity interactions (Gagné 2020; Tashi 2023), what is most evident is that the dichotomy between Culture and Nature is not universally recognized. Instead, many societies enfold aspects of the natural environment into their social lives and organizations in ways that blur the distinction altogether.

Using the analytical tools that this discipline affords us, the values that this approach to conservation engenders appear different in kind from those that local and Indigenous people attach to aspects of the natural environment. Considering animals to be your kin relations—in the way that the Achuar of the Amazon regard certain game animals—assigns value to those animals. But it bears explicit mention that because of the parallel in relationships, the value is *exactly the same kind* as that given to, for example, one's human brother-in-law. Moreover, what is expressed is what is meant; there are no metaphors at work here. When the Achuar describe jaguars as their brothers-in-law, they are not saying that jaguars are *like* their brothers-in-law or that it is *as if* jaguars are their brothers-in-law. The statement is to be understood as it is expressed: my relationship to jaguar is the same as my relationship to my human brother-in-law. Moreover, this assertion can only adequately be comprehended according to the cosmology, social structure, and culture of the Achuar.

In starkly different terms, the value assigned to Nature in the context of bioconservation firstly relies on the methods of empirical science. The focus on identifying, quantifying, and homogenizing Nature through models such as ecosystem services (see Box 9a) underscores how value is created in this context by combining scientific assessments and valuation. As Ruffo and Kareiva explain in a guest editorial in *Frontiers in Ecology*, "the idea of 'ecosystem services'—identifying and quantifying the resources and processes that nature provides for people—gives us a framework to measure nature's contribution to human well-being, and to understand the cost of its loss" (2009).

The theory behind ecosystem services developed in Western-dominated institutions during the 1970s–1990s. In 1996, Maurice Strong, Secretary General at the 1972 UN Conference on the Human Environment in Stockholm and the 1992 Earth Summit in Rio and the first UNEP Executive Director, gave a lecture to the Korea Institute of International Economic Policy. He stated: "In addressing the challenge of achieving global sustainability, we must apply the basic principles of business. This means running 'Earth Incorporated' with a depreciation, amortization, and maintenance account" (cited in Sullivan 2011). In this articulation, he placed "Earth Incorporated" as a service provider and argued that it could be regarded as the largest corporation on [its own] planet. Subsequent iterations of ecological-systems services have modified this stance, but service commodification remains its core principle.

Academics and others have taken two divergent paths in thinking through the quantification of Nature's services. The first, which is highly critical of this approach, holds that assigning value to Earth Systems as a service provider places them within the same structures—and problems—as global capital. As such, the ecological-systems services model represents another strand of neoliberal reform and becomes susceptible to *laissez-faire* capitalist norms that shape governance, property, and money flows and, ultimately, the values and decisions affecting complex socio-ecological systems around the globe. Moreover, these norms will replicate the inherent inequalities of global trade within local environmental decision-making despite language to the contrary. Another more pragmatic path recognizes the limitations implicit in regarding Nature as a service provider but notes that the approach mitigates rampant exploitation while also providing a common (globalized) ground for collaboration among nation-states and a more equitable distribution of costs and benefits. It takes the view that some action to protect the Earth's resources and the local Indigenous peoples who depend on them, especially as it acknowledges Indigenous values, is better than none. These paths, representing two opposite ends of a spectrum against and for ecosystem services, are considered in turn.

It is no coincidence that the rise of ecosystem services within conservation practice occurred during the same decades that neoliberalism was transforming the world economy through trade liberalization and globalization (Harvey 2005, 1–4). Neoliberalism's core theory holds that power and capital centralization should be removed from the hands of states. The first step in any neoliberal process is, therefore, deregulation. As many observers have noted, however, this deregulation coincides with another kind of coding, the reregulation of power and capital in the hands of corporations and the supposedly free market. Various effects flow from this central shift. Castree (2008) notes that it transforms previously untradable things into tradable commodities. In the sphere of conservation, Igoe and Brockington (2007) unpack neoliberalism's practical effects on conservation through a series of promises:

> Increased democracy and participation by dismantling restrictive state structures and practices ... to protect rural communities by guaranteeing their property rights and helping them enter into conservation-oriented business ventures ... to promote green business practices, by demonstrating to corporations that green is also profitable ... [through ecotourism] to promote environmental consciousness for western consumers by encouraging them to fall in love with the environment through direct connections to it.
>
> *(2007, 434)*

Castree also points to multiple instances in which those promises have not been fulfilled. One reason that McAfee (1999) gives for this failure is the pervasive assumption that "conservation can be achieved without addressing the difficult and systemic inequities and power relationships that are inextricably linked to so many of our global environmental problems today."

Hoping to bypass the difficult, sometimes intractable, barriers to conservation, advocates for a neoliberal approach to conservation turn to the power of money to achieve their aims. They hope a universal desire for money and its inherent homogenous quality will provide a common and globalized ground that enables environmental conservation. At the core of the ecosystems approach is the logic of capitalism. In practical terms, conservation's move toward neoliberalism, framed by deregulation and reregulation, has been marked by increased direct involvement of non-state actors, NGOs, and for-profit businesses. These two organizational classes are also increasingly working together. One indicator of this growing collaboration has been the number of chief executive officers of large corporations who sit on the boards of conservation NGOs, particularly in the United States (Dorsey 2005), where many development projects originate. While this collaboration can make it easier for conservation solutions to find needed financial backing, it inevitably brings complications, and it is more difficult for NGOs to take a stand on environmental issues that contradict corporate interests. Igoe and Brockington (2007) note, as well, that the hope that money can flow as a universal equalizer to solve conservation problems tends to overlook the fact that the networks created around neoliberal conservation are often constrained and highly exclusive: "flows of money within them—be they investments, multi-lateral loans, or conservation funding—tend to stay within them."

Despite these trenchant critiques regarding ecological systems services, the Asian Highlands' many environmental issues, geopolitical constraints (see Chapter 8), and diverse governance regimes in its various states (see Chapter 9) make it a region in which it is difficult to establish environmental justice and borderland conservation. Those who argue for the more pragmatic embrace of the neoliberal approach to conservation (particularly through the ecosystems services approach) say that it provides several avenues for greater cooperation in this context. The framework of ecosystem services has been used in much bioconservation scholarship in India, but it is yet to be applied in any systemic way through government or NGO programs, although many international NGOs and bioconservation scholars are calling for its use (Srivathsa et al. 2023). Bhutan's conservation policies are often assessed according to ecosystem services, and there is some evidence that this approach has influenced the state's Gross National Happiness policy (Harris and Roach 2021, 251–282). In China, a form of ecosystem service is found in the national Payment for Ecosystem Services (PES) schemes, proposed in 2016 through two policies by the central government (Pan et al. 2017). One policy in particular, Guidelines for Facilitating Lateral Mechanism of Eco-compensation between Upper and Down Stream of Cross-Province Rivers, targets watershed conservation. Challenges remain with implementing effective cross-provincial schemes, but initial studies report some success in monitoring water pollution and improving water quality in certain provinces such as Anhui and Zhejiang (Pan et al. 2017: 204–205).

Some also argue for extending the ecosystem services approach to the international level. The region's various actors could establish a common ground (and language) by settling on terms assigned by the values of capital and money, using these to conduct inter-state and inter-community negotiations. This would extend the state policy of compensating citizens monetarily for ecological loss into the international arena. For example, suppose a dam in China or India affects the fish population or sediment flow downstream. In that case, the upper riparian countries could negotiate compensation or mitigation measures that lead to more equitable sharing of benefits and costs. These kinds of payments are part of the water-sharing agreements in the Syr Darya catchment in Central Asia (Wegerich et al. 2015) but have not yet been used in the Brahmaputra catchment.

Implicit in the use of monetary exchange to compensate for water inequities in this model is the hope that a more holistic view of river systems management can be achieved, one that encourages nations to consider a catchment in its entirety and understand how different factors, including climate change, can affect its health and productivity. On this basis, potential conflicts related to environmental degradation or loss of livelihoods can be identified before they escalate, and destructive activities such as the over-extraction of water, unregulated fishing or sand mining, and pollution can be identified and paid for through fines. Increased communication among nations could also promote sustainable practices, such as Integrated Water Resources Management (IWRM; see Chapters 5 and 9), the restoration of riparian habitats, and the creation of sustainable fisheries that benefit all. The emphasis on benefits can also take on the language of natural defense mechanisms and reduce the need for costly infrastructures such as dams or embankments that can create transboundary tensions. It also offers a method to realize the opportunity costs of climate resilience; healthy ecosystems are more able to adapt to climate change impacts such as droughts, floods, and landslides (see Chapters 4, 5, and Box 4a). Finally, a leveled and more equitable playing field in which communication is more open could also encourage flows of knowledge and scientific research across nations and toward their populations, creating more public awareness and community buy-in.

Despite this potential, the Asian Highlands' complex geopolitics and multiple political systems place severe limitations on capital flows through the mountains, which does and will limit the use of the ecosystem services approach. National and subnational states often prioritize industrialization-oriented policies such as state-endorsed hydroelectric power generation and agricultural intensification, and these imperatives are frequently misaligned with the sustainable management of ecosystem services at local scales. Moreover, cultural and livelihood variations, especially among Indigenous communities in the upper Brahmaputra Valley, significantly differentiate ecosystem service valuation. For instance, traditional agricultural systems, intrinsically linked to natural flood pulses, face irreversible alterations due to extensive water infrastructure development. Simultaneously, marginalizing traditional knowledge about environmental hazards in national decision-making processes exacerbates potential risks associated with infrastructural projects.

There are no straightforward solutions to safeguard the biodiversity and environmental vulnerability of Asian Highlands' river systems, nor can we draw simple conclusions about how best to mitigate continued human impacts. The values assigned to Nature through the cost/benefit quantification of ecosystem services offer a way for nation-states to communicate with each other and negotiate for a commonly agreed and more equitable distribution of services provided by the river systems. They could also incorporate norms based on LEK into their negotiations, which would measure and cost TEK in the overall framework. On the other hand, the disembedding processes resulting from this inclusion are a cause for concern. They not only inevitably break down into discrete parts a complex socio-ecological system that must be understood holistically but are also subject to the rearranged power hierarchies and relations that always result from the interplay between human discourse and social organization.

10.5 Conclusion

This chapter has outlined how discourse and forms of social organization as human activities have impacted the material conditions and processes of Asian Highlands' river systems. These factors take many expressions, from direct impacts on riverine species and their habitats and associated efforts to protect and conserve them to the indirect effects on humans and other species that occur through resettlement to the cultural revival of environmental motifs. All these actions have implications for the river catchments, which, as socio-ecological systems, also implicate the local and Indigenous populations living within them.

By focusing specifically on environmental conservation NGOs, the chapter distinguishes the activities of this broad set of institutions with the riparian management activities of sovereign states, outlined in the previous chapter. The cooperative model suggested by NGOs is undoubtedly a step toward greater participation and inclusivity of Asian Highland communities and their local experiences. In some instances, such as with black-necked crane conservation in Bhutan, this model enables expressions of cultural evolution. And the uptake of TEK approaches to develop LEK as a situated and experientially based form of understanding and interacting with the natural environment has seen some success in other ways (see Box 10a). Nonetheless, there are limits to this model of cooperation, especially where some national contexts (Lowland governments, for example) subsume LEK under more dominant manifestations of ecological knowledge. Prioritizing a dominant form of ecological knowledge, underpinned by a conceptual dichotomy between "nature" as wilderness-biodiversity and "culture" as human activity, reveals how these different sets of knowledges are often incommensurable with each other. A comparative anthropological perspective allows us to appreciate the cultural specificity of the Nature/Culture divide, which is not universal but more dominant in Western societies than in most other areas of the world.

The chapter also considered how the Nature/Culture divide underpinned the valuation of Nature in the ecosystem services approach. It explained the positives and negatives of approaching Nature as a service provider within the context of international economic structures that trend toward neoliberalism. It concludes that within this framework, there is no benefit without cost; the distribution can be configured and changed, but the formulation remains, so all actions will be influenced by ongoing structures of hierarchy and capitalism. Additionally, continuing questions remain about the possibility of other avenues and alternative solutions to the pressing issue of anthropogenic actions on complex socio-ecological systems.

BOX 10a FISH CONSERVATION IN YUNNAN

In Yunnan, China, the NGO Protection Association of Endemic Fish Species of Shangri-la (PAEFSS) works with local governments, residents, research institutes, firms, small-business owners, volunteers, fishermen, monks, and other groups to conserve freshwater biodiversity, control illegal fishing, restore banks, and facilitate local recreational fishery in the Dri Chu Catchment. It was established to deal with two primary challenges to fish management: fish release and illegal fishing.

Fish release into Asian Highland rivers is a multipurpose, multicultural, and multijurisdictional phenomenon. State-sanctioned release of captive-bred fish into rivers is intended to restock rivers for conservation and economic reasons. The captive breeding and release of endemic fish species is used widely to conserve, reintroduce, and supplement populations of imperiled freshwater species (Rytwinski et al. 2021). Chinese authorities regularly use the captive breeding of fish to restore freshwater biodiversity loss caused by dam construction (Xu and Pittock 2019), but the cost of fish captive breeding and the scale needed to undo the dams' damage means these programs rarely achieve their aims. Hydropower dams have depleted the numbers of many endangered fish species, and the central and local governments, working with hydropower developers, could only restore a few of the catchment's iconic species.

(Continued)

While the government programs are underused, increasing numbers of Buddhists are releasing fish into Highland rivers for a very different reason, namely as part of the *tsetar* (life-release) ritual. When this practice is conducted without coordination, various exotic fish that may harm endemic fish populations are released. Until recently, *tsetar* of fish was only practiced occasionally but has become increasingly widespread, with local Tibetans purchasing live fish from markets and releasing them into rivers (see Box 6a). The practitioner's ability to save numerous fish lives makes this a more meritorious action than releasing the life of one larger animal. Since the improvement of Tibet's economy in the 2000s and the expansion of the intra-China fish trade, the number of exotic fish released into Tibetan rivers has increased exponentially. *Tsetar* is particularly popular during religious festivals such as the Saga Dawa Festival in Lhasa, which celebrates the month of the Buddha's birth in May–June. Many of the 70,000 people at this festival release live fish and other aquatic species into the Kyi Chu and surrounding rivers, particularly Chinese softshell turtle (*Trionychidae*) and American bullfrog (*Lithobates catesbeianus*), neither of which are indigenous to the Plateau's river system. A 2014 survey found that *tsetar* practitioners had released 14 introduced fish species into Tibetan rivers, lakes, and wetlands and that this activity has endangered local species.

By engaging with fish conservation and Buddhist stakeholders, PAEFSS developed a program that uses *tsetar* to complement the government's limited fish release programs in the Dri Chu Catchment. They began by explaining the impact of exotic fish release on local river ecosystems to socially influential monks from almost all local monasteries. They then collaborated with these

FIGURE 10A.1 Monks conduct a *tsetar* (life-release) ceremony for endemic fish into the Dri Chu. Photography, Hongzhang Xu 2019.

(*Continued*)

temples to develop *tsetar* programs that released captivity-bred endemic fish (see Figure 10a.1). The temples provided the funds for this program, typically 1–2 million yuan per month, and PAEFSS has used these funds to purchase endemic larval fish from breeding organizations, transport these larval fish to rivers, and provide buckets and speakers through which the monks can say prayers to aid the monks' *tsetar* rituals.

The success of this fish release program in the Dri Chu Catchment showcases a model that aligns religious practices with ecological goals. As PAEFSS contemplates scaling up this program, it will be crucial to prioritize *tsertar*'s Buddhist framework, which necessitates consistent dialogue, participatory decision-making, and a genuine respect for religious autonomy. While the unique religious and cultural context might make national expansion challenging in China, countries such as Bhutan, which has a strong Buddhist heritage and environmental consciousness, could be more receptive and suitable to such initiatives.

The other major issue that PAEFSS deals with is illegal fishing. Illegal, unreported, and unregulated fishing is a global issue causing perverse environmental and social impacts, with global losses of 10–23.5 billion US dollars per year (Ma et al. 2018). Restrained resources and a lack of public awareness and coordination between agencies make managing illegal fishing on rivers in remote and rural regions more challenging than in urban and other more populated areas. In these circumstances, strict conservation strategies such as fishing bans often encourage more illegal fishing (Shalehin et al. 2022). This global trend plays out within the Yangzi Catchment. Fishing was banned on the Yangzi and its tributaries in 2020 for ten years to recover freshwater biodiversity (Yang and Duan 2022). Yet Illegal fishing continues in remote and rural areas, particularly at night when it is difficult to detect (Yang and Duan 2022).

To better manage illegal fishing in the Yunnan reaches of the Dri Chu Catchment, PAEFSS has adopted a subsidy-driven and decentralized reporting system. In short, it pays people to inform authorities of illegal fishing activities by providing subsidies to incentivize accurate and comprehensive reporting on illegal fishing activities. This data helps authorities make informed decisions to address illegal fishing. Those who report larger and more severe illegal fishing activities get more money. Mr. Yang from PAEFSS told two of this book's authors, Xu and Pittock, that the highest subsidy they gave to anyone for information about illegal fishing was around 8,000 yuan. They only offer this amount to people who report large-scale operations, like a family-owned electrofishing team that kills more than 15,000 fish daily. Yang also explained that those who reported illegal fishing were typically recreational fishing people or villagers with land near the river.

PAEFSS connects potential reporters through WeChat (a Chinese message service) groups, which enables reporters to provide regular updates on fishing activities in their region. PAEFSS also gives participants free jackets (see

(*Continued*)

FIGURE 10A.2 *Left*, A free jacket printed with PAEFSS's working WeChat Group QR code. *Top right*, Education poster on the conservation of endemic and endangered fish species in the Dri Chu. *Bottom right*, An illegal fisherperson on the Dri Chu, caught after community reporting. Photographs, Hongzhang Xu 2019.

Figure 10a.2, left) emblazoned with a QR code generator that links to the WeChat groups (see Figure 10a.2, top right). This encourages more potential supporters to become reporters and transparency and accountability in the payment process. The system implicitly holds everyone accountable; reporters know their observations are visible to all, and PAEFSS is similarly obligated to act upon the information they provide. After identifying illegal activities, PAEFSS will contact local governmental agencies, such as police stations, to direct them to illegal fishing.

This reporting not only directly reduces illegal fishing but also raises public awareness of its impact on local communities and the environment. Moreover, it makes local people active participants in preserving the biodiversity of their regions. PAEFSS has capitalized on this greater awareness by creating educational posters to introduce the endemic and endangered fish species in the Jinsha River and explains the impacts of killing them (see Figure 10a.2, bottom right). This, in turn, helps mobilize more support for action against illegal fishing.

11

CONCLUDING REFLECTIONS

Braiding Flows

The Asian Highlands' presence in the center of the large Eurasian continent shapes movement around them. This movement, these flows, are multiple and interconnected. The Highlands drive climate systems that evaporate water from the ocean, direct clouds back to them, and cause precipitation pathways down through large river systems. They shape biological flows. Flora and fauna use the rivers as conduits up to and down from the Highlands. Birds take a more direct route, making seasonal flights over the mountains from one wetland to another and taking advantage of the diverse temperatures the Highlands provide. The Highlands have shaped human flows. Humans have followed Highland rivers to mountain passes on transhumant and longer-term migrations since at least the start of the Holocene (the past 11,700 years). They continue to travel across the mountains for pastures, trade, or pilgrimage wherever and whenever the region's complicated geopolitics allows. The Highlands themselves continue to flow, too. The Indian tectonic plate is still pushing under Eurasia, constantly reshaping rocks, rivers, and ecologies, redirecting flows.

The book's river stories also flow through channels that converge, augment, dissipate, and diverge. As Chapter 1 (our Introduction) explained, the braided sections of the Dri Chu and the upper Brahmaputra have been our model for the multidisciplinary and interdisciplinary research that produced this book. Sometimes, the various disciplines have flowed in separate channels as multidisciplinary research; at other times, they have converged into interdisciplinary research. Each discipline represented in the book—anthropology, climatology, conservation ecology, cultural and environmental history, geography, geology, geomorphology, geopolitics, political ecology, and religious studies—has contributed distinctive perspectives to the

DOI: 10.4324/9781003392033-11

book's chapters and their accompanying boxes. The juxtaposition between disciplinary approaches is most apparent in the chapters with accompanying boxes, where ideas flow through discrete channels. Chapter 2, for example, takes us through the epic story of the Highlands' geological and geomorphological formation, while its box tells some of the Highlanders' cosmological origin stories. Chapter 5 explains the rivers' ecologies and the conservation regimes that operate within them, and its box explains the relationship between the *lu* and water.

At other points in the book, co-researching and co-writing brought about interdisciplinarity, where disciplinary channels converged and integrated disciplines to produce enhanced insights. These points have multiple valences. At its most integrated, the writing is co-informed from the start, and the result is a piece that highlights different perspectives while offering fresh details. Box 6a, which tells the story of the Highland's fish from the perspective of political ecology and environmental history, is an example. At other times, the writing started from the perspective of one discipline but integrated whole subsections from other perspectives. Chapter 10 is an example of this. It began as a chapter written from an anthropological perspective but was heavily modified by insights from other disciplines.

Working across disciplines also forced us to learn new terms and modify those we used. From geomorphology, we learned to fix altitude in time as well as place. History taught us that "modern" is not always a synonym of "contemporary." This process was neither properly multidisciplinary nor interdisciplinary but something in-between, a form of disciplinary space-making. Once again, it evokes the interplay within a braided river; any channel's morphology is affected by the presence of the river's other channels, even when they run separately.

Crucially, these multidisciplinary and interdisciplinary processes have given us a better understanding of how flows work in the Asian Highlands. We have learned (and the book reflects) how flows interact with "natural," "cultural," and geopolitical borders. The Highlands' deep history—its geomorphological formation—created "natural borders" to movement over tens of millions of years. They directed climate-system flows, creating circular monsoonal precipitation on one side of the mountains and rain shadows on the other. The precipitation that landed in the mountains—rain and snow—flowed through valleys, creating river systems that began co-creating the landscape. Together, these mountains and river systems directed biota flows, creating variant evolutionary trajectories on either side of impassable peaks. And they have slowed—but not stopped—animal and human movement, as curious and adaptative quadrupeds and bipeds found different ways to engage with the "natural borders" around them.

Even though they found multiple ways to cross the Highland's mountains and rivers, the mountains' human communities still do not take this process for granted. Historically, Highlanders saw mountain passes as places of great metaphysical power and risk. Crossing them required appropriate incantations and offerings to a corresponding deity. Even now, as they sit in buses, trains, and planes that traverse passes, many Highlanders will chant prayers and, where possible, throw paper adorned with *lungta* (wind horses) to spread printed blessings. They approach river crossings with similar caution. Highlanders attach *lungta* to bridges and say prayers before crossing rivers or passing springs, lakes, and other waterbodies where *lu* and other powerful beings live.

In stark contrast to this millennia-long history of human movement across passes and rivers, since the 1960s, the region's governments have worked hard to block movement across the Highlands. Paradoxically, they have situated militarized blockades at border posts on passes and simultaneously developed narratives describing these new, lineal, geopolitical borders as eternal and "natural." The upper Dri Chu marks the border between Ganzi Tibetan Autonomous Prefecture in Sichuan Province and Chamdo Prefecture in the Tibetan Autonomous Region. The lower Dri Chu divides Ganzi from the Dechen Tibetan Autonomous Prefecture in Yunnan Province. South of the Himalaya, borders run along watersheds: Sikkim's borders are the Teesta's eastern, northern, and western watersheds, and much of the Bhutan-Arunachal Pradesh border runs along the Kameng-Drangme Chu (Manas) watershed. The Highlands' most contested, militarized, and—in terms of flows—directional border, the Line of Actual Control between China and India, traces the highest peaks between the Yarlung Tsangpo and the Brahmaputra. While Lowland governments, armies, and many geopolitical thinkers present this as a "natural" border, it was never a barrier for Highlanders. It is a Lowland creation, drawn by the British on a map in 1913 and only enforced after the 1962 Sino-Indian War (see Chapter 8).

Despite their best efforts, the region's governments have been unable to stop or even completely regulate Highland flows. Rivers remain notorious border flouters, and while yaks and dogs are sometimes stuck on either side of borders with their human companions, snow leopards, and black-neck cranes cross them without sanction. Occasionally, geomorphological flows even move geopolitical borders. The Nepal–China border transects the Earth's highest peak, Chomolungma (Mount Everest). In 2020, both governments conducted a survey to resolve their dispute over its height. They discovered it reached 8848.86 meters above sea level, just under a meter higher than earlier estimates. Earth scientists speculated that the 2015 Nepal earthquake had increased the mountain's height, and cartographers had to recalibrate the entire Central Himalayan section of the border. This example not only shows how geomorphology shifts borders but also how Earth science data—geodetic surveys, hydrology, and mathematics—are increasingly shaping them.

That said, geomorphological data destabilizes rather than fixes our view of the Highlands. It demonstrates large processes that are in continual movement but operate beyond the geographic and temporal scales of human lives, so we are not always aware of them. The Indian tectonic plate continues to slide under Eurasia, pushing the Highlands centimeters higher each year. The rains and rivers then erode much of this growth, transporting sediment across Asia's river plains and out into the Indian and Pacific Oceans. Throughout their deep history, Asian Highland rivers have intercepted tributaries and been intercepted by more aggressive streams cutting into mountain terrain. This river capture and other river drainage reorganizations have probably affected the genetics and diversity of the fish in the Highland catchments. However, knowledge of this topic is limited (see Chapter 2). Significant changes by river capture and catchment area increase and decrease divided fish migration and could have affected the Highlands' aquatic biota in the Highlands. Ecologically, each river's character—low water temperature, low food availability, rapid currents—has led to the evolution of bottom-dwelling, slow-growing, later-maturing, long-lived fish species in their upper reaches. These fish species' character has, in turn, helped shape the Highlanders' relationship with fish (see Box 6a), as their high levels of toxicity and comparatively small—and slow-replenishing—populations aligned with religious proscriptions to discourage fish eating across much of the region.

This interaction between religious belief and biophysical character is a good example of how the Highlands' environmental conditions influenced but in no way determined human experiences there. The Highlands' specific environmental conditions necessitated human populations' social organization and cultural practices, but this exchange was neither inevitable nor unidirectional. Rather than the one-way flow of environmental determinism, there were multi-directional flows of interdependent causes and effects. These causes and effects generated feedback loops that created further cycles. While mountains and rivers have created conditions for life over millions of years, the fish, flora, fauna, and people of the Asian Highlands have evolved and adapted with them. What is more, the conditions themselves keep changing, as fish, flora, fauna, people, and climactic factors influence them. The system's complexity is inestimable.

11.1 Humans and Rivers as Complex, Interconnected Systems

As the various chapters and accompanying boxes in this book have demonstrated, the Asian Highland rivers on which this book has focused, the upper Brahmaputra and Dri Chu (Yangzi), are remarkable and important in distinct ways. Yet, they and all the Earth's rivers share one characteristic: they are systems, not singularities. As discussed in Chapter 1 and revisited in Chapters 6 and 8, these rivers have single names on most maps because

Western geographical conventions insist rivers are singular entities that run from sources through courses to outlets on lakes, seas, or other sea-bound rivers. These naming conventions—calling rivers Brahmaputra, Yangzi, Nile, Amazon, or Mississippi—obfuscate the reality of the thing itself: rivers are systems. They enfold numerous streams and branches of flowing sky, surface, and subterranean water. They interact with the valleys, wetlands, and plains surrounding them. They are formed by slow time movements in the Earth's crust, shaped by growing mountains, and transformed by rapid events such as floods and landslides. And they are interconnected with numerous organisms—including humans—that depend on and impact their flows.

Looking back at the various writings of the book, the framework, "rivers as systems," provides some coherence to the different braids of our collective river thinking. This idea is not a grand synthesis of our views (see Chapter 1), nor does our use of it imply that we have consistently applied this framework to each chapter. Rather, this framework elaborates on our shared conceptual common ground. Rivers are systems, and, more specifically, they are complex systems. Approaching rivers as complex systems means considering them integrally composed of constituent parts, including humans. It resonates with a *general* definition of systems thinking, where any identifiable event or relation exists as a constituent part in a larger context. Critically, the parts and contexts inform each other; they are mutually constitutive.

Systems thinking has been used—and interpreted with different emphases—by many disciplines in various real-life situations, from mathematics and information technology to education and psychology (see Chapter 1). Because of its multidisciplinary and interdisciplinary nature, this book does not ascribe to any one specific disciplinary interpretation or definition of systems thinking. Instead, what we suggest as a coherent framework relies on the conceptual ground of general systems thinking. What we propose then as general systems thinking relies on key components: constituent parts in a larger context where the parts and contexts inform each other; mutual constitution of parts and contexts in a mechanism of feedback loops that are akin to "circular causal paths"; and general hierarchies of systems (comprising both parts and contexts) that link direct and indirectly with each other through circular causal paths. The result is a complex systems framework.

The relevance of this framework is most evident in Chapters 2–5, which approach the rivers primarily from the Planetary Rivers perspective. These chapters describe how the Asian Highlands rose to their great elevation and gave birth to large rivers whose tributaries, flow quantities, and directions are constantly reorganizing, shaping precipitation pathways and the ecosystems that develop within them. It also created a cryosphere at its higher altitudes, comprised of permafrost, seasonal and perennial snow, and glaciers that played a constitutive role in the Asian Highlands' climate variability, shaping the river systems' human and other biotic constitutive parts. Within this framework of complex systems, the explication of cause and effect is never linear but inherently circular so that outcomes, whether produced systemically or by human anthropogenic activity, feed back into the system and affect its many parts. The sheer scale of the Asian Highlands system, comprising (in Hutchison's terms) *biogeochemical* and *biodemographic* components (see Chapter 1 and Tan 2014), is impossible to comprehend fully.

Harrell's recent sweeping work on an ecological history of modern China (2023) demonstrates how complex systems necessarily include humans. Indeed, when we consider anthropogenic activities of various levels of magnitude—or scales—such as those we examine from the Social and Regulated Rivers perspectives, we begin to appreciate how much human activity has impacted the Asian Highlands over time and across scales. The Social Rivers perspective describes a socio-ecological system that developed over millennia. The Regulated Rivers approach describes and analyses a system of state, corporate, and non-governmental human activities that are simultaneously terraforming river systems while seeking to conserve parts of them. How we read these disturbances in terms of the systems' persistence is critical: can they adapt to the disturbances and maintain themselves, albeit in a changed way?

Moreover, by explicitly considering these complex systems as socio-ecological systems, we ascribe to the view that these are "neither humans embedded in an ecological system nor ecosystems embedded in human systems, but rather a different thing altogether" (Walker et al. cited in Harrell 2023, 16). As Harrell notes from Abel (2007), these systems consist of not only material and energy flows but also ideas, power, and social relations (Harrell 2023, 16). As we detail in Chapters 6, 8, and 10, these systems also involve trade and capital flows.

As the systems approach is not mechanistic, it assumes new flows are always possible. It thus avoids the conclusions of High Modernism, which ascribes to the "technological lock-in" (Wasson et al. 2020) and the tendency to implement "a fix to fix the fix" (Harrell 2023, 2). Despite the possibility of new flows, the corollary question of timescale is ever-present: how quickly can those new flows be created considering the intense onslaught of human anthropogenic activities? In a complex systems framework, we agree with Harrell that there are "both spatial and temporal components; the relations between spatial parts change over time, and the temporal changes over time work differently in different spaces" (Harrell 2023, 16). We offer these insights with some caution since a complex systems framework, which has not been applied or tested consistently through our chapters, can only be understood retrospectively. However, future research could develop a semi-predictive model.

11.2 *Tendrel* (Interdependent Origination) and the Limits of System Thinking

We also want to be careful not to take the idea of complex systems too far; it is one of the book's themes, not its entire framing. In the book, we have placed this way of understanding reality in conversation with others more important to Highland communities. In Chapters 1 and 7, we introduced the cherished Buddhist concept of *tendrel*, or interdependent origination, *pratityasamutpada* in Sanskrit. *Tendrel* is a key Buddhist doctrine shared by all its schools but interpreted in various ways by Buddhist philosophers (Kardas 2015) and communities (see Chapter 7). Its basic principle is that all phenomena arise in dependence on each other. The Mahayana form of Buddhism practiced in the Highlands interpreted this interdependent origination to mean that all phenomena were empty of independent and, therefore, ultimate existence. This definition makes it a synonym of *tongpanyi* (emptiness) but approaches this ultimate reality from another perspective. Viewing *tongpanyi* or emptiness as *tendrel* avoids the trap of nihilism. Things do exist; they just do not exist as we assume. They are conventional, momentary, and dependent on causes and conditions that are, in turn, the workings of *karma*, cause and effect.

As other thinkers such as Joanne Macy (1991) have noted, there are resonances between the idea of *tendrel* and the interdependence evident in the systems thinking that underpins ecology and, particularly, the holistic concepts of deep ecology. However, the Mahayana Buddhist understanding of reality is quite distinct from systems thinking, ecology, or any type of empirical science because it is not grounded in materialism. In Mahayana Buddhist thought, reality does not arise from matter but from the non-inherently existent mind. Through *tendrel*'s lens, things arise and fall because of *karma; karma* arises from mental intention, and the mind, ultimately, does not exist. The two knowledge lineages started from different positions, developed through different histories, and operate with different intentions. To synthesize them would be to distort them, and, given the power differentials between supporters of materialism and *tendrel, tendrel* would be more distorted by the interaction than materialism (see Chapter 1 and Stengers 2018). So, rather than attempting to synthesize them, we merely note their agreement on the general existence of interdependence.

11.3 Generated Insights

Writing this book involved discussions that, while grounded in disciplinary perspectives, generated insights beyond them. These insights came in three main kinds: (1) clearly generative connections where the synergy among disciplines enabled new ideas and further possibilities for collaborative thinking; (2) potential synergies that could be developed through further multidisciplinary or interdisciplinary investigations; and (3) limited synergies that demonstrate disjuncture between the various disciplines' starting points, frameworks, and goals.

11.3.1 Generative Connections

Our reflections on the many flows that traverse the Asian Highlands is one example of the generative connections from this multidisciplinary and interdisciplinary research. The idea of rivers and humans as complex systems and its juxtaposition—not synergy—with the Buddhist concept *tendrel* is another. There were many more, but we will highlight two here: the interdependence of Highlands and Lowlands and the pervasive role of Gravitational Potential Energy (GPE) in the region's histories and present.

We introduced the Highland–Lowland distinction in Chapter 1 as a heuristic device to give the reader a sense of the book's setting, but it has proved a fecund framing throughout the book, informing multiple disciplinary approaches. One element of this Highland–Lowland distinction is its physicality: the region's geomorphology has produced steep inclines, and their altitudinal change produces proximate changes in climate, hydrology, and ecology. Human societies within these ecotones co-evolved with them, growing rice and raising cows at low altitudes, millet in mid-altitudes, barley higher up, herding yaks in even higher regions, and leaving the mountain peaks to the gods.

More recently, the Highland–Lowland distinction has also shaped the region's geopolitics and economies. Since two lowland states, India and China, territorialized the Highlands in the mid-twentieth century (see Chapter 8), they have dominated their political economies and ecologies, intensifying governance of and extraction from the region. In a seemingly odd dichotomy, they have increased both the exploitation and conservation of Highland environments at the expense of their local and Indigenous peoples (see Chapter 10). Their competition for territory and the unresolved borders has intensified the construction of road and railway development across the Highlands, and these projects have then driven further development. They focus their river management practices on the extraction of water and hydroelectricity for Lowland populations. Technocratic conservation, growth in tourism and recreation facilities for visiting Lowlanders, and an emerging desire to sequester carbon have all led to more reserves in the Highlands.

One of the benefits of working across disciplines as we engaged with this Highland–Lowland distinction was that while we could see it, we could also provide each other with examples of connections that subverted it. Rivers are the most apparent Highland–Lowland connector, but so are the climatic systems that drag water up to the mountains from evaporating seas. Historic trade and pilgrimage networks transect the Highland–Lowland divide, as do the contemporary hard infrastructure such as roads, trains, and electrical "powersheds" (electrical networks that direct energy flows), along with the administrative zones and education systems that lowland states use to govern the highlands.

The other intensively generative insight our research produced was perhaps more unexpected: the role GPE has played and continues to play throughout the Asian Highlands (see Box 2a). More GPE is stored on the Tibetan Plateau's edges than anywhere else. It dissipates to low-altitude areas of low GPE through downhill flows, such as rivers, landslides, avalanches, glacial lake outburst floods (GLOFs) and landslide lake outburst floods (LLOFs). In these downward flows, the Plateau's GPE is transformed into kinetic energy, which is measured as stream power in rivers.

Refracting the idea of GPE through multiple disciplinary lenses led us to several conceptual breakthroughs. By enlivening the Highland's metabolism, this kinetic energy has enabled much of the region's multi-scaler *more-than-human* history. Geomorphologically, the downward flow of water has carved out the river valleys in which most of the Highlands' plants, animals, and humans live. Hydrologically, GPE has provided the running water that humans and animals need for drinking (see Chapter 7) and humans used to irrigate their crops (see Chapter 6). The presence of GPE also made the Plateau's edge an inherently risky place (see Box 4a). Large floods, earthquakes, and landslides shaped and reshaped the Highlands' politics and economies (see Chapters 6 and 8).

As noted in Chapter 1, the Highlands most influential belief systems—Indigenous, Buddhist, and *Mentsi* (medicine and calculation, including geomancy)—evolved with these forces (see also Chapters 6 and 7 and Boxes 1a, 5b, and 7a). At the village level, these belief systems were entwined with memories of previous disasters, leading Highlanders to utilize their space wisely (see Chapter 7 and Box 9b). These localized knowledge lineages led villagers to build houses from earthquake-resistant materials away from flood and landslide paths. Their relationship with these spaces stands in contrast to Lowland administrators' technocratic approach to Highland development. These relationships, along with livelihood opportunities provided by tourist and pilgrimage routes, direct development to valley floors, which are floodplains and sites of landslides, and, therefore, require increasingly intrusive engineering interventions in an unrealistic effort to reduce risks.

These twin capacities of GPE to enliven and destroy have shaped much of the region's history and present political-ecological tensions. The push to regulate the Highlands' rivers is another example of this. Currently, it is focused on extracting this energy, primarily by constructing large hydropower dams (see Chapters 8, 9, and Box 9b). However, as noted in Box 4a, the rivers' dynamic histories are etched into the mountains' geoarchives, and they carry dire warnings about the construction of these large dams. The presence of GPE may be the primary reason hydropower dams are built on the Tibetan Plateau's edge, but a complex relationship of elements determines its intensity, and these elements are continually self-regulating, trying to reach stability through rock uplift, valley incision, and landslides. A shift in one constitutive part results in effects that are greater than a similar shift in a less risky place. Human interventions, such as the building of dams, create a significant shift in the self-regulation of the region by exponentially multiplying the potential effects within the larger system. This dynamic means that the site and size of hydropower projects matter intensely. Those engaging in infrastructure planning should take a deep-time perspective before rushing to build—geologically speaking—short-term projects that can have disastrous outcomes.

Our multifaceted approach to GPE also enabled us to assess social and political responses to it critically. We noted, for example, that the Lowland-based governments who have administered the rivers' catchments since the mid-twentieth century have used their co-emergent energy potential and geomorphological dynamism—along with the emerging threats from climate change—to justify greater controls in the mountains. In the Himalaya, multiple governments focused on the disproven theory of Himalayan Environmental Degradation, painting the region as a "hazardscape" (Chakraborty 2021) and increasing both their governance of and extraction from the region. The Chinese state deployed similar language in the Sangjiangyuan region (see Chapter 9), describing it as "degraded" before intensifying its regulation of Highland landscape and water courses. This approach also ignores and attempts to erase the multiple, innovative, and less impactful ways local communities have lived and worked with this energy for millennia. In cases like this, with such profound practical ramifications, we need to choose the concepts and words we use to describe places carefully. Representing the Asian Highlands as a site of high GPE, with all that this entails, may be a better way to understand its dynamic elemental processes than focusing on its hazards. At least it grants space for governments to learn from community solutions to mountain dynamics rather than engaging in hubristic attempts to control them.

11.3.2 Further Potential Synergies

While collaborating on this book, we also discovered synergies that need further exploration. These include (1) the study of aquatic biota to better understand river network reorganizations in deep time; (2) the historical and contemporary socio-ecological effects of atmospheric rivers; and (3) the multidisciplinary analysis of environmental justice in the Highlands.

The first two are discrete phenomena that particular disciplines could investigate. Studying aquatic biota and river networks would require collaboration between biologists and geomorphologists. It would build on the available evidence that the timescales of aquatic species evolution are comparable with those of long-term geomorphic change (Stokes and Perron 2020; Craw et al. 2016). These studies provide possible clues relevant to the Asian Highlands, as exemplified by Ma et al. (2015), who argue from molecular data that the glyptosternoid fishes in the Tibetan Plateau's southeastern rivers can largely be explained by their drainage networks' major reorganization.

Other changes evident in the Asian Highlands' geomorphic processes may also have changed habitat connectivity, producing new species, new corridors for dispersal, and barriers for aquatic organisms. Modeling by Stokes and Perron (2020) suggests that the catchment area is a primary control of species richness, and river capture increases species richness temporarily. Elevated speciation and extinction rates can provide a persistent biological record of river network reorganization, as evidenced by data from Aotearoa (New Zealand), where fish genomes have retained signs of tectonic landscape development. A correlation between geologic age and DNA sequence divergence shows that "... landscape evolution has controlled ongoing biological diversification over the past 25 million years" (Craw et al. 2016). Further testing this idea in the Asian Highlands would require collaboration between fish biologists and geomorphologists, but its results would produce important insights into the Highlands' formation and its aquatic biota.

Another area of further synergies is the study of atmospheric rivers. As we learned in Chapter 3, atmospheric rivers are relatively long, narrow regions in the atmosphere that transport water vapor from tropical oceans to temperate zones. These ephemeral rivers transport immense amounts of water and, when they occur, often cause floods, landslides, and other hazards. As monsoonal Asia's climate warms, it produces an increasing number of these rivers in the sky. A multidisciplinary study of historical and contemporary social and physical atmospheric rivers would produce numerous intensely generative insights. Following the rivers would make connections between Asia's Highlands and its two oceans, the Indian and Pacific, regions that are, like the Highlands, already experiencing the deleterious impacts of climate heating. This topic could also produce multiple methodological insights as Asia's Highlands and Indo-Pacific Islands contain geoarchives within ice cores and pollen deposits that could be further juxtaposed with their written and oral history archives.

All the book's disciplines could also analyze more generalized topics. One potential topic is environmental justice failures in the Asian Highlands. The book's last three chapters tangentially addressed this issue. Chapter 8 traced the mid-twentieth century territorialization of the upper Brahmaputra and Dri Chu catchments and the concurrent minoritization of its local and Indigenous peoples. Chapters 9 and 10 outline the impact of this territorialization on the Highlands and the governance, extraction, and conservation programs that followed it. Each of these disciplinary approaches notes that the construction of dams and other river-system infrastructure disproportionately impacts local and Indigenous populations. It is unevenly situated on their lands and excessively affects their knowledge lineages, lifeways and livelihoods. Sometimes, as the Rong (Lepcha) anti-dam campaigners insist (see Box 9b), one project threatens an entire community and its culture.

This book outlined how the Chinese state operates a technocratic governance system that applies centralized rules to river management. This form of governance has led them to fully exploit some rivers' hydropower potential and leave others for biodiversity conservation. Competition between India's major political parties and the states in its federal union has led its multiple governments to the *ad hoc* development of hydropower projects without proper environmental or cultural appraisals. By contrast, Bhutan has far more explicit policies and practices for the sustainable development of its mountain rivers, policies that place great emphasis on sustaining cultural landscapes. Further study of this topic would investigate more thoroughly how contemporary sovereign states' hydraulic missions—the impetus to maximize and distribute available energy and water—clash with emplaced, relational knowledge lineages, and how each sovereign state's history, politics, geopolitics, and culture determines its approach to riverine resource management.

By tracing the Asian Highlands' territorialization and the cascading effects that flowed from it, all the book's disciplines have also described and analyzed various facets of the region's complicated international hydropolitics. The Earth sciences describe the fundamental geomorphological, climatic, and hydrological systems within which these politics operate. The geopolitical (or hydropolitical) ecological lens that our co-authors Dechen Palmo and Alexander Davis use in Chapter 8 and their other research (Davis 2023) reveals how these international tensions play out in the contemporary setting. Further integration of these approaches would be insightful. Rivers such as the Yarlung Tsangpo and many other Brahmaputra tributaries arise on the Tibetan Plateau within China's borders and descend through India and Bhutan to Bangladesh. As they descend, they supply water, food, and other ecosystem services (ESS) to ecosystems and communities. Upstream activities, such as dam construction in China or India, affect downstream water flow and livelihoods, agriculture, and biodiversity in lower riparian countries like Bangladesh. Once again, this disparity highlights environmental injustice, but this time between sovereign states, who have unequal power and control over the rivers, and this equates with their power and control over resources. Historical readings of these issues suggest the hydropolitics' contingency, and many of the tensions between the region's sovereign states reflect, intensify, or mitigate their historical relations. Investigating these tensions from a multidisciplinary approach would produce many further insights.

11.3.3 Limited Synergies

To effectively pursue these more general topics from a multidisciplinary and possibly interdisciplinary approach, we would also need to be cognizant of the limitations of this kind of work. We have already encountered these limitations within this study as we have negotiated incompatibilities between our different disciplines' starting points and frameworks. Two areas of limited synergy that we encountered were the valuing of ecosystems and the benefits of offering policy suggestions.

Chapter 10 traced the limited synergies between our approaches to ecosystem valuation. It analyzes two bioconservation approaches to human–environment interactions, ESS, traditional ecological knowledge (TEK), and local ecological knowledge (LEK) from conservation biology and anthropology perspectives. It explains that environmental economists developed ESS by identifying ecosystems' services and calculating their economic value in the 1980s and 1990s. Calculating ESS, they theorized, would reduce natural resource exploitation and compensate environments and communities for degradation.

It also traces the emergence of TEK, the term anthropologists developed to describe local, often Indigenous communities' ways of knowing and organizing their environments. Early descriptions of TEK included the complexity of emplaced human–environment experiences. Later, however, it was abstracted (Dwyer 2005) and distilled (Nadasdy 1999) to make it more compatible with bioconservation's language and knowledge systems. It took decades for TEK to gain any recognition within the field of bioconservation, but in recent years, some conservation experts have begun using the language of collaboration, embracing this abstracted and distilled version of TEK. Others are still resisting, framing TEK as the qualitative, experiential, and non-codified knowledge of Indigenous and local communities, which cannot be incorporated into the science of bioconservation as it is based on quantitative data with objective-based outcomes. This form of knowledge is also called LEK. The division between LEK and TEK reflects the different knowledge lineages we explored in Chapter 1: (1) the enumerated sciences that take a phenomenological approach to knowledge; (2) the experiential emplaced knowledge lineages; and (3) the humanities and social science disciplines that study these experiential systems qualitatively.

ESS adds another degree of complexity to this quantitative–experiential–qualitative triangle by introducing *value*. As described in Chapter 10, ESS has been variously successful and problematic. It undoubtedly values the knowledge and experiences of local and Indigenous peoples and aims to recompense them for maintaining their environments, but it still operates within the logic of capitalism and imposes the values of neoliberal institutions. The "success" of an ESS approach depends on how rigorously its principles are applied. When this is done well, technocrats must explicitly identify cultural values and TEK. When it is not, it merely uses different language to perpetuate the inequalities between dominant and subservient positions.

Our primary insight into this debate is that starting points and framing matter. If the project intends to prioritize local and Indigenous practices and the framing is comparative, which is anthropology's approach, the debate about whether TEK or ESS is better placed to advance local and Indigenous peoples is moot. As we have explained repeatedly in this book, the knowledge lineages of local and Indigenous peoples and those of environmental conservation are irreducibly different, and the power imbalances between the two groups are severe. From this perspective, any bioconservation intervention, whether reframing their belief system through TEK or applying ESS, does epistemic violence to them. If the debate starts from a pragmatic, solutions-based perspective of conservation biology, by contrast, the answers become more complex, and there are various possible outcomes. In a best-case scenario, each community would assess the governance structures, risks, and benefits they live with and make collective decisions based on their circumstances. The specificities of the Asian Highlands' complex politics and geopolitics must also be considered when deciding whether these best-case scenarios are likely. Its governance structures are tilted toward Lowland majoritarianism, and Highland communities get little say in the decisions made about their lands. This leads those who are in favor of ESS to ask: What is the counterfactual to not using it? If these systems are not used, is it likely that the Chinese and Indian states will revert control of these resources to Indigenous and local communities? Probably not. In lieu of this, they argue, ESS may provide the best justification for contemporary states to recognize and sustain Indigenous environmental care.

The complexity of human activities in the Highlands—states, NGOs, experts, and multiple communities—along with our various disciplinary and personal backgrounds, also meant that we could not agree on whether this book should offer policy suggestions. Some disciplines, particularly conservation biology, political ecology, and geopolitical analysis, tend to require their researchers to produce policy and other recommendations. Others, such as anthropology, history, and religious studies, do not. There are also non-disciplinary, ethical reasons to make and not make recommendations. There are, for example, good reasons why a group of people working in Global North institutions, only one of whom is from the Highlands, should refrain from offering advice to local and Indigenous communities. Such advice-giving perpetuates colonial paradigms and solution cycles, in which one interventionist solution creates the causes for new "solutions." But what if the advice arises obviously from the data we have examined, such as the imperative to shift toward off-river pumped hydro? Where does the presentation of information become an insistence on solutions? And does it matter if that information/advice is aimed at governments rather than local communities? We have no definitive answer for this, so to continue our commitment to braided research, we have compiled a list of suggestions from those authors who wanted to contribute policy suggestions as the book's last box, which accompanies these Concluding Reflections.

BOX 11a POLICY SUGGESTIONS

- The Asian Highlands' abundance of GPE can support the region's energy transition, but it needs to be produced in much less hazardous and less environmentally and socially impactful ways. The best means of doing this is through off-river, pumped-storage hydropower stations that back up solar and wind energy generators. Smaller projects such as this would also involve less risky tunneling and displacement.
- Governments should recognize and support Highland communities. This recognition should enable them to choose their own development pathways. This autonomy could be seen in the more systematic application of embedded knowledge, such as avoiding floodplain construction and building with earthquake-proof materials.
- Governments should support Highlanders' livelihoods through payments for carbon sequestration, water catchment services, and tourist recreational services. Rather than extracting from these communities, governments should support their efforts to manage cultural landscapes and conserve biodiversity.
- Both governments and conservationists should use local geographic and species naming conventions to describe prioritized and non-prioritized places and species. This is particularly important for species that are endemic to the region. It will draw attention to and generate support for their conservation. Using local naming practices enforces state and substate recognition of cultural differences. This awareness among government workers and conservationists would support cultural and environmental justice efforts.
- Adhering to international conventions, such as the Ramsar Convention on Wetlands, can provide a framework for conservation strategies and actions that often align with the desires of traditional communities and mitigate the over-exploitation of highland resources. These codes give primacy to principles, such as conserving each species and representative areas of each type of ecosystem.

11.4 Multiple Voices, Multiple Scales

Part of the reason it was difficult to agree on the inclusion or exclusion of suggestions was the complexity of the river systems' temporal and spatial scales. Throughout the book, we have tried to indicate its multiple and enormously variant time scales through the turtle timelines that introduce each chapter. These range from tens of millions of years to now and reflect the influence of the ideas of "deep time" and the separation of planetary, Earth, and human histories on our temporal discourses (Fildani 2022). Our multiple temporal scales also intersect various spatial scales. The deep-time geological cycle created the mountains, and the hydrological cycle connects them to the ocean; water arises from the hydrosphere's oceans, evaporates through the atmosphere, lands as snow in the cryosphere, and melts back into the hydrosphere's rivers. Among these hydro-geological interactions, diverse species-made habitats, societies, and cultures evolved, historical empires and kingdoms came and went, present-day sovereign states continue to govern territories and fix borders, and NGOs operate with and around governments' development and bioconservation agendas.

Specific contexts within these multiple temporal and spatial scales sometimes merge time and space. Bioconservation, for example, emerged from a history of discursive ideas in Enlightenment Europe and flowed into another context, namely the management of present-day Asian Highlands' river systems. In this way, contexts—as scales—exist within contexts, an example being where a solution, such as ESS, might work or make sense at one level, namely a national context, but not so much at another level, namely a subnational context. Multiple contexts thus denote the multiple scales this book has attempted to narrate.

Achieving a coherent yet multiple narrative has been key: multiple narratives have manifested from different disciplinary perspectives, yet we have also attempted to give multiple "voices" to the book itself. It was important to co-write with Indigenous academics from communities across the catchments. If we had our time again, we would put better geopolitical-proof strategies into our research plans and grant these voices a leading role. It was also challenging to think with—and against—stark dichotomies, for example, abstract large-scale realities of geology and hydrology versus the on-the-ground realities of anthropology or the specificities of philological textual analysis.

The voices of the book's co-authors were the braided river's divergent channels in our collective river thinking, and, given our different starting points, perspectives, and communication styles, it was undoubtedly a challenge to merge our thoughts. Nonetheless, our collective goodwill and focus on the flows between scales and topics allowed most dichotomies to become distinctions. Flows were also crucial in our group discussions, which were generative because we let the process dictate what happened next. Going with the discursive flow required a commitment to engage with and listen to each other, even when the conversation showed limited synergies. At those times, we could only listen and allow the differences to co-exist. Often, however, the conversations mirrored the free-flowing course of a braided river, and after one divergence, there was another convergence.

APPENDIX

River Names

On maps/ in text	*Tibetan/ Dzongkha name*	*Tibetan/ Dzongkha spelling*	*International name*	*Chinese name*
YARLUNG TSANGPO (Brahmaputra) CATCHMENT				
Yarlung Tsangpo	Yarlung Tsangpo Tachog Khabap	Yar klung gtsang po Rta mchog kha ’bab		Yalu Cangbu Jiang 雅鲁藏布江
Brahmaputra			Brahmaputra	Bulamaputela He 布拉马普特拉河
Myang Chu[a]	Nyang Chu	Myang chu		Niyang Qu尼洋曲
Kyi Chu (Lhasa)	Kyi Chu	Skyid chu	Lhasa	Lasa He 拉萨河
Tolung Chu	Tolung Chu	Stod lung chu		Donglu Qu洞庐曲
Yarlung Chu	Yarlung Chu	Yar klung chu		Yalong He 亚龙河
Nyang Chu	Nyang Chu	Nyang chu		Nyang Qu尼洋曲
Parlung Tsangpo	Parlung Tsangpo	Spa lung gtsang po		Palong Zangbu 帕隆藏布
Dibang			Dibang	Dangba Qu 丹巴曲
Dzayul Chu (Lohit)	Dzayul Chu	Rdza yul chu	Lohit	Zayuqu 察隅曲
Chayul Chu (Subansiri)	Chayul Chu	Bya yul chu	Subansiri	Subanxili He 苏班西里河
Mangde Chu	Mangde Chu	Mang sde chu		Mande Qu 曼德曲

(*Continued*)

(Continued)

On maps/ in text	*Tibetan/ Dzongkha name*	*Tibetan/ Dzongkha spelling*	*International name*	*Chinese name*
Chamkhar Chu	Chamkhar Chu	Cham dkar chu		Chamukaer Qu 查姆卡尔曲
Drangme Chu (Manas)	Drangme Chu	Grangs med chu	Manas	Delangmu Qu 德朗木曲
Puna Tsangchu (Sunkosh)	Puna Tsangchu	Spu na gtsang chu	Sunkosh	Zhalong Qu 扎龙曲
Wang Chu (Raidak)	Wang Chu	Dbang chu	Raidak	Wang Qu 旺曲
Amo Chu (Torsa)	Amo Chu	A mo chu	Torsa	Tuo'ersa He 托尔萨河 Tisita He 提斯塔河
Teesta	(Also known as Tista)			
DRI CHU (Yangzi) CATCHMENT				
Dri Chu (Yangzi)	Dri Chu	'Bri chu	Yangzi / Yangtze	Yangzi Jiang 扬子江
Nyak Chu (middle Yalong)	Nyak Chu	Nyag chu	Yalong	Yalong Jiang 雅砻江
Dza Chu (upper Yalong)	Dza Chu	Rdza chu	Yalong	Yalong Jiang 雅砻江
Eastern Gyelmo Ngul Chu (Dadu)	Gyelmo Ngul Chu	Rgyal mo dngul chu	Dadu	Dadu He 大渡河
Min			Min	Min Jiang 岷江
She Chu (Xian Shui)	She Chu	Shas Chu Shing Chu		Xianshui He 鲜水河
Other Rivers				
Western Gyelmo Ngul Chu (Salween)	Gyelmo Ngul Chu	Rgyal mo dngul chu	Salween	Nu Jiang 怒江
Da Chu (Mekong)	Da Chu	Zla Chu (Sometimes Rdza Chu)	Mekong	Lancang 澜沧江
Sengge Tsangpo (Indus)	Sengge Tsangpo Sengge Khabap	Seng ge gtsang po Seng ge kha 'bab	Indus	Yindu He 印度河

(Continued)

(Continued)

On maps/ in text	*Tibetan/ Dzongkha name*	*Tibetan/ Dzongkha spelling*	*International name*	*Chinese name*
Langchen Tsangpo (Sutlej)	Langchen Tsangpo Langchen Khabap	Glang chen gtsang po Glang chen kha 'bab	Sutlej	Satelaijie He 萨特莱杰河
Macha Tsangpo (Karnali)	Macha Tsangpo Macha Khabap	Rma bya gtsang po Rma bya kha 'bab	Karnali	Jiagela He 加格拉河
Ganga		Gangga	Ganges	Heng He 恒河
Ma Chu (Huang He)	Ma Chu	Rma Chu	Yellow	Huang He 黄河
Ayeyarwady (Also known as Irrawaddy)				Yiluowadi Jiang 伊洛瓦底江

[a] This river's name is often written phonetically as Nyang Chu. Following its Tibetan spelling, we have chosen to write it as Myang Chu to differentiate it from the other similar-sounding river that enters the Yarlung Tsangpo in Kongpo.

BIBLIOGRAPHY

Abel, Thomas. 2007. "World-Systems as Complex Human Ecosystems." In *The World System and the Earth System*, edited by Alf Hornborg, and Carole Crumley, 56–73. Walnut Creek: Left Coast Press.

Abell, Jordan T., Gisela Winckler, Robert F. Anderson, et al. 2021. "Poleward and Weakened Westerlies during Pliocene Warmth." *Nature* 589 (7840): 70–75.

Abell, Robin, Michele L. Thieme, Carmen Revenga, et al. 2008. "Freshwater Ecoregions of the World: A New Map of Biogeographic Units for Freshwater Biodiversity Conservation." *BioScience* 58 (5): 403–414.

Abram, Nerilie J., Nicky M. Wright, Bethany Ellis, et al. 2020. "Coupling of Indo-Pacific Climate Variability over the Last Millennium." *Nature* 579 (7799): 385–392.

Acheampong, Michael. 2020. "'Critical Ecosystems' as a Concept in Political Ecology–Developing a Comprehensive Analytical Framework." *Journal of Political Ecology* 27 (1): 190–212.

Afroz, Rounak, and Md Ataur Rahman. 2013. "Transboundary River Water for Ganges and Teesta Rivers in Bangladesh: An Assessment." *Global Science and Technology Journal* 1 (1): 100–111.

Agrawal, Arun. 2002. "Indigenous Knowledge and the Politics of Classification." *International Social Science Journal* 54 (173): 287–297.

Agrawal, Arun. 2005. "Environmentality: Community, Intimate Government, and the Making of Environmental Subjects in Kumaon, India." *Current Anthropology* 46 (2): 161–190.

Ahmad, Zahiruddin (tr.), 1995. *A History of Tibet by Ngags Dbang blo Bzang rgyam mtsho, fifth Dalai Lama of Tibet*. Bloomington: Indiana University Oriental Series, 1995.

Akhtar, M. P., Nayan Sharma, and C. S. P. Ojha. 2011. "Braiding Process and Bank Erosion in the Brahmaputra River." *International Journal of Sediment Research* 26 (4): 431–444.

Alam, Firoz, Quamrul Alam, Suman Reza, et al. 2017. "Sourcing Green Power in Bhutan: A Review." *Energy Procedia* 110: 586–591.

Alessandri, Andrea, Franco Catalano, Matteo De Felice, et al. 2021. "Varying Snow and Vegetation Signatures of Surface-Albedo Feedback on the Northern Hemisphere Land Warming." *Environmental Research Letters* 16 (3): 034023.

Allen, D. J., S. Molur, and B. A. Daniel. 2010. *The Status and Distribution of Freshwater Biodiversity in the Eastern Himalaya*. Cambridge: IUCN.

Allouche, Jeremy. 2020. "Nationalism, Legitimacy and Hegemony in Transboundary Water Interactions." *Water Alternatives* 13 (2): 286–301.

Altner (Lange), Diana. 2009. "Fishery in Southern and Central Tibet: An Economic Niche Is Going to Disappear." In *Contemporary Visions in Tibetan Studies: Proceedings of the International Seminar of Young Tibetologists*, edited by Brandon Dotson, Kalsang Norbu Gurung, Georgios Halkias, et al., 45–67. Chicago: Serindia Publications.

Altner (Lange), Diana. 2010. "From Water Radish to Fish Restaurant: Recent Developments of Fisheries in Central Tibet." In *AAS Working Papers in Social Anthropology*, edited by Andre Gingrich, and Guntram Hazod, Vol. 18, 1–11. Vienna: Austrian Academy of Sciences.

Amrith, Sunil. 2018. *Unruly Waters: How Rains, Rivers, Coasts, and Seas Have Shaped Asia's History*. New York: Basic Books.

An, Baosheng, Weicai Wang, Wei Yang, et al. 2022. "Process, Mechanisms, and Early Warning of Glacier Collapse-Induced River Blocking Disasters in the Yarlung Tsangpo Grand Canyon, Southeastern Tibetan Plateau." *Science of the Total Environment* 816: 151652.

An, Wenling, Shugui Hou, Wangbin Zhang, et al. 2016. "Possible Recent Warming Hiatus on the Northwestern Tibetan Plateau Derived from Ice Core Records." *Scientific Reports* 6 (1): 32813.

Angchok, Dorjey, and Premlata Singh. 2006. "Traditional Irrigation and Water Distribution System in Ladakh." *Indian Journal of Traditional Knowledge* 5 (3): 397–402.

Anisimov, Oleg, and Reneva Svetlana. 2006. "Permafrost and Changing Climate: The Russian Perspective." *Ambio* 35 (4): 169–175. doi: 10.1579/0044-7447.

Anonymous. 2015. "Tibetans and Fish – Part I: From Taboo to Eating Fish." *High Peaks Pure Earth*, January 15. Accessed 27 May 2023. https://highpeakspureearth.com/tibetans-and-fish-part-i-from-taboo-to-eating-fish/.

Ardussi, John A. 1977a. "The Quest for the Brahmaputra River and Its Course According to Tibetan Sources." *The Tibet Journal* 2 (1): 35–49.

Ardussi, John A. 1977b. "Bhutan before the British: A Historical Study." PhD diss., The Australian National University.

Argand, E. 1924. "La tectonique de l'Asie." Proceedings of the 13th International Geological Congress, Brussels 1: 181–371.

Armijo, Rolando, Paul Tapponnier, J. L. Mercier, et al. 1986. "Quaternary Extension in Southern Tibet: Field Observations and Tectonic Implications." *Journal of Geophysical Research: Solid Earth* 91 (B14): 13803–13872.

Arora, Vibha. 2009. "'They Are All Set to Dam(n) Our Future': Contested Development through Hydel Power in Democratic Sikkim." *Sociological Bulletin* 58 (1): 94–114.

Arthington, Angela H., C. M. Finlayson, and Jamie Pittock. 2018. "Freshwater Ecological Principles." In *Freshwater Ecosystems in Protected Areas: Conservation and Management*, edited by C. Max Finlayson, Angela H. Arthington, and Jamie Pittock, 34–53. London: Routledge.

Artyushkov, E. V. 1973. "Stresses in the Lithosphere Caused by Crustal Thickness Inhomogeneities." *Journal of Geophysical Research* 78 (32): 7675–7708.

Asthana, Vandana, and A. C. Shukla. 2014. *Water Security in India: Hope, Despair, and the Challenges of Human Development*. London: Bloomsbury Publishing.

Avouac, J. P. 2015. "Mountain Building: From Earthquakes to Geologic Deformation." In *Treatise on Geophysics (Second Edition)*, edited by Gerald Schubert, 381–432. Oxford: Elsevier.

Baghel, Ravi, and Lea Stepan. 2017. "Contextualizing the Anthropocene: The Cultures, Pracices and Politics of Water Knowledge in Asia." In *Water, Knowledge and the Environment in Asia: Epistemologies, Practices and Locales*, edited by Ravi Baghel, Leea Stephan, and Joseph Hill, 1–18. London: Taylor and Francis.

Bajracharya, S., and B. Shrestha. 2011. *The Status of Glaciers in the Hindu Kush-Himalayas*. Kathmandu: International centre for integrated mountain development (ICIMOD).

Barnett, Jon, Sarah Rogers, Michael Webber, et al. 2015. "Sustainability: Transfer Project Cannot Meet China's Water Needs." *Nature* 527 (7578): 295–297.

Barros, A. P., M. Joshi, J. Putkonen, et al. 2000. "A Study of the 1999 Monsoon Rainfall in a Mountainous Region in Central Nepal Using TRMM Products and Rain Gauge Observations." *Geophysical Research Letters* 27 (22): 3683–3686.

Barstow, Geoffrey. 2017. *Food of Sinful Demons: Meat, Vegetarianism, and the Limits of Buddhism in Tibet*. New York: Columbia University Press.

Barua, Anamika, Sumit Vij, and Mirza Zulfiqur Rahman. 2018. "Powering or Sharing Water in the Brahmaputra River Basin." *International Journal of Water Resources Development* 34 (5): 829–843.

Baruah, Sanjib. 2020. *In the Name of the Nation: India and Its Northeast*. Stanford: Stanford University Press.

Baten, Mohammed Abdul, and Rashed Al Mahmud Titumir. 2016. "Environmental Challenges of Trans-Boundary Water Resources Management: The Case of Bangladesh." *Sustainable Water Resources Management* 2 (1): 13–27.

Baviskar, Amita. 1997. *In the Belly of the River: Tribal Conflicts over Development in the Narmada Valley*. Delhi: Oxford University Press.

Bej, Sourina. 2020. "Locked in Hydropolitics: Understanding the Local Protests and Differing Dynamics in Tawang." In *Tawang, Monpas and Tibetan Buddhism in Transition: Life and Society along the India-China Borderland*, edited by M. Mayilvaganan, Nasima Khatoon, and Sourina Bej, 135–151. Singapore: Springer.

Bell, Charles. 1992. *Tibet Past and Present*. Delhi: Motilal Banarsidass.

Bellezza, John Vincent. 1997. *Divine Dyads Ancient Civilization in Tibet*. Dharamsala: Library of Tibetan Works and Archives.

Bentinck, A. 1913. "The Abor Expedition: Geographical Results." *The Geographical Journal* 41 (2): 97–109.

Bentley, Jenny. 2021. "Protectors of the Land and Water: Citizenship, Territory and Vulnerability among the Lepcha in Sikkim and West Bengal." *Asian Ethnicity* 22 (2): 330–352.

Berkes, Fikret, Johan Colding, and Carl Folke. 2000. "Rediscovery of Traditional Ecological Knowledge as Adaptive Management." *Ecological Applications* 10 (5): 1251–1262.

Berzin, Alexander. 2011. "Buddhist Cosmology in Abhidharma and Kalachakra." *Study Buddhism*. https://studybuddhism.com/en/advanced-studies/abhidharma-tenet-systems/time-the-universe/buddhist-cosmology-in-abhidharma-and-kalachakra.

Best, Jim. 2019. "Anthropogenic Stresses on the World's Big Rivers." *Nature Geoscience* 12 (1): 7–21.

Bhatt, Jay P., and Maharaj K. Pandit. 2016. "Endangered Golden mahseer Tor putitora Hamilton: A Review of Natural History." *Reviews in Fish Biology and Fisheries* 26 (1): 25–38.

Bhattacharjya, B. K., U. Bhaumik, and A. P. Sharma. 2017. "Fish Habitat and Fisheries of Brahmaputra River in Assam, India." *Aquatic Ecosystem Health & Management* 20 (1–2): 102–115.

Bhattacharya, Gourab, Delores M. Robinson, and Matthew M. Wielicki. 2020. "Detrital Zircon Provenance of the Indus Group, Ladakh, NW India: Implications for the Timing of the India-Asia Collision and Other Syn-Orogenic Processes." *GSA Bulletin* 133 (5–6): 1007–1020.

Bhutia, Kalzang Dorjee. 2021. "Trees as Village Protectors, Guru Rinpoche's Wayfinders and Adopted Family Members: Arboreal Imagination, Agency and Relationality in Sikkim." *Worldviews: Global Religions, Culture, and Ecology* 25 (2): 151–170.

Biba, Sebastian. 2014. "Desecuritization in China's Behavior towards Its Transboundary Rivers: The Mekong River, the Brahmaputra River, and the Irtysh and Ili Rivers." *Journal of Contemporary China* 23 (85): 21–43.

Bilham, Roger, and Walter Szeliga. 2008. "Interaction between the Himalaya and the Flexed Indian Plate—Spatial Fluctuations in Seismic Hazard in India in the Past Millennium?" In *2008 Seismic Engineering Conference Commemorating the 1908 Messina and Reggio Calabria Earthquake*, edited by A. Santini, and N. Moraci, Vol. 1020, 224–231. Reggio Calabria: American Institute of Physics.

Billé, Franck. 2014. "Territorial Phantom Pains (and Other Cartographic Anxieties)." *Environment and Planning D: Society and Space* 32 (1): 163–178.

Billé, Franck. 2016. "Introduction to 'Cartographic Anxieties'." *Cross-Currents: East Asian History and Culture Review* 21: 1–18.

Biswas, Soutik. 2021. "'Hanging' Glacier Broke Off to Trigger India flood." In *BBC News*, February 10.

Bizzi, S., and D. N. Lerner. 2015. "The Use of Stream Power as an Indicator of Channel Sensitivity to Erosion and Deposition Processes." *River Research and Applications* 31 (1): 16–27.

Blisniuk, Peter M., Bradley R. Hacker, Johannes Glodny, et al. 2001. "Normal Faulting in Central Tibet since at least 13.5 Myr Ago." *Nature* 412 (6847): 628–632.

Blo bzang lhun sgrub rdo rje. 2018. *Mtsho bod mtho sgang gi ri chu'i rig gnas: Gnyan po gyu rtse'i gnas yig*. Pe cin: Krung go'i bod rig pa dpe skrun khang.

Boelens, Rutgerd, Arturo Escobar, Karen Bakker, et al. 2022. "Riverhood: Political Ecologies of Socionature Commoning and Translocal Struggles for Water Justice." *The Journal of Peasant Studies* 50 (3): 1125–1156.

Bolch, T., A. Kulkarni, A. Kääb, et al. 2012. "The State and Fate of Himalayan Glaciers." *Science* 336 (6079): 310–314.

Bollinger, L., F. Perrier, J.-P. Avouac, et al. 2007. "Seasonal Modulation of Seismicity in the Himalaya of Nepal." *Geophysical Research Letters* 34 (8): L08304.

Bookhagen, Bodo, and Douglas W. Burbank. 2006. "Topography, Relief, and TRMM-Derived Rainfall Variations along the Himalaya." *Geophysical Research Letters* 33 (8): L08405.

Bookhagen, Bodo, and Douglas W. Burbank. 2010. "Toward a Complete Himalayan Hydrological Budget: Spatiotemporal Distribution of Snowmelt and Rainfall and Their Impact on River Discharge." *Journal of Geophysical Research: Earth Surface* 115 (F3): F03019.

Borgohain, Biraj, George Mathew, Naveen Chauhan, et al. 2020. "Evidence of Episodically Accelerated Denudation on the Namche Barwa massif (Eastern Himalayan syntaxis) by megafloods." *Quaternary Science Reviews* 245: 106410.

Bracciali, Laura, Yani Najman, Randall R. Parrish, et al. 2015. "The Brahmaputra Tale of Tectonics and Erosion: Early Miocene River Capture in the Eastern Himalaya." *Earth and Planetary Science Letters* 415: 25–37.

Braulik, Gill T., Masood Arshad, Uzma Noureen, et al. 2014. "Habitat Fragmentation and Species Extirpation in Freshwater Ecosystems; Causes of Range Decline of the Indus River Dolphin (Platanista gangetica minor)." *PLOS ONE* 9 (7): e101657.

Braun, J., D. R. Burbidge, F. N. Gesto, et al. 2009. "Constraints on the Current Rate of Deformation and Surface Uplift of the Australian Continent from a New Seismic Database and low-T Thermochronological Data." *Australian Journal of Earth Sciences* 56 (2): 99–110.

Bsod nams rgyal mtshan. 1973. *rGyal rabs gsal ba'i me long*. Tibetan Bonpo Monastic Centre.

Buckley, Brendan M., Kevin J. Anchukaitis, Daniel Penny, et al. 2010. "Climate as a Contributing Factor in the Demise of Angkor, Cambodia." *Proceedings of the National Academy of Sciences* 107 (15): 6748–6752.

Buckley, Tim, and Kashish Shah. 2019. *Pumped Hydro Storage in India*. Sydney: Institute for Energy Economics and Financial Analysis (IEEFA).

Buell, Lawrence. 2023. *Henry David Thoreau: Thinking Disobediently*. Oxford: Oxford University Press.

Buffetrille, Katia. 1994. "The Blue Lake of A mdo and Its Island: Legends and Pilgrimage Guide." *The Tibet Journal* 19 (4): 2–22.

Buffetrille, Katia. 2013. "Reflections on Pilgrimages to Sacred Mountains, Lakes and Caves." In *Pilgrimage in Tibet*, edited by Alex McKay, 18–34. Abingdon: Routledge.

Burchell, Graham, Colin Gordon, and Peter Miller, eds. 1991. *The Foucault Effect: Studies in Governmentality*. Chicago: University of Chicago Press.

Burrard, Sidney Gerald, and Henry Hubert Hayden. 1907. *A Sketch of the Geography and Geology of the Himalaya Mountains and Tibet*. Calcutta: Superintendent Government printing.

Burton, T. M., and R. W. Tiner. 2009. "Ecology of Wetlands." In *Encyclopedia of Inland Waters*, edited by Gene E. Likens, 507–515. Amsterdam: Elsevier.

Byang chub rgyal mtshan, Ta'i si tu. 1989. *Ta si byang chub rgyal mtshan gyi Bka' chems mthong ba don ldan*. Lha sa: Bod ljongs mi dmangs dpe skrun khang.

Byrnes, Corey. 2018. *Fixing Landscape: A Techno-Poetic History of China's Three Gorges*. New York: Columbia University Press.

Cai, Bing. 1992. "Xizang Autonomous Region Conducting Whole-Region Fishery Resources Survey." *China Fisheries* 8: 19.

Cai, D., L. Han, H. Zhou, et al. 2007. "Mitochondrial DNA Sequence Polymorphism Analysis of Ancient Sheep from the Taosi and Erlitou Sites." *Keji Kaogu* 2: 35–40.

Cai, D. X., X. Y. Wang, G. W. Li, et al. 2021. "Late Cenozoic Denudation and Topographic Evolution History of the Lhasa River Drainage in Southern Tibetan Plateau: Insights from Inverse Thermal History Modeling." *Frontiers in Earth Science* 9: 636459.

Cai, Wenju, Lixin Wu, Matthieu Lengaigne, et al. 2019. "Pantropical Climate Interactions." *Science* 363 (6430): eaav4236.

Cannon, Forest, Leila M. V. Carvalho, Charles Jones, et al. 2015. "Multi-Annual Variations in Winter Westerly Disturbance Activity Affecting the Himalaya." *Climate Dynamics* 44 (1): 441–455.

Cao, Jianjun, Jan F. Adamowski, Ravinesh C. Deo, et al. 2019. "Grassland Degradation on the Qinghai–Tibetan Plateau: Reevaluation of Causative Factors." *Rangeland Ecology and Management* 72 (6): 988–995.

Cao, Zetao, Nan Zhuotong, Hu, Jianan et al. 2023. "A new 2010 permafrost distribution map over the Qinghai–Tibet Plateau based on subregion survey maps: a benchmark for regional permafrost modeling." *Earth System Science Data Discussions* 15: 3905–3930.

Carrivick, Jonathan L., and Fiona S. Tweed. 2013. "Proglacial Lakes: Character, Behaviour and Geological Importance." *Quaternary Science Reviews* 78: 34–52.

Castelltort, Sébastien, and Guy Simpson. 2006. "Growing Mountain Ranges and Quenched River Networks." *Comptes Rendus Geoscience* 338 (16): 1184–1193.

Castree, Noel. 2008. "Neoliberalising Nature: Processes, Effects, and Evaluations." *Environment and Planning A* 40 (1): 153–173.

CBD. 2009. *Ecosystem Approach*. Bonn: Convention on Biological Diversity.

CBD. 2010. *X/28. Inland Waters Biodiversity*. Montreal: Convention on Biological Diversity.

CBD. 2022. *Kunming–Montreal Global Biodiversity Framework*. Montreal: Convention on Biological Diversity.

CBD, and RAMSAR. 2011. The Convention on Biological Diversity (CBD) and the Ramsar Convention on Wetlands (Ramsar): 5th Joint Work Plan (JWP) 2011–2020. Bonn and Gland: Convention on Biological Diversity and Ramsar Convention on Wetlands.

Central Water Commission. 2019. *National Register of Large Dams -2019*. New Delhi: Government of India, Central Water Commission, Central Dam Safety Organization.

Chakrabarty, Dipesh. 2021. *The Climate of History in a Planetary Age*. Chicago: University of Chicago Press.

Chakraborty, Ritodhi, Mabel D. Gergan, Pasang Y. Sherpa, et al. 2021. "A Plural Climate Studies Framework for the Himalayas." *Current Opinion in Environmental Sustainability* 51: 42–54.

Chang, Edmund Z. 2000. "Geology and Tectonics of the Songpan-Ganzi Fold Belt, Southwestern China." *International Geology Review* 42 (9): 813–831.

Chauhan, Chetan, and Pramod Giri. 2023. "Washing Away of Dam Brings Focus on Sikkim's Hydel Projects." *Hindustan Times*, October 6.

Chaskar, Ketan, Suraj Bhosale, D. K. Pandey, et al. 2023. "Jurassic Geoheritage and Geotourism Potential in the Spiti Valley of India: A Thing of the Past, Shielded for the Future." *Geoheritage* 15 (4): 108.

Chawla, Swati. 2023. "Fashioning a 'Buddhist'Himalayan Cartography: Sikkim Darbar and the Cabinet Mission Plan." *India Quarterly* 79 (1): 29–44.

Che, Ping, and Jianghu Lan. 2021. "Climate Change along the Silk Road and Its Influence on Scythian Cultural Expansion and Rise of the Mongol Empire." *Sustainability* 13 (5): 2530.

Chen, F., Zhang J. et al. 2020. "Climate Change, Vegetation History, and Landscape Responses on the Tibetan Plateau during the Holocene: A Comprehensive Review." *Quaternary Science Reviews* 243: 106444.

Chen, Fahu, Frido Welker, Chuan-Chou Shen, et al. 2019. "A Late Middle Pleistocene Denisovan Mandible from the Tibetan Plateau." *Nature* 569 (7756): 409–412.

Chen, Fahu, Huan Xia, Zhenxiu Jia, et al. 2022. "Earliest Hand- and Footprint Art Indicates that Denisovans May Have Occupied the Interior of the High-Altitude Tibetan Plateau since 200 Thousand Years Ago." *Science China Earth Sciences* 65 (4): 769–772.

Chen, Huai, Gang Yang, Changhui Peng, et al. 2014a. "The Carbon Stock of Alpine Peatlands on the Qinghai–Tibetan Plateau during the Holocene and Their Future Fate." *Quaternary Science Reviews* 95: 151–158.

Chen, Jianhui, Fahu Chen, Song Feng, et al. 2015. "Hydroclimatic Changes in China and Surroundings during the Medieval Climate Anomaly and Little Ice Age: Spatial Patterns and Possible Mechanisms." *Quaternary Science Reviews* 107: 98–111.

Chen, Peng, Bing Yan, and Yuan Liu. 2021. "Active Strike-Slip Faulting and Systematic Deflection of Drainage Systems along the Altyn Tagh Fault, Northern Tibetan Plateau." *Remote Sensing* 13 (16): 3109.

Chen, Wang-Ping, Chu-Yung Chen, and John L. Nábelek. 1999. "Present-Day Deformation of the Qaidam Basin with Implications for Intra-Continental Tectonics." *Tectonophysics* 305 (1): 165–181.

Chen, Y., and W. Cao. 2000. "Schizothoracinae." In *Fauna Sinica. Osteichthyes. Cypriniformes (III)*, edited by Peiqi Yue, et al., 273–390. Beijing: Science Press.

Chen, Yiyu, Beate Jessel, Bojie Fu, et al., eds. 2014b. *Ecosystem Services and Management Strategy in China*. Heidelberg: Springer.

Cheng, Guodong, and Tonghua Wu. 2007. "Responses of Permafrost to Climate Change and Their Environmental Significance, Qinghai–Tibet Plateau." *Journal of Geophysical Research: Earth Surface* 112 (F2): F02S03.

Chettri, Kul Bahadur. 2021. "Current Status of Rainbow Trout Culture in Sikkim: A Sustainable Farming System in the Hills." *Indian Journal of Hill Farming* 34 (1): 126–133.

Chevalier, M.-L., P. Tapponnier, J. van der Woerd, et al. 2020. "Late Quaternary Extension Rates across the Northern Half of the Yadong-Gulu Rift: Implication for East-West Extension in Southern Tibet." *Journal of Geophysical Research: Solid Earth* 125 (7): e2019JB019106.

Chinese Academy of Sciences. 2015. "Scientific Assessment of Environmental Change on the Tibetan Plateau (in Chinese)." *Chinese Academy of Sciences*. https://www.cas.cn/print/index.shtml?docurl=//www.cas.cn/yw/201511/t20151117_4465636.shtml.

Chopra, Ashish, and Parthasarathy Ramachandran. 2021. "Understanding Water Institutions and Their Impact on the Performance of the Water Sector in India." *Water Policy* 23 (2): 466–486.

Choudhury, S. Dutta. 1996. *Arunachal Pradesh District Gazetteers: East Kameng, West Kameng and the Tawang Districts*. Arunachal Pradesh: Government of Arunachal Pradesh.

Chowdhury, Arnab Roy, and Sanjay Kumar Rajhans. 2021. "Revisiting Wittfogel: 'Hydraulic Society' in Colonial India and Its Post-Colonial Legacies in Hydropower Management." In *Theory, Policy, Practice: Development and Discontents in India*, edited by Suman Nath, and Debraj Bhattacharya, 123–140. Abingdon: Routledge.

Chowdhury, Nasima Tanveer. 2009. "Water Management in Bangladesh: An Analytical Review." *Water Policy* 12 (1): 32–51.

Cioc, Mark. 2002. *The Rhine: An Eco-Biography, 1815–2000*. Seattle: University of Washington Press.

Clark, M. K., L. H. Royden, K. X. Whipple, et al. 2006. "Use of a Regional, Relict Landscape to Measure Vertical Deformation of the Eastern Tibetan Plateau." *Journal of Geophysical Research: Earth Surface* 111 (F3): F03002.

Clark, Marin K., John W. M. Bush, and Leigh H. Royden. 2005. "Dynamic Topography Produced by Lower Crustal Flow against Rheological Strength Heterogeneities Bordering the Tibetan Plateau." *Geophysical Journal International* 162 (2): 575–590.

Clark, Marin Kristen, and Leigh Handy Royden. 2000. "Topographic Ooze: Building the Eastern Margin of Tibet by Lower Crustal Flow." *Geology* 28 (8): 703–706.

Clift, Peter D., and Jerzy Blusztajn. 2005. "Reorganization of the Western Himalayan River System after Five Million Years Ago." *Nature* 438 (7070): 1001–1003.

Coen, Deborah R., and Fredrik Albritton Jonsson. 2022. "Between History and Earth System Science." *Isis* 113 (2): 407–416.

Coggins, Chris. 2019. "Sacred Watersheds and the Fate of the Village Body Politic in Tibetan and Han Communities under China's Ecological Civilization." *Religions* 10 (11): 600.

Colby, Michael E. 1991. "Environmental Management in Development: The Evolution of Paradigms." *Ecological Economics* 3 (3): 193–213.

Conklin, Harold C. 1957. *Hanunoo Agriculture. A Report on an Integral System of Shifting Cultivation in the Philippines*. Rome: Food and Agricultural Organization of the United Nations.

Conrad, Clinton P. 2000. "Convective Instability of Thickening Mantle Lithosphere." *Geophysical Journal International* 143 (1): 52–70.

Cook, Edward R., Kevin J. Anchukaitis, Brendan M. Buckley, et al. 2010. "Asian Monsoon Failure and Megadrought during the Last Millennium." *Science* 328 (5977): 486–489.

Cook, Edward R., Yochanan Kushnir, Jason E. Smerdon, et al. 2019. "A Euro-Mediterranean Tree-Ring Reconstruction of the Winter NAO Index since 910 C.E." *Climate Dynamics* 53 (3): 1567–1580.

Copley, Alex. 2008. "Kinematics and Dynamics of the Southeastern Margin of the Tibetan Plateau." *Geophysical Journal International* 174 (3): 1081–1100.

Courtney, Chris. 2018. *The Nature of Disaster in China: The 1931 Yangzi River Flood*. Cambridge: Cambridge University Press.

Craw, Dave, Phaedra Upton, Christopher P. Burridge, et al. 2016. "Rapid Biological Speciation Driven by Tectonic Evolution in New Zealand." *Nature Geoscience* 9 (2): 140–144.

Crawford, Stephen S., and Andrew M. Muir. 2008. "Global Introductions of Salmon and Trout in the Genus Oncorhynchus: 1870–2007." *Reviews in Fish Biology and Fisheries* 18 (3): 313–344.

Crow-Miller, Britt, Michael Webber, and Sarah Rogers. 2017. "The Technopolitics of Big Infrastructure and the Chinese Water Machine." *Water Alternatives* 10 (2): 233–249.

Cullather, Nick. 2010. *The Hungry World: America's Cold War Battle against Poverty in Asia*. Cambridge, MA: Harvard University Press.

Curio, Julia, Fabien Maussion, and D. Scherer. 2015. "A 12-Year High-Resolution Climatology of Atmospheric Water Transport over the Tibetan Plateau." *Earth System Dynamics* 6 (1): 109–124.

D'Alpoim Guedes, Jade, and Mark Aldenderfer. 2020. "The Archaeology of the Early Tibetan Plateau: New Research on the Initial Peopling through the Early Bronze Age." *Journal of Archaeological Research* 28 (3): 339–392.

D'Souza, Radha. 2002. "At the Confluence of Law and Geography: Contextualising Inter-State Water Disputes in India." *Geoforum* 33 (2): 255–269.

D'Souza, Radha. 2004. "The Democracy-Development Tension in Dam Projects: The Long Hand of the Law." *Political Geography* 23 (6): 701–730.

D'Souza, Rohan. 2008. "Framing India's Hydraulic Crises." *Monthly Review: An Independent Socialist Magazine* 60 (3): 112–113.

D'Souza, Rohan. 2019. "Should Clean Energy Be Politics as Usual? Reflections on India's Energy Transition." *Georgetown Journal of Asian Affairs* 4 (2): 38–44.

D'Souza, Rohan. 2023. "Nostalgia to Melancholy: Towards a History of the Environmental History of South Asia in the Anthropocene." *Comparativ* 32 (6): 691–703.

Da Cunha, Dilip. 2018. *The Invention of Rivers: Alexander's Eye and Ganga's Descent*. Philadelphia: University of Pennsylvania Press.

Dahlquist, Maxwell P., and A. Joshua West. 2021. "Central Himalayan Rivers Record the Topographic Signature of Erosion by Glacial Lake Outburst Floods. Earth Surface Dynamics." *Earth Surface Dynamics Discussions*.

Dai, Jin-Gen, Matthew Fox, Xu Han, et al. 2021. "Two Stages of Accelerated Exhumation in the Middle Reach of the Yarlung River, Southern Tibet since the Mid-Miocene." *Tectonics* 40 (6): e2020TC006618.

Dangar, Swarup, Akarsh Asoka, and Vimal Mishra. 2021. "Causes and Implications of Groundwater Depletion in India: A Review." *Journal of Hydrology 596*: 126103.

Dätwyler, C., M. Grosjean, N. J. Steiger, et al. 2020. "Teleconnections and Relationship between the El Niño–Southern Oscillation (ENSO) and the Southern Annular Mode (SAM) in Reconstructions and Models over the Past Millennium" *Climate of the Past* 16 (2): 743–756.

Davis, Alexander E. 2023. *The Geopolitics of Melting Mountains: An International Political Ecology of the Himalaya*. Singapore: Palgrave Macmillan.

DeCelles, P. G., P. Kapp, J. Quade, et al. 2011. "Oligocene–Miocene Kailas Basin, Southwestern Tibet: Record of Postcollisional Upper-Plate Extension in the Indus-Yarlung Suture Zone." *GSA Bulletin* 123 (7–8): 1337–1362.

De Groot, Rudolf, Mishka Stuip, Max Finlayson, et al. 2006. "Valuing Wetlands: Guidance for Valuing the Benefits Derived from Wetland Ecosystem Services." *EconPapers*, H039735.

Dendup, Pema, Leki Wangdi, Yenten Jamtsho, et al. 2021. "Bird Diversity and Conservation Threats in Jigme Dorji National Park, Bhutan." *Global Ecology and Conservation* 30: e01771.

Deng, Haijun, N. C. Pepin, and Yaning Chen. 2017. "Changes of Snowfall under Warming in the Tibetan Plateau." *Journal of Geophysical Research: Atmospheres* 122 (14): 7323–7341.

Descola, Philippe. 1994. *In the Society of Nature: A Native Ecology in Amazonia*. Cambridge: Cambridge University Press.

Dettinger, Michael D., F. Martin Ralph, and Jonathan J. Rutz. 2018. "Empirical Return Periods of the Most Intense Vapor Transports during Historical Atmospheric River Landfalls on the U.S. West Coast." *Journal of Hydrometeorology* 19 (8): 1363–1377.

Devoie, Élise G., James R. Craig, Ryan F. Connon, et al. 2019. "Taliks: A Tipping Point in Discontinuous Permafrost Degradation in Peatlands." *Water Resources Research* 55 (11): 9838–9857.

Dharmadhikary, Shripad. 2005. *Unravelling Bhakra: Assessing the Temple of Resurgent India*. Badwani: Manthan Adhyayan Kendra.

Diemberger, Hildegard. 2021. "When *lha lu* Spirits Suffer and Sometimes Fight Back: Tibetan Cosmopolitics at a Time of Environmental Threats and Climate Change." In *Cosmopolitical Ecologies across Asia*, edited by Riamsara Kuyakanon, Hildegard Diemberger, and David Sneath, 19–42. Abingdon: Routledge.

Ding, Lin, Paul Kapp, Fulong Cai, et al. 2022. "Timing and Mechanisms of Tibetan Plateau Uplift." *Nature Reviews Earth & Environment* 3 (10): 652–667.

Doblas-Reyes, F. J., A. A. Sorensson, M. Almazroui, et al. 2021. "Linking Global to Regional Climate Change." In *Climate Change 2021: The Physical Science Basis. Contribution of Working Group I to the Sixth Assessment Report of the Intergovernmental Panel on Climate Change*, edited by Valérie Masson-Delmotte, Panmao Zhai, Anna Pirani, et al., 1363–1512. Cambridge: Cambridge University Press.

Dong, Wenhao, Yanluan Lin, Jonathon S. Wright, et al. 2016. "Summer Rainfall over the Southwestern Tibetan Plateau Controlled by Deep Convection over the Indian Subcontinent." *Nature Communications* 7 (1): 10925.

Dong, Zhibao, Guangyin Hu, Guangqiang Qian, et al. 2017. "High-Altitude Aeolian Research on the Tibetan Plateau." *Reviews of Geophysics* 55 (4): 864–901.

Dong, Zhibao, X. Wang, and L. Liu. 2000. "Wind Erosion in Arid and Semiarid China: An Overview." *Journal of Soil and Water Conservation* 55 (4): 439–444.

Dorji, Gyeltshen, and Ugyen Dorji. 2021. "Habitat Use, Feeding Behaviour and Conservation Threats of Black-Necked Cranes at Gangtey-Phobji Valley, Wangdue Phodrang, Bhutan." *International Journal of Environment* 10 (2): 64–82.

Dorji, Yeshey, ed. 2016. *Water: Securing Bhutan's Future*. Thimphu: Asian Development Bank, and National Environment Commission, Royal Government of Bhutan.

Dorsey, Michael. 2005. "Conservation, Collusion and Capital." *Anthropology News* 46 (7): 45–46.

Dotson, Brandon. 2008. "Complementarity and Opposition in Early Tibetan Ritual." *Journal of the American Oriental Society* 128 (1): 41–67.

Dotson, Brandon. 2011. "Theorising the King: Implicit and Explicit Sources for the Study of Tibetan Sacred Kingship." *Revue d'Etudes Tibétaines* 21: 83–103.

Dotson, Brandon. 2013. "The Victory Banquet: The Old Tibetan Chronicle and the Rise of Tibetan Historical Narrative." Unpublished Habilitation script, Ludwig Maximilian University of Munich.

Dou, X., X. Fan, A. P. Yunus, et al. 2021. "Spatio-Temporal Evolution of Glacial Lakes in the Tibetan Plateau over the Past 30 Years." *Earth System Science Data Discussions* 2021: 1–31.

Dou, Xiangyang, Xuanmei Fan, Xin Wang, et al. 2023. "Spatio-Temporal Evolution of Glacial Lakes in the Tibetan Plateau over the Past 30 Years." *Remote Sensing* 15 (2): 416.

Du, Heqiang, Xian Xue, Tao Wang, et al. 2022. "Modeling Dust Emission in Alpine Regions with Low Air Temperature and Low Air Pressure – A Case Study on the Qinghai-Tibetan Plateau (QTP)." *Geoderma* 422: 115930.

Dudgeon, David. 2019. "Multiple Threats Imperil Freshwater Biodiversity in the Anthropocene." *Current Biology* 29 (19): R960–R967.
Dudley, Nigel. 2008. *Guidelines for Applying Protected Area Management Categories*. Gland: IUCN.
DWMNFGA. 2021. *National Report to Ramsar COP14 The People's Republic of China*. Beijing: Department of Wetland Management, National Forestry and Grassland Administration.
Dwyer, Peter D. 2005. "Ethnoclassification, Ethnoecology, and the Imagination." *Journal de la Société des Océanistes* 120–121: 11–25.
Edney, Matthew H. 1997. *Mapping an Empire: The Geographical Construction of British India, 1765–1843*. Chicago: University of Chicago Press.
Eizenhöfer, Paul R., Nadine McQuarrie, Eitan Shelef, et al. 2019. "Landscape Response to Lateral Advection in Convergent Orogens over Geologic Time Scales." *Journal of Geophysical Research: Earth Surface* 124 (8): 2056–2078.
Ekvall, Robert B., and James F. Downs. 1963. "Notes on Water Utilization and Rule in the Sakya Domain–Tibet." *The Journal of Asian Studies* 22 (3): 293–303.
Elias, Scott. 2021. "Changes in Ocean Circulation Patterns." In *Threats to the Arctic*, edited by S. Elias, 27–44. Amsterdam: Elsevier.
England, Philip, and Peter Molnar. 2022. "Changes in Plate Motions Caused by Increases in Gravitational Potential Energy of Mountain Belts." *Geochemistry, Geophysics, Geosystems* 23 (10): e2022GC010389.
Erlewein, Alexander. 2014. "The Promotion of Dams through the Clean Development Mechanism: Between Sustainable Climate Protection and Carbon Colonialism." In *Large Dams in Asia: Contested Environments between Technological Hydroscapes and Social Resistance*, edited by Marcus Nüsser, 149–168. Dordrecht: Springer.
Escobar, Arturo. 1998. "Whose Knowledge, Whose nature? Biodiversity, Conservation, and the Political Ecology of Social Movements." *Journal of Political Ecology* 5 (1): 53–82.
Estrada, Alejandro, Paul A. Garber, Sidney Gouveia, et al. 2022. "Global Importance of Indigenous Peoples, Their Lands, and Knowledge Systems for Saving the World's Primates from Extinction." *Science Advances* 8 (32): eabn2927.
Fan, Ka-wai. 2023. "The Little Ice Age and the Fall of the Ming Dynasty: A Review." *Climate* 11 (3): 71.
Feng, Xiu, Bing Li, Yifeng Chen, et al. 2023. "Species-Level Monitoring of Rare and Invasive Fishes using eDNA Metabarcoding in the Middle and Upper Yarlung Zangbo River, Tibet." *Water Biology and Security* 2 (1): 100089.
Feng, Yan, Wenling Wang, and Jiang Liu. 2019. "Dilemmas in and Pathways to Transboundary Water Cooperation between China and India on the Yaluzangbu-Brahmaputra River." *Water* 11 (10): 2096.
Ferris, Nick. 2022. "Weekly Data: Why Hydropower Is Key to India's Net-Zero Plans." *Energy Monitor*, March 28. https://www.energymonitor.ai/renewables/weekly-data-why-hydropower-is-key-to-indias-net-zero-plans/?cf-view&cf-closed.
Fielding, Eric, Bryan Isacks, Muawia Barazangi, et al. 1994. "How Flat Is Tibet?" *Geology* 22 (2): 163–167.
Fildani, Andrea. 2022. "The Revolutionary Impact of the Deep Time Concept: Geology's Modernity and Societal Implications." *The Sedimentary Record* 20 (1): 1–4.
Finlayson, C. M., and J. Pittock. 2018. "Climate Change and the Management of Freshwater Protected Areas." In *Freshwater Ecosystems in Protected*

Areas: Conservation and Management, edited by C. Max Finlayson, Angela H. Arthington, and Jamie Pittock, 242–255. London: Routledge.
Finnegan, Noah J., Gerard Roe, David R. Montgomery, et al. 2005. "Controls on the Channel Width of Rivers: Implications for Modeling Fluvial Incision of Bedrock." *Geology* 33 (3): 229–232.
FitzHerbert, Solomon George. 2017. "Tibetan Buddhism and the Gesar Epic." In *Oxford Research Encyclopedia of Religion*. https://oxfordre.com/religion/view/10.1093/acrefore/9780199340378.001.0001/acrefore-9780199340378-e-560.
Foucault, Michel. 2003. *'Society Must Be Defended': Lectures at the Collège de France, 1975-1976*. Vol. 1. Macmillan.
Fox, Jefferson. 2002. "Siam Mapped and Mapping in Cambodia: Boundaries, Sovereignty, and Indigenous Conceptions of Space." *Society & Natural Resources* 15 (1): 65–78.
Fox, M., A. Carter, and J. G. Dai. 2020. "How Continuous Are the "Relict" Landscapes of Southeastern Tibet?" *Frontiers in Earth Science* 8: 587597.
Frank, Mark E. 2016. "Planting and Its Discontents: Or How Nomads Produced Spaces of Resistance in China's Erstwhile Xikang Province." *Resilience: A Journal of the Environmental Humanities* 3: 112–141.
Frank, Mark E. 2019. "Wheat Dreams: Scientific Interventions at Chinese Model Farms in Kham, 1937–1949." In *Frontier Tibet: Patterns of Change in the Sino-Tibetan Borderlands*, edited by Stéphane Gros, 217–254. Amsterdam: Amsterdam University Press.
Frey, Karen E., and James W. McClelland. 2009. "Impacts of Permafrost Degradation on Arctic River Biogeochemistry." *Hydrological Processes* 23 (1): 169–182.
Gabrielli, Paolo, Anna Wegner, M. Roxana Sierra-Hernández, et al. 2020. "Early Atmospheric Contamination on the Top of the Himalayas since the Onset of the European Industrial Revolution." *Proceedings of the National Academy of Sciences* 117 (8): 3967–3973.
Gadgil, Madhav, and Ramachandra Guha. 1993. *This Fissured Land: An Ecological History of India*. Berkeley: University of California Press.
Gaerrang, Kabzung. 2017. "Tibetan Buddhism, Wetland Transformation, and Environmentalism in Tibetan Pastoral Areas of Western China." *Conservation and Society* 15 (1): 14–23.
Gagné, Karine. 2020. "The Materiality of Ethics: Perspectives on Water and Reciprocity in a Himalayan Anthropocene." *WIREs Water* 7 (4): e1444.
Gain, Animesh K., M. Shahjahan Mondal, and Rezaur Rahman. 2017. "From Flood Control to Water Management: A Journey of Bangladesh towards Integrated Water Resources Management." *Water* 9 (1): 55.
Gamble, Ruth. 2018. *Reincarnation in Tibetan Buddhism: The Third Karmapa and the Invention of a Tradition*. New York: Oxford University Press.
Gamble, Ruth. 2021. "Muddying the Waters: The Invention and Enclosure of Tibet's Wetlands." In *Environmental Humanities in the New Himalayas: Symbiotic Indigeneity, Commoning, Sustainability*, edited by Dan Smyer Yü, and Erik de Maaker, 115–131. London: Routledge.
Gamble, Ruth. 2023. "Climing the Great Gorge: The Many Discoveries of the Yarlung Tsangpo Gorge." In *Storying Multipolar Climes of the Himalaya, Andes and Arctic*, edited by Dan Smyer Yü, and Jelle J. P. Wouters, 203–222. Abingdon: Routledge.
Gamble, Ruth, and Alexander E. Davis. 2024. "Bordering Ladakh, Again: From Ecological Flows to Cartographic Competition." *Asian Studies Review* 48 (1): 179–199.

Gamble, Ruth, John Powers, and Paul Hackett. 2022. "Central Tibetan Famines 1280–1400: When Premodern Climate Change and Bad Governance Starved Tibet." *Bulletin of the School of Oriental and African Studies* 85 (2): 215–233.

Gan, Weijun, Peter Molnar, Peizhen Zhang, et al. 2021. "Initiation of Clockwise Rotation and Eastward Transport of Southeastern Tibet Inferred from Deflected Fault Traces and GPS Observations." *GSA Bulletin* 134 (5–6): 1129–1142.

Gao, Yan. 2022. *Yangzi Waters: Transforming the Water Regime of the Jianghan Plain in Late Imperial China*. Leiden: Brill.

Gao, Jing, You He, Valerie Masson-Delmotte, et al. 2018. "ENSO Effects on Annual Variations of Summer Precipitation Stable Isotopes in Lhasa, Southern Tibetan Plateau." *Journal of Climate* 31 (3): 1173–1182.

Gao, Mingxing, Ming Hao, Gerold Zeilinger, et al. 2023. "Recent Uplift Characteristics of the Southeast Tibetan Plateau, an Analysis Based on Fluvial Indices." *Remote Sensing* 15 (2): 433.

García-Quijano, Carlos G. 2007. "Fishers' Knowledge of Marine Species Assemblages: Bridging between Scientific and Local Ecological Knowledge in Southeastern Puerto Rico." *American Anthropologist* 109 (3): 529–536.

Gardelle, Julie, Yves Arnaud, and Etienne Berthier. 2011. "Contrasted Evolution of Glacial Lakes along the Hindu Kush Himalaya Mountain Range between 1990 and 2009." *Global and Planetary Change* 75 (1): 47–55.

Gardner, Alexander. 2006. "The Twenty-Five Great Sites of Khams: A Narrative Map of an Imperiled Place." In *Studies in the History of Eastern Tibet*, edited by Wim van Spengen, and Lama Jabb, 97–132. Leiden: Brill.

Gardner, Kyle J. 2021. *The Frontier Complex: Geopolitics and the Making of the India–China Border, 1846–1962*. Cambridge: Cambridge University Press.

Garikipati, Supriya. 2002. "Resettlement Sites of Narmada Valley Project: Some Revealing Insights." *Economic and Political Weekly* 37 (23): 2257–2266.

Gelek, Surkhang Wangchen. 1986. "Government, Monastic and Private Taxation in Tibet." *The Tibet Journal* 11 (1): 21–40.

Gerbeaux, P., C. M. Finlayson, and A. A. van Dam. 2016. "Wetland Classification: Overview." In *The Wetland Book I: Structure and Function, Management and Methods*, edited by C. Max Finlayson, Mark Everard, Kenneth Irvine, et al., 1–8. Dordrecht: Springer Netherlands.

Gergan, Mabel Denzin. 2020. "Disastrous Hydropower, Uneven Regional Development, and Decolonization in India's Eastern Himalayan Borderlands." *Political Geography* 80: 102175.

Ghosh, Attreyee, William E. Holt, and Lucy M. Flesch. 2009. "Contribution of Gravitational Potential Energy Differences to the Global Stress Field." *Geophysical Journal International* 179 (2): 787–812.

Ghosh, Kausik, and Poulomi Chakraborty. 2021. "Assessing the Benefits, Threats and Conservation of Reservoir-Based Wetlands in the Eastern Himalayan River Basin." In *Wetlands Conservation: Current Challenges and Future Strategies*, edited by Sanjeev Sharma, and Pardeep Singh, 140–161. Hoboken: Wiley-Blackwell.

Ghosh, Rupa, Pradeep Srivastava, U. K. Shukla, et al. 2019. "100 kyr Sedimentary Record of Marginal Gangetic Plain: Implications for Forebulge Tectonics." *Palaeogeography, Palaeoclimatology, Palaeoecology* 520: 78–95.

Giersch, C. Patterson. 2010. "Across Zomia with Merchants, Monks, and Musk: Process Geographies, Trade Networks, and the Inner-East–Southeast Asian Borderlands." *Journal of Global History* 5 (2): 215–239.

Giersch, C. Patterson. 2019. "The Origins of Disempowered Development in the Tibetan Borderlands." In *Frontier Tibet: Patterns of Change in the Sino-Tibetan Borderlands*, edited by Stéphane Gros, 255–280. Amsterdam: Amsterdam University Press.

Gilchrist, A. R., M. A. Summerfield, and H. A. P. Cockburn. 1994. "Landscape Dissection, Isostatic Uplift, and the Morphologic Development of Orogens." *Geology* 22 (11): 963–966.

Gilfillan, Daniel, and Jamie Pittock. 2022a. *APEC Workshop on the Use of Pumped Storage Hydropower to Enable Greater Renewable Energy Use and Reliable Electricity Supply*. Singapore: Asia-Pacific Economic Cooperation (APEC).

Gilfillan, Daniel, and Jamie Pittock. 2022b. "Pumped Storage Hydropower for Sustainable and Low-Carbon Electricity Grids in Pacific Rim Economies." *Energies* 15 (9): 3139.

Gilmartin, David. 2020. *Blood and Water: The Indus River Basin in Modern History*. Oakland: University of California Press.

Gimeno, Luis, Raquel Nieto, Marta Vázquez, et al. 2014. "Atmospheric Rivers: A Mini-Review." *Frontiers in Earth Science* 2: 2.

Glaskin, Katie. 2012. "Anatomies of Relatedness: Considering Personhood in Aboriginal Australia." *American Anthropologist* 114 (2): 297–308.

Godard, V., J. Lavé, J. Carcaillet, et al. 2010. "Spatial Distribution of Denudation in Eastern Tibet and Regressive Erosion of Plateau Margins." *Tectonophysics* 491 (1): 253–274.

Gohain, Swargajyoti. 2020a. *Imagined Geographies in the Indo-Tibetan Borderlands: Culture, Politics, Place*. Amsterdam: Amsterdam University Press.

Gohain, Swargajyoti. 2020b. "Producing Monyul as Buffer: Spatial Politics in a Colonial Frontier." *Modern Asian Studies* 54 (2): 432–470.

Gong, Li. 2007. *Huanghe shihua*. Beijing: Zhongguo dabaike quanshu chubanshe.

Gouin, Margaret. 2012. *Tibetan Rituals of Death: Buddhist Funerary Practices*. London: Routledge.

Govin, Gwladys, Peter van der Beek, Yani Najman, et al. 2020. "Early Onset and Late Acceleration of Rapid Exhumation in the Namche Barwa Syntaxis, Eastern Himalaya." *Geology* 48 (12): 1139–1143.

Graham, N. E., C. M. Ammann, D. Fleitmann, et al. 2011. "Support for Global Climate Reorganization during the 'Medieval Climate Anomaly'." *Climate Dynamics* 37 (5): 1217–1245.

Gray, Donald Maurice, and David Harold Male. 1981. *Handbook of Snow: Principles, Processes, Management & Use*. Oxford: Pergamon Press.

Gregory, John Walter, and C. J. Gregory. 1925. "V. The Geology and Physical Geography of Chinese Tibet, and Its Relations to the Mountain System of South-Eastern Asia, from Observations Made during the Percy Sladen Expedition, 1922." *Philosophical Transactions of the Royal Society of London. Series B, Containing Papers of a Biological Character* 213 (402–410): 171–298.

Gros, Stéphane. 2019. "Frontier (of) Experience: Introction and Prolegomenon." In *Frontier Tibet: Patterns of Change in the Sino-Tibetan Borderlands*, edited by Stéphane Gros, 41–84. Amsterdam: Amsterdam University Press.

Gros, Stéphane. 2023. "The Sino-Tibetan Borderlands." In *Oxford Research Encyclopedia of Asian History*. https://oxfordre.com/asianhistory/view/10.1093/acrefore/9780190277727.001.0001/acrefore-9780190277727-e-727.

Grothmann, Kerstin. 2012. "Population History and Identity in the Hidden Land of Pemakö." *Journal of Bhutan Studies* 26 (2): 21–52.

Gruber, S. 2012. "Derivation and Analysis of a High-Resolution Estimate of Global Permafrost Zonation." *The Cryosphere* 6 (1): 221–233.

Gruber, Stephan, Renate Fleiner, Emilie Guegan, et al. 2017. "Inferring Permafrost and Permafrost Thaw in the Mountains of the Hindu Kush Himalaya Region." *The Cryosphere* 11 (1): 81–99.

Gtsug lag phreng ba (Dpa' bo). 1986. *Chos 'byung mkhas pa'i dga' ston*. 2 vols. Pe cin: Mi rigs dpe skrun khang.

Guan, Chong, Maodu Yan, Weilin Zhang, et al. 2021. "Paleomagnetic and Chronologic Data Bearing on the Permian/Triassic Boundary Position of Qamdo in the Eastern Qiantang Terrane: Implications for the Closure of the Paleo-Tethys." *Geophysical Research Letters* 48 (6): e2020GL092059.

Guillot, S., F. Goussin, L. Airaghi, et al. 2019. "How and When Did the Tibetan Plateau Grow?" *Russian Geology and Geophysics* 60 (9): 957–977.

Guillot, Stéphane, and Anne Replumaz. 2013. "Importance of Continental Subductions for the Growth of the Tibetan Plateau." *Bulletin de la Société Géologique de France* 184 (3): 199–223.

Gulati, Ashok, and Pritha Banerjee. 2016. "Emerging Water Crisis in India: Key Issues and Way Forward." *Indian Journal of Economics* 96: 681–704.

Guo, Yongqiang, Chun Chang Huang, Jiangli Pang, et al. 2015. "Investigating Extreme Flood Response to Holocene Palaeoclimate in the Chinese Monsoonal Zone: A Palaeoflood Case Study from the Hanjiang River." *Geomorphology* 238: 187–197.

Gupta, Avijit. 2008. *Large Rivers: Geomorphology and Management*. Hoboken: John Wiley & Sons.

Gupta, Nishikant, Varun Tiwari, Mark Everard, et al. 2020. "Assessing the Distribution Pattern of Otters in Four Rivers of the Indian Himalayan Biodiversity Hotspot." *Aquatic Conservation: Marine and Freshwater Ecosystems* 30 (3): 601–610.

Gupta, S. K., and R. D. Deshpande. 2004. "Water for India in 2050: First-Order Assessment of Available Options." *Current Science* 86 (9): 1216–1224.

Gupta, Sonika. 2021. "Frontiers in Flux: Indo-Tibetan Border: 1946–1948." *India Quarterly* 77 (1): 42–58.

Guthman, J. 1997. "Representing Crisis: The Theory of Himalayan Environmental Degradation and the Project of Development in Post-Rana Nepal." *Development and Change*, 28 (1): 45–69.

Guyot-Réchard, Bérénice. 2016. *Shadow States: India, China and the Himalayas, 1910–1962*. Cambridge: Cambridge University Press.

Gyatso, Janet. 1980. "The Teachings of Thang-Stong rGyal-po." In *Tibetan Studies in Honour of Hugh Richardson*, edited by Michael Aris, and Aung San Suu Kyi, 111–119. Warminster: Aris and Phillips.

Haarh, Erik. 1969. *The Yar-luṅ Dynasty*. København: G.E.C. GAD's Forlag.

Hadi, Tahmina. 2019. "An Analysis of Water Policies and Strategies of Bangladesh in the Context of Climate Change." *Asia-Pacific Journal of Rural Development* 29 (1): 111–123.

Hagg, Wilfried, Stefan Ram, Alexander Klaus, et al. 2021. "Hazard Assessment for a Glacier Lake Outburst Flood in the Mo Chu River Basin, Bhutan." *Applied Sciences* 11 (20): 9463.

Haider, Vicky L., István Dunkl, Hilmar von Eynatten, et al. 2013. "Cretaceous to Cenozoic Evolution of the Northern Lhasa Terrane and the Early Paleogene Development of Peneplains at Nam Co, Tibetan Plateau." *Journal of Asian Earth Sciences* 70–71: 79–98.

Haldar, Partha, and Nipu Modak. 2022. "Potential of Pumped Hydro Storage as an Electrical Energy Storage in India." *Journal of the Institution of Engineers (India): Series C* 103 (2): 181–187.

Hallet, B., and P. Molnar. 2001. "Distorted Drainage Basins as Markers of Crustal Strain East of the Himalaya." *Journal of Geophysical Research: Solid Earth* 106 (B7): 13697–13709.

Hao, Zhixin, Maowei Wu, Yang Liu, et al. 2020. "Multi-Scale Temperature Variations and Their Regional Differences in China during the Medieval Climate Anomaly." *Journal of Geographical Sciences* 30 (1): 119–130.

Harman, Jay R. 1987. "Westerlies, Middle-Latitude West Winds." In *The Encyclopedia of Climatology*, edited by J. E. Oliver, and Rhodes W. Fairbridge, 922–928. New York: Springer.

Harrell, Stevan. 2023. *An Ecological History of Modern China.* Seattle: University of Washington Press.

Harris, Jonathan M., and Brian Roach. 2021. *Environmental and Natural Resource Economics: A Contemporary Approach.* London: Routledge.

Harris, Tina. 2013. *Geographical Diversions: Tibetan Trade, Global Transactions.* Athens: The University of Georgia Press.

Harris, Tina. 2017. "The Mobile and the Material in the Himalayan Borderlands." In *Making Relations across China's Borders*, edited by Juan Zhang, and Martin Saxer, 145–163. Amsterdam: Amsterdam University Press.

Harvey, David. 2005. *A Brief History of Neoliberalism.* New York: Oxford University Press.

Hatton MacDonald, Darla, Mark D. Morrison, and Mary B. Barnes. 2010. "Willingness to Pay and Willingness to Accept Compensation for Changes in Urban Water Customer Service Standards." *Water Resources Management* 24: 3145–3158.

Hayes, Jack Patrick. 2013. "From Great Green Walls to Deadly Mires: Wetlands as Military Environments and Ecosystems in Chinese History." *Water History* 5 (1): 7–26.

Hazod, Guntram. 2007. "The Grave on the 'Cool Plain'. On the Identification of 'Tibet's First Tomb' in Nga-ra-thang of 'Phyong-po." In *Pramāṇakīrtiḥ. Papers Dedicated to Ernst Steinkellner on the Occasion of His 70th Birthday, Part 1*, edited by Birgit Kellner, Helmut Krasser, Horst Lasic, et al., 259–283. Wien: WSTB.

He, Chuansong, Shuwen Dong, and Yanghua Wang. 2019. "Lithospheric Delamination and Upwelling Asthenosphere in the Longmenshan Area: Insight from Teleseismic P-wave Tomography." *Scientific Reports* 9 (1): 6967.

He, Dekui, and Yifeng Chen. 2006. "Biogeography and Molecular Phylogeny of the Genus Schizothorax (Teleostei: Cyprinidae) in China Inferred from Cytochrome *b* Sequences." *Journal of Biogeography* 33 (8): 1448–1460.

He, Dekui, and Yifeng Chen. 2009. "Phylogeography of Schizothorax o'connori (Cyprinidae: Schizothoracinae) in the Yarlung Tsangpo River, Tibet." *Hydrobiologia* 635 (1): 251–262.

Hedin, Sven. 1909. *Trans-Himalaya: Discoveries and Adventures in Tibet.* New York: MacMillan Company.

Helman-Ważny, Agnieszka, and Charles Ramble. 2020. *The Mardzong Manuscripts: Codicological and Historical Studies of an Archaeological Find in Mustang, Nepal.* Leiden: Brill.

Hetzel, Ralf, István Dunkl, Vicky Haider, et al. 2011. "Peneplain Formation in Southern Tibet Predates the India–Asia Collision and Plateau Uplift." *Geology* 39 (10): 983–986.

Hinkel, Kenneth M., Yongwei Sheng, John D. Lenters, et al. 2012. "Thermokarst Lakes on the Arctic Coastal Plain of Alaska: Geomorphic Controls on Bathymetry." *Permafrost and Periglacial Processes* 23 (3): 218–230.

Hinzman, Larry D., Neil D. Bettez, W. Robert Bolton, et al. 2005. "Evidence and Implications of Recent Climate Change in Northern Alaska and Other Arctic Regions." *Climatic Change* 72 (3): 251–298.

Ho, Selina. 2016. "'Big Brother, Little Brothers': Comparing China's and India's Transboundary River Policies." *Water Policy* 18 (S1): 32–49.

Hobbs, R. J., Higgs, E., Harris, J. A. 2009. "Novel Ecosystems: Implications for Conservation and Restoration." *Trends in Ecology & Evolution* 24 (11): 599–605.

Hodges, K. V., J. M. Hurtado, and K. X. Whipple. 2001. "Southward Extrusion of Tibetan Crust and Its Effect on Himalayan Tectonics." *Tectonics* 20 (6): 799–809.

Hou, Juzhi, Kejia Ji, Erlei Zhu, et al. 2023. "Climate Change Fostered Rise and Fall of the Tibetan Empire during 600–800 AD." *Science Bulletin* 68 (11): 1187–1194.

Houde, Nicolas. 2007. "The Six Faces of Traditional Ecological Knowledge: Challenges and Opportunities for Canadian Co-Management Arrangements." *Ecology and Society* 12 (2): 34.

Hu, Fangyang, Fu-Yuan Wu, Mihai N. Ducea, et al. 2022. "Does Large-Scale Crustal Flow Shape the Eastern Margin of the Tibetan Plateau? Insights from Episodic Magmatism of Gongga–Zheduo Granitic Massif." *Geophysical Research Letters* 49 (12): e2022GL098756.

Hu, Shuai, Tianjun Zhou, and Bo Wu. 2021. "Impact of Developing ENSO on Tibetan Plateau Summer Rainfall." *Journal of Climate* 34 (9): 3385–3400.

Huang, Liangliang, and Jianhua Li. 2016. "Status of Freshwater Fish Biodiversity in the Yangtze River Basin, China." In *Aquatic Biodiversity Conservation and Ecosystem Services*, edited by Shin-ichi Nakano, Tetsukazu Yahara, and Tohru Nakashizuka, 13–30. Singapore: Springer.

Huber, Amelie. 2019. "Hydropower in the Himalayan Hazardscape: Strategic Ignorance and the Production of Unequal Risk." *Water* 11 (3): 414.

Huber, Toni. 2004. "Territorial Control by 'Sealing' (rgya sdom-pa): A Religio-Political Practice in Tibet." *Zentralasiatische Studien* 33: 127–152.

Huber, Toni. 2008. *The Holy Land Reborn: Pilgrimage and the Tibetan Reinvention of Buddhist India.* Chicago: University of Chicago Press.

Huber, Toni. 2012. "The Changing Role of Hunting and Wildlife in Pastoral Communities of Northern Tibet." In *Pastoral Practices in High Asia: Agency of 'Development' Effected by Modernisation, Resettlement and Transformation*, edited by Hermann Kreutzmann, 195–215. Dordrecht: Springer.

Hummel, Siegbert, and G. Vogliotti. 2000. "On the Origin of the Irrigation Technique in Tibet." *The Tibet Journal* 25 (3): 8–13.

Huo, Bin, Cong Xin Xie, Charles Paul Madenjian, et al. 2014. "Feeding Habits of an Endemic Fish, Oxygymnocypris Stewartii, in the Yarlung Zangbo River in Tibet, China." *Environmental Biology of Fishes* 97 (11): 1279–1293.

Husson, Laurent, Matthias Bernet, Stéphane Guillot, et al. 2014. "Dynamic Ups and Downs of the Himalaya." *Geology* 42 (10): 839–842.

Hutchinson, G. Evelyn. 1948. "Circular Causal Systems in Ecology." *Annals of the New York Academy of Sciences* 50 (4): 221–246.

Hydropower Sustainability Secretariat. 2023. *Hydropower Sustainability Standard*. London: Hydropower Sustainability Council.

Iaffaldano, Giampiero. 2021. "Has the Tibetan Plateau Risen in the Early/Mid-Miocene? Constraints from Plate-Motion Reconstructions and Seismicity of the Indian Ocean Lithosphere." *Geophysical Journal International* 225 (2): 1349–1358.

Iaffaldano, Giampiero, Laurent Husson, and Hans-Peter Bunge. 2011. "Monsoon Speeds Up Indian Plate Motion." *Earth and Planetary Science Letters* 304 (3): 503–510.

Igoe, Jim, and Dan Brockington. 2007. "Neoliberal Conservation: A Brief Introduction." *Conservation and Society* 5 (4): 432–449.

Imaeda, Yoshiro, Tsuguhito Takeuchi, Izumi Hoshi, et al. 2007. *Tibetan Documents from Dunhuang: Kept at the Bibliothèque Nationale de France and the British Library*. Tokyo: Research Institute for Languages and Cultures of Asia and Africa, Tokyo University of Foreign Studies.

Immerzeel, Walter W., Ludovicus P. H. van Beek, and Marc F. P. Bierkens. 2010. "Climate Change Will Affect the Asian Water Towers." *Science* 328 (5984): 1382–1385.

Indian Space Applications Centre (ISRO). 2009. *National Wetland Atals: Arunachal Pradesh*. Ahmedabad: Space Applications Centre, Indian Space Research Organisation.

Indian Space Applications Centre (ISRO). 2012. *National Wetland Atlas: High Altitude Lakes of India*. Ahmedabad: Space Applications Centre.

IRENA. 2019. *Renewables Readiness Assessment: Kingdom of Bhutan*. Abu Dhabi: International Renewable Energy Agency.

Jacoby, Sarah H. 2015. *Love and Liberation: Autobiographical Writings of the Tibetan Buddhist Visionary Sera Khandro*. New York: Columbia University Press.

Jamieson, Rebecca A., Martyn J. Unsworth, Nigel B. W. Harris, et al. 2011. "Crustal Melting and the Flow of Mountains." *Elements* 7 (4): 253–260.

Ji, Zhenming, Shichang Kang, Qianggong Zhang, et al. 2016. "Investigation of Mineral Aerosols Radiative Effects over High Mountain Asia in 1990–2009 Using a Regional Climate Model." *Atmospheric Research* 178–179: 484–496.

Jiang, Wei, Yihe Lü, Yuanxin Liu, et al. 2020b. "Ecosystem Service Value of the Qinghai–Tibet Plateau Significantly Increased during 25 Years." *Ecosystem Services* 44: 101146.

Jiang, Xingwen, Tuantuan Zhang, Chi-Yung Tam, et al. 2019. "Impacts of ENSO and IOD on Snow Depth over the Tibetan Plateau: Roles of Convections over the Western North Pacific and Indian Ocean." *Journal of Geophysical Research: Atmospheres* 124 (22): 11961–11975.

Jodha, N. S. 1998. "Poverty and Environmental Resource Degradation: An Alternative Explanation and Possible Solutions." *Economic and Political Weekly* 33 (36/37): 2384–2390.

Johns, Jason, and Jyotsna Rani Nag. 2021. "Muchalinda Buddha: An Interdisciplinary Approach to Reinterpret the Depiction of the Buddha with Muchalinda Naga." *Journal of Archaeological Studies in India* 1 (1): 140–157.

Jolivet, M., S. Dominguez, J. Charreau, et al. 2010. "Mesozoic and Cenozoic Tectonic History of the Central Chinese Tian Shan: Reactivated Tectonic Structures and Active Deformation." *Tectonics* 29 (6): TC6019.

Jones, Craig H., Jeffrey R. Unruh, and Leslie J. Sonder. 1996. "The Role of Gravitational Potential Energy in Active Deformation in the Southwestern United States." *Nature* 381 (6577): 37–41.

Junk, Wolfgang J., Shuqing An, C. M. Finlayson, et al. 2013. "Current State of Knowledge Regarding the World's Wetlands and Their Future under Global Climate Change: A Synthesis." *Aquatic Sciences* 75 (1): 151–167.

Kale, Vishwas S. 2003. "Geomorphic Effects of Monsoon Floods on Indian Rivers." *Natural Hazards* 28 (1): 65–84.

Kamae, Youichi, Wei Mei, Shang-Ping Xie, et al. 2017. "Atmospheric Rivers over the Northwestern Pacific: Climatology and Interannual Variability." *Journal of Climate* 30 (15): 5605–5619.

Kandel, Pratikshya, Nakul Chettri, Sunita Chaudhary, et al. 2021. "Ecosystem Services Research Trends in the Water Tower of Asia: A Bibliometric Analysis from the Hindu Kush Himalaya." *Ecological Indicators* 121: 107152.

Kang, Shichang, Qianggong Zhang, Yun Qian, et al. 2019. "Linking Atmospheric Pollution to Cryospheric Change in the Third Pole Region: Current Progress and Future Prospects." *National Science Review* 6 (4): 796–809.

Kapp, Paul, and Peter G. DeCelles. 2019. "Mesozoic–Cenozoic Geological Evolution of the Himalayan–Tibetan Orogen and Working Tectonic Hypotheses." *American Journal of Science* 319 (3): 159–254.

Kapp, Paul, Jon D. Pelletier, Alexander Rohrmann, et al. 2011. "Wind Erosion in the Qaidam Basin, Central Asia: Implications for Tectonics, Paleoclimate, and the Source of the Loess Plateau." *GSA Today* 21 (4/5): 4–10.

Kardas, Goran. 2015. "From Etymology to Ontology: Vasubandhu and Candrakīrti on Various Interpretations of Pratītyasamutpāda." *Asian Philosophy* 25 (3): 293–317.

Karkee, Rajendra, and Jude Comfort. 2016. "NGOs, Foreign Aid, and Development in Nepal." *Frontiers in Public Health* 4: 177.

Karmay, Samten Gyaltsen. 1998. *The Arrow and the Spindle: Studies in History, Myths, Rituals and Beliefs in Tibet*. Vol. 3. Delhi: Mandala Book Point.

Kaspari, S., T. H. Painter, M. Gysel, et al. 2014. "Seasonal and Elevational Variations of Black Carbon and Dust in Snow and Ice in the Solu-Khumbu, Nepal and Estimated Radiative Forcings." *Atmospheric Chemistry and Physics* 14 (15): 8089–8103.

Khalid, Iram. 2020. "Interstate Water Politics: An Analysis of Indian Case Water Governance in India." *Pakistan Social Sciences Review* 4 (2): 1–16.

Khan, Shariq, and Susan Fenton. 2023. "Bhutan to Add More Solar Capacity to Reduce Reliance on Imports - Minister." *Reuters*, February 6. Accessed 14 November 2023. https://www.reuters.com/business/energy/bhutan-add-more-solar-capacity-reduce-reliance-imports-minister-2023-02-06/.

Khatiwara, Sunita, and Karma Choden Bhutia. 2020. "A Distribution Survey for Otters in Sikkim, India." *IUCN Otter Specialist Group Bulletin* 37 (4): 212–218.

Khetsun, Tubten. 2007. *Memories of Life in Lhasa under Chinese Rule*. New York: Columbia University Press.

Kingdon-Ward, Francis. 1926. *The Riddle of the Tsangpo Gorges*. London: Edward Arnold.

Kirby, Eric, Kelin X. Whipple, Wenqing Tang, et al. 2003. "Distribution of Active Rock Uplift along the Eastern Margin of the Tibetan Plateau: Inferences from Bedrock Channel Longitudinal Profiles." *Journal of Geophysical Research: Solid Earth* 108, 2217, (B4).

Klemme, A., T. Warneke, H. Bovensmann, et al. 2023. "Sediment Transport in Indian Rivers High Enough to Impact Satellite Gravimetry." *Hydrology and Earth System Sciences Discussions* 2023: 1–15.

Kocurek, Jakub. 2013. "Tree Beings in Tibet: Contemporary Popular Concepts of Klu and Gnyan as a Result of Ecological Change." *Journal of Ethnology and Folkloristics* 7 (1): 19–30.

Kok, Jasper F., Trude Storelvmo, Vlassis A. Karydis, et al. 2023. "Mineral Dust Aerosol Impacts on Global Climate and Climate Change." *Nature Reviews Earth & Environment* 4 (2): 71–86.

Kornman, Robin, Sangye Khandro, Lama Chonam, et al. 2012. *The Epic of Gesar of Ling*. Boston: Shambhala.

Korup, Oliver, and David R. Montgomery. 2008. "Tibetan Plateau River Incision Inhibited by Glacial Stabilization of the Tsangpo Gorge." *Nature* 455 (7214): 786–789.

Kothari, Uma, ed. 2019. *A Radical History of Development Studies: Individuals, Institutions and Ideologies*. London: Zed Books.

Kotlia, B. S., J. Sanwal, and S. K. Bhattacharya. 2008. "Climatic Record between ca. 31 and 22 ka BP in East-Central Uttarakhand Himalaya, India." *Himalayan Geology* 29 (1): 25–33.

Kraaijenbrink, Philip D. A., Emmy E. Stigter, Tandong Yao, et al. 2021. "Climate Change Decisive for Asia's Snow Meltwater Supply." *Nature Climate Change* 11 (7): 591–597.

Krasting, John P., Anthony J. Broccoli, Keith W. Dixon, et al. 2013. "Future Changes in Northern Hemisphere Snowfall." *Journal of Climate* 26 (20): 7813–7828.

Krishna, K. S., M. Ismaiel, K. Srinivas, et al. 2016. "Sediment Pathways and Emergence of Himalayan Source Material in the Bay of Bengal." *Current Science* 110 (3): 363–372.

Krishna, Sankaran. 1994. "Cartographic Anxiety: Mapping the Body Politic in India." *Alternatives: Global, Local, Political* 19 (4): 507–521.

Kuhle, Matthias. 1995. "Glacial Isostatic Uplift of Tibet as a Consequence of a Former Ice Sheet." *GeoJournal* 37 (4): 431–449.

Kumar, M. Dinesh. 2018. *Water Policy Science and Politics: An Indian Perspective*. Amsterdam: Elsevier.

Kutzbach, J. E., W. L. Prell, and Wm F. Ruddiman. 1993. "Sensitivity of Eurasian Climate to Surface Uplift of the Tibetan Plateau." *The Journal of Geology* 101 (2): 177–190.

Kuyakanon, Riamsara, and Dorji Gyeltshen. 2017. "Propiating the Tsen, Sealing the Mountain: Community Mountain-Closure Ritual and Practice in Eastern Bhutan." *Himalaya* 37 (1): 8–25.

Kyab, Lhamo. 2023. *Bod mdo dbus mtho sgang gi gtsang po*. Dharamsala: Lhamo Kyab.

Lakshmi, D. Dhana, and A. N. V. Satyanarayana. 2019. "Influence of Atmospheric Rivers in the Occurrence of Devastating Flood Associated with Extreme Precipitation Events over Chennai using Different Reanalysis Data Sets." *Atmospheric Research* 215: 12–36.

Lange, Diana. 2014. "'The Government Forced Us to Send the Boat in the Middle of the Night.' Water Transport in Pre-1959 Tibet." *The Tibet Journal* 39 (1): 75–93.

Lange, Diana. 2020. *An Atlas of the Himalayas by a 19th Century Tibetan Lama: A Journey of Discovery*. Leiden: Brill.

Lange, Diana. 2023. "'My Karma Selected Me to Become a Ferryman': The Role of Waterways and Watercraft in the Corvée Tax System in Pre-1959 Tibet." In *Taxation in Tibetan Societies: Rules, Practices and Discourses*, edited by Alice Travers, Peter Schwieger, and Charles Ramble, 197–210. Leiden: Brill.

Larsen, Isaac J., and David R. Montgomery. 2012. "Landslide Erosion Coupled to Tectonics and River Incision." *Nature Geoscience* 5 (7): 468–473.

Laskowski, Andrew K., Paul Kapp, Lin Ding, et al. 2017. "Tectonic evolution of the Yarlung suture zone, Lopu Range region, southern Tibet." *Tectonics* 36 (1): 108–136.

Latour, Bruno. 1993. *We Have Never Been Modern* (translated by Catherine Porter). Harvard, MA: Harvard University Press.

Lavers, David A., and Gabriele Villarini. 2013. "The Nexus between Atmospheric Rivers and Extreme Precipitation across Europe." *Geophysical Research Letters* 40 (12): 3259–3264.

Lauer, Matthew and Shankar Aswani Canela. 2009. "Indigenous Ecological Knowledge as Situated Practices: Understanding Fishers' Knowledge in the Western Solomon Islands." *American Anthropologist* 111 (3): 317–329.

Lazcano, Santiago. 2005. "Ethnohistoric Notes on the Ancient Tibetan Kingdom of sPo bo and Its Influence on the Eastern Himalayas." *Revue d'Etudes Tibétaines* 7: 41–63.

Leibold, James. 2006. "Competing Narratives of Racial Unity in Republican China: From the Yellow Emperor to Peking Man." *Modern China* 32 (2): 181–220.

Lenard, Sebastien J. P., Jérôme Lavé, Christian France-Lanord, et al. 2020. "Steady Erosion Rates in the Himalayas through Late Cenozoic Climatic Changes." *Nature Geoscience* 13 (6): 448–452.

Lepcha, Kachyo. 2020. "The Teesta Hydro Power Projects: A Historical Analysis of the Protest Movement in North Sikkim (1964–2011)." PhD diss., Sikkim University.

Lepcha, Rongnyoo, and Mongfing Lepcha. 2021. "Painting the Genesis of the Lepcha: A World Emerging from Water Spirits." In *Environmental Humanities in the New Himalayas: Symbiotic Indigeneity, Commoning, Sustainability*, edited by Dan Smyer Yü, and Erik de Maaker, 99–114. London: Routledge.

Lévi-Strauss, Claude. 2021. *Wild Thought: A New Translation of La Pensée Sauvage*. Translated and edited by Jeffrey Mehlman, and John Leavitt. Chicago: University of Chicago Press.

Lévy-Bruhl, Lucien. 1985. *How Natives Think*. Princeton: Princeton University Press.

Lhaden, Yangyel. 2023. "DoE Aims to Add 500 Megawatts Renewable Energy to the Energy Mix." *Kuensel*, September 23.

Lhamo, Phurpa, and Indraman Chhetri. 2022. "Assessing Water Resource Management in Bhutan from a Whole-of-Government Approach: A Perception Study." *Bhutan Journal of Management* 2 (1): 154–182.

Li, B., J. Li, and Z. Cui, eds. 1991. *Quaternary glacial distribution map of Qinghai–Xizang (Tibet) Plateau 1:3,000,000*. Quaternary Glacier, and Environmental Research Center, Lanzhou University.

Li, Chunhong, Fengge Su, Daqing Yang, et al. 2018. "Spatiotemporal Variation of Snow Cover over the Tibetan Plateau Based on MODIS Snow Product, 2001–2014." *International Journal of Climatology* 38 (2): 708–728.

Li, Dongfeng, Xixi Lu, Desmond E. Walling, et al. 2022. "High Mountain Asia Hydropower Systems Threatened by Climate-Driven Landscape Instability." *Nature Geoscience* 15 (7): 520–530.

Li, He, Pan Liu, Shenglian Guo, et al. 2020a. "Climatic Control of Upper Yangtze River Flood Hazard Diminished by Reservoir Groups." *Environmental Research Letters* 15 (12): 124013.

Li, Jinbao, Jiangfeng Shi, David D. Zhang, et al. 2017a. "Moisture Increase in Response to High-Altitude Warming Evidenced by Tree-Rings on the Southeastern Tibetan Plateau." *Climate Dynamics* 48 (1): 649–660.

Li, Ni, Yong-Wei Zhao, Liu-Yi Zhang, et al. 2020b. "The Quaternary Eruptive Sequence of the Tengchong Volcanic Group, Southwestern China." *Lithos* 354–355: 105173.

Li, Shu-Feng, Paul J. Valdes, Alex Farnsworth, et al. 2021. "Orographic Evolution of Northern Tibet Shaped Vegetation and Plant Diversity in Eastern Asia." *Science Advances* 7 (5): eabc7741.

Li, Xiaofei, Shichang Kang, Xiaobo He, et al. 2017b. "Light-Absorbing Impurities Accelerate Glacier Melt in the Central Tibetan Plateau." *Science of the Total Environment* 587–588: 482–490.

Li, Yingkui, Dewen Li, Gengnian Liu, et al. 2014. "Patterns of Landscape Evolution on the Central and Northern Tibetan Plateau Investigated Using In-Situ Produced 10Be Concentrations from River Sediments." *Earth and Planetary Science Letters* 398: 77–89.

Li, Yongxian. 2007. "Animal Bones and Economy at the Site of Karuo: An Opinion on Prehistoric Agriculture in the Hengduan Mountain Chain." *Sichuan Cultural Relics* 5: 50–56.

Liang, Shiming, Weijun Gan, Chuanzheng Shen, et al. 2013. "Three-Dimensional Velocity Field of Present-Day Crustal Motion of the Tibetan Plateau Derived from GPS Measurements." *Journal of Geophysical Research: Solid Earth* 118 (10): 5722–5732.

Liang, Wendong, Eduardo Garzanti, Xiumian Hu, et al. 2022. "Tracing Erosion Patterns in South Tibet: Balancing Sediment Supply to the Yarlung Tsangpo from the Himalaya versus Lhasa Block." *Basin Research* 34 (1): 411–439.

Liang, Yu-Hsuan, Sun-Lin Chung, Dunyi Liu, et al. 2008. "Detrital Zircon Evidence from Burma for Reorganization of the Eastern Himalayan River System." *American Journal of Science* 308 (4): 618–638.

Lieberman, Victor. 2003. *Strange Parallels: Southeast Asia in Global Context, c.800–1830*. Cambridge: Cambridge University Press.

Lieberman, Victor, and Brendan Buckley. 2012. "The Impact of Climate on Southeast Asia, circa 950–1820: New Findings." *Modern Asian Studies* 46 (5): 1049–1096.

Linderholm, Hans W., Andrea Seim, Tinghai Ou, et al. 2013. "Exploring Teleconnections between the Summer NAO (SNAO) and Climate in East Asia over the Last Four Centuries – A Tree-Ring Perspective." *Dendrochronologia* 31 (4): 297–310.

Linton, Jamie, and Jessica Budds. 2014. "The Hydrosocial Cycle: Defining and Mobilizing a Relational-Dialectical Approach to Water." *Geoforum* 57: 170–180.

Liu, Fei, Mingzheng Li, Jian Wang, et al. 2021a. "Species Composition and Longitudinal Patterns of Fish Assemblages in the Middle and Lower Yarlung Zangbo River, Tibetan Plateau, China." *Ecological Indicators* 125: 107542.

Liu, Han, Chuanbo Guo, Xiao Qu, et al. 2020b. "Fish Diversity, Endemism, Threats, and Conservation in the Jinsha River Basin (Upper Yangtze River), China." *North American Journal of Fisheries Management* 41 (4): 967–984.

Liu, Huancai, Keqin Duan, Man Li, et al. 2015a. "Impact of the North Atlantic Oscillation on the Dipole Oscillation of Summer Precipitation over the Central and Eastern Tibetan Plateau." *International Journal of Climatology* 35 (15): 4539–4546.

Liu, Jie, Richard I. Milne, Guang-Fu Zhu, et al. 2022. "Name and Scale Matter: Clarifying the Geography of Tibetan Plateau and Adjacent Mountain Regions." *Global and Planetary Change* 215: 103893.

Liu, Kam-biu, Zuju Yao, and Lonnie G. Thompson. 1998. "A Pollen Record of Holocene Climatic Changes from the Dunde Ice Cap, Qinghai–Tibetan Plateau." *Geology* 26 (2): 135–138.

Liu, S., J.-M. Nocquet, X. Xu, et al. 2023. "Present-Day Uplift of the East Kunlun Shan, Northern Tibetan Plateau." Paper presented at EGU General Assembly 2022, Vienna, May 23–27.

Liu, Shengsheng, Shunwu Zhou, Ping Wu, et al. 2021b. "Response of Winter Precipitation in Eastern Tibetan Plateau to Arctic Oscillation." *Acta Meteorologica Sinica* 79 (4): 558–569.

Liu, Weiming, Paul A. Carling, Kaiheng Hu, et al. 2019. "Outburst Floods in China: A Review." *Earth-Science Reviews* 197: 102895.

Liu, X., Bai, A. and Liu, C., 2009. "Diurnal Variations of Summertime Precipitation Over the Tibetan Plateau in Relation to Orographically-induced Regional Circulations." *Environmental Research Letters* 4 (4): 045203.

Liu, Xiaodong, and Baode Chen. 2000. "Climatic Warming in the Tibetan Plateau during Recent Decades." *International Journal of Climatology* 20 (14): 1729–1742.

Liu, Xiaodong, and Zhi-Yong Yin. 2001. "Spatial and Temporal Variation of Summer Precipitation over the Eastern Tibetan Plateau and the North Atlantic Oscillation." *Journal of Climate* 14 (13): 28962–909.

Liu, Yang. 2015b. "Transboundary Water Cooperation on the Yarlung Zangbo/Brahmaputra – A Legal Analysis of Riparian State Practice." *Water International* 40 (2): 354–374.

Liu, Yong, Yunsheng Wang, Liangshuai Wei, et al. 2020a. "River Capture in the Middle Reaches of the Palaeo-Yarlung Zangbo River." *Royal Society Open Science* 7 (4): 191753.

Liu-Zeng, J., P. Tapponnier, Y. Gaudemer, et al. 2008. "Quantifying Landscape Differences across the Tibetan Plateau: Implications for Topographic Relief Evolution." *Journal of Geophysical Research: Earth Surface* 113 (F4): F04018.

Lo chen 'gyur med bde chen. 1982. *Thang stong rgyal po'i rnam thar.* Khreng tu 'u: Si khron mi rigs dpe skrun khang.

Loiko, Sergey V., Oleg S. Pokrovsky, Tatiana V. Raudina, et al. 2017. "Abrupt Permafrost Collapse Enhances Organic Carbon, CO2, Nutrient and Metal Release into Surface Waters." *Chemical Geology* 471: 153–165.

Lu, Jun-Mei, Jian-Hua Ju, Seong-Joong Kim, et al. 2008. "Arctic Oscillation and the Autumn/Winter Snow Depth over the Tibetan Plateau." *Journal of Geophysical Research: Atmospheres* 113 (D14): D14117.

Ludden, David. 2003. "Presidential Address: Maps in the Mind and the Mobility of Asia." *The Journal of Asian Studies* 62 (4): 1057–1078.

Lupker, M., J. Lavé, C. France-Lanord, et al. 2017. "^{10}Be Systematics in the Tsangpo-Brahmaputra Catchment: The Cosmogenic Nuclide Legacy of the Eastern Himalayan Syntaxis." *Earth Surface Dynamics* 5, 429–449.

Lyons, N. J., P. Val, J. S. Albert, et al. 2020. "Topographic Controls on Divide Migration, Stream Capture, and Diversification in Riverine Life." *Earth Surface Dynamics* 8 (4): 893–912.

Ma, Jing, Arjen Y. Hoekstra, Hao Wang, et al. 2006. "Virtual versus Real Water Transfers within China." *Philosophical Transactions of the Royal Society B: Biological Sciences* 361 (1469): 835–842.

Ma, Qianqian, Maierdang Keyimu, Xiangyi Li, et al. 2023. "Climate and Elevation Control Snow Depth and Snow Phenology on the Tibetan Plateau." *Journal of Hydrology* 617: 128938.

Ma, Tingting, Wei Zhou, Like Chen, et al. 2018. "Concerns about the Future of Chinese Fisheries Based on Illegal, Unreported and Unregulated Fishing on the Hanjiang River." *Fisheries Research* 199: 212–217.

Ma, X., Kang, J., Chen, W., et al. 2015. "Biogeographic History and High-Elevation Adaptations Inferred from the Mitochondrial Genome of Glyptosternoid Fishes (Sisoridae, Siluriformes) from the Southeastern Tibetan Plateau." *BMC Evolutionary Biology* 15: 233.

McAfee, Kathleen. 1999. "Selling Nature to Save It? Biodiversity and Green Developmentalism." *Environment and Planning D: Society and Space* 17 (2): 133–154.

MacAllister, D. J., G. Krishan, M. Basharat, et al. 2022. "A Century of Groundwater Accumulation in Pakistan and Northwest India." *Nature Geoscience* 15 (5): 390–396.

McCarthy, Susan K. 2022. "Liberating Party Animals: Cultural Governance and 'Life Release' Rituals in China." *The China Journal* 88: 55–77.

MacCormack, Ian. 2018. "Buddhism and State in Seventeenth-Century Tibet: Cosmology and Theology in the Works of Sangyé Gyatso." PhD diss., Harvard University.

McCracken, Melissa, and Aaron T. Wolf. 2019. "Updating the Register of International River Basins of the World." *International Journal of Water Resources Development* 35 (5): 732–782.

Mace, Georgina M. 2014. "Whose Conservation?" *Science* 345 (6204): 1558–1560.

McGlade, Ciara L. O., James W. E. Dickey, Richard Kennedy, et al. 2022. "Behavioural Traits of Rainbow Trout and Brown Trout May Help Explain Their Differing Invasion Success and Impacts." *Scientific Reports* 12 (1): 1757.

McGranahan, Carole, 2020. *Arrested Histories: Tibet, the CIA, and Memories of a Forgotten War*. Durham: Duke University Press.

McKay, Alex. 1997. *Tibet and the British Raj: The Frontier Cadre, 1904–1947*. London: RoutledgeCurzon.

McKay, Alex. 2016. *Kailas Histories: Renunciate Traditions and the Construction of Himalayan Sacred Geography*. Leiden: Brill.

McKay, Alex. 2021. *The Mandala Kingdom: A Political History of Sikkim*. Gangtok: Rachna Books.

Macy, Joanna. 1991. *Mutual Causality in Buddhism and General Systems Theory: The Dharma of Natural System*. Albany: State University of New York Press.

Manish, Kumar, Maharaj K. Pandit, Yasmeen Telwala, et al. 2017. "Elevational Plant Species Richness Patterns and Their Drivers across Non-Endemics, Endemics and

Growth Forms in the Eastern Himalaya." *Journal of Plant Research* 130 (5): 829–844.

Mann, Michael E. 2012. *The Hockey Stick and the Climate Wars: Dispatches from the Front Lines*. New York: Columbia University Press.

Mao, Dehua, Zongming Wang, Baojia Du, et al. 2020. "National Wetland Mapping in China: A New Product Resulting from Object-Based and Hierarchical Classification of Landsat 8 OLI Images." *ISPRS Journal of Photogrammetry and Remote Sensing* 164: 11–25.

Mao, Peini, Yongqiang Guo, and Tao Liu. 2023. "Holocene Extreme Palaeofloods Recorded by Slackwater Deposits along the Jiacha Gorge of the Yarlung Tsangpo River Valley, Southern Tibetan Plateau." *CATENA* 231: 107360.

Mao, Rui, DaoYi Gong, YaPing Shao, et al. 2013. "Numerical Analysis for Contribution of the Tibetan Plateau to Dust Aerosols in the Atmosphere over the East Asia." *Science China Earth Sciences* 56 (2): 301–310.

Maurer, Petra. 2009. *Die Grundlagen der tibetischen Geomantie dargestellt anhand des 32. Kapitels des Vaidurya dkar po von sde srid Sangs rgyas rgya mtsho (1653–1705)*. Halle: IITBS.

Maurer, Petra. 2019. "When the Tiger Meets Yul 'Khor Srung, or How to Protect a Construction Site." *Études mongoles et sibériennes, centrasiatiques et tibétaines* 50: https://doi.org/10.4000/emscat.3950.

Maurer, Petra. 2024. "Adapting Construction to Suit the Environment: Living with Deities and Demons in Traditional Tibet." In *Kucha and Beyond, Divine and Human Landscapes from Central Asia to the Himalaya*, edited by Eli Franco, Charles Ramble, and Monika Zin. New Delhi: Dev Publishers.

Mekonnen, Mesfin M., and Arjen Y. Hoekstra. 2016. "Four Billion People Facing Severe Water Scarcity." *Science Advances* 2 (2): e1500323.

Meng, Xiangang, Dagang Zhu, Zhaogang Shao, et al. 2008. "Late Cenozoic Stratigraphy and Paleomagnetic Chronology of the Zanda Basin, Tibet, and Records of the Uplift of the Qinghai-Tibet Plateau." *Acta Geologica Sinica - English Edition* 82 (1): 63–72.

Menon, Shruti. 2022. "Climate Change: What Emission Cuts Has India Promised?" *BBC News*, November 7. https://www.bbc.com/news/world-asia-india-58922398.

Merchant, Carolyn. 2003. "Shades of Darkness: Race and Environmental History." *Environmental History* 8 (3): 380–394.

Meyer, M.C., Aldenderfer, M.S., Wang, Z., et al. 2017. "Permanent Human Occupation of the Central Tibetan Plateau in the Early Holocene." *Science* 355 (6320): 64–67.

Miao, Yunfa, Xiaomin Fang, Jimin Sun, et al. 2022. "A New Biologic Paleoaltimetry Indicating Late Miocene Rapid Uplift of Northern Tibet Plateau." *Science* 378 (6624): 1074–1079.

Miehe, Georg, Sabine Miehe, Frank Schlütz, et al. 2006. "Palaeoecological and Experimental Evidence of Former Forests and Woodlands in the Treeless Desert Pastures of Southern Tibet (Lhasa, A.R. Xizang, China)." *Palaeogeography, Palaeoclimatology, Palaeoecology* 242 (1): 54–67.

Millennium Ecosystem Assessment. 2005. *Ecosystems and Human Well-Being: Synthesis*. Washington, DC: Island Press.

Miller, Manjari Chatterjee. 2020. *Wronged by Empire: Post-Imperial Ideology and Foreign Policy in India and China*. Stanford: Stanford University Press.

Ming, Jing, Yaqiang Wang, Zhencai Du, et al. 2015. "Widespread Albedo Decreasing and Induced Melting of Himalayan Snow and Ice in the Early 21st Century." *PLOS One* 10 (6): e0126235.

Ministry of Ecology and Environment. 2016. *The List of China's Nature Reserves.* Beijing: Ministry of Ecology and Environment of People's Republic of China.

Ministry of Environment Forests & Climate Change. 2021. *National Report to Ramsar CoP14 India.* Gland: Ramsar Convention on Wetlands.

Ministry of Power. 2022. *Policy on Hydro Power Development.* New Delhi: Ministry of Power, Government of India.

Ministry of Water Resources. 1999. *National Water Policy.* Dhaka: Ministry of Water Resources, Government of the People's Republic of Bangladesh.

Mirumachi, Naho, Mark Zeitoun, and Jeroen Warner. 2013. "Transboundary Water Interactions and the UN Watercourses Convention: Allocating Waters and Implementing Principles." In *The UN Watercourses Convention in Force: Strengthening International Law for Transboundary Water Management*, edited by Flavia Rocha Loures, and Alistair Rieu-Clarke, 352–364. Abingdon: Routledge.

Mischke, Steffen, Zhencheng Sun, Ulrike Herzschuh, et al. 2010. "An Ostracod-Inferred Large Middle Pleistocene Freshwater Lake in the Presently Hyper-Arid Qaidam Basin (NW China)." *Quaternary International* 218 (1): 74–85.

Mishra, Deepak K. 2023. "Ecological Ruptures in the Eastern Himalayas: The Political Economy of Hydropower Development in Arunachal Pradesh." In *Capital and Ecology: Developmentalism, Subjectivity and the Alternative Life-Worlds*, edited by Rakhee Bhattacharya, and G. Amarjit Sharma, 97–119. London: Routledge.

Mitchell, F. J. 1930. "How Trout Were Introduced into Kashmir." *Journal of the Bombay Natural History Society* 34: 491–501.

Mitsuishi, Mayumi, Simon R. Wallis, Mutsuki Aoya, et al. 2012. "E–W Extension at 19Ma in the Kung Co area, S. Tibet: Evidence for Contemporaneous E–W and N–S Extension in the Himalayan Orogen." *Earth and Planetary Science Letters* 325–326: 10–20.

Mkhas pa lde'u jo sras. 1987. *Mkhas pa lde'us mdzad pa'i rgya bod kyi chos 'byung rgyas pa.* Lhasa: Bod ljongs mi dmangs dpe skrun khang.

Mölg, Thomas, Fabien Maussion, and Dieter Scherer. 2014. "Mid-Latitude Westerlies as a Driver of Glacier Variability in Monsoonal High Asia." *Nature Climate Change* 4 (1): 68–73.

Molle, François, Peter P. Mollinga, and Philippus Wester. 2009. "Hydraulic Bureaucracies and the Hydraulic Mission: Flows of Water, Flows of Power." *Water Alternatives* 2 (3): 328–349.

Molnar, Peter. 2022. "The Rise of the Tibetan Plateau and Removal of Some Its Mantle Lithosphere in Light of Recent Observations." *Himalayan Geology* 43 (1B): 123–139.

Molnar, Peter, and Paul Tapponnier. 1975. "Cenozoic Tectonics of Asia: Effects of a Continental Collision." *Science* 189 (4201): 419–426.

Montgomery, David R., Bernard Hallet, Liu Yuping, et al. 2004. "Evidence for Holocene Megafloods Down the Tsangpo River Gorge, Southeastern Tibet." *Quaternary Research* 62 (2): 201–207.

Moore, Scott M. 2018. *Subnational Hydropolitics: Conflict, Cooperation, and Institution-Building in Shared River Basins.* New York: Oxford University Press.

Morgan, Ruth. 2017. "The Anthropocene as Hydro-Social Cycle: Histories of Water and Technology for the Age of Humans." *Icon* 23: 36–54.

Mostern, Ruth. 2021. *The Yellow River: A Natural and Unnatural History*. New Haven: Yale University Press.

Mueggler, Erik. 2011. *The Paper Road: Archive and Experience in the Botanical Exploration of West China and Tibet*. Berkeley: University of California Press.

Mullard, Saul. 2011. *Opening the Hidden Land: State Formation and the Construction of Sikkimese History*. Leiden: Brill.

Mumford, Stan Royal. 1989. *Himalayan Dialogue: Tibetan Lamas and Gurung Shamans in Nepal*. Madison: University of Wisconsin Press.

Myers, Norman, Russell A. Mittermeier, Cristina G. Mittermeier, et al. 2000. "Biodiversity Hotspots for Conservation Priorities." *Nature* 403 (6772): 853–858.

Nadasdy, Paul. 1999. "The Politics of Tek: Power and the "Integration" of Knowledge." *Arctic Anthropology* 36 (1–2): 1–18.

Nanson, G. C., and J. C. Croke. 1992. "A Genetic Classification of Floodplains." *Geomorphology* 4 (6): 459–486.

National Environment Commission. 2016. *National Integrated Water Resources Management Plan 2016*. Thimphu: National Environment Commission, Royal Government of Bhutan.

National Energy Administration (NEA). 2021. *Pump Storage Hydropower Development Plan (2021–2035)*. Beijing: National Energy Administration, People's Republic of China.

National Environment Commission. 2020. *Climate Change Policy of the Kingdom of Bhutan 2020*. Thimphu: National Environment Commission, Royal Government of Bhutan.

National Environment Commission. 2015. *Bhutan Water Vision and Bhutan Water Policy*. Thimphu: National Environment Commission.

Nayak, M. A., Azam, M. F., and Lyngwa, R. V. 2021. "ERA5-based Database of Atmospheric Rivers over Himalayas," *Earth System Science Data Discussions* [preprint], https://doi.org/10.5194/essd-2020-397.

Nazarea, Virginia D. 2006. "Local Knowledge and Memory in Biodiversity Conservation." *Annual Review of Anthropology* 35 (1): 317–335.

Nebesky-Wojkowitz, René de. 1996. *Oracles and Demons of Tibet: The Cult and Iconography of the Tibetan Protective Deities (reprint)*. Kathmandu: Book Faith India.

Olson, Richard G. 2015. *Scientism and Technocracy in the Twentieth Century: The Legacy of Scientific Management*. London: Lexington Books.

Orlove, Ben. 2016. "Two Days in the Life of a River: Glacier Floods in Bhutan." *Anthropologica* 58 (2016): 227–242.

Ortlieb, Luc. 2000. "The Documented Historical Record of El Nino Events in Peru: An Update of the Quinn Record (Sixteenth through Nineteenth Centuries)." In *El Niño and the Southern Oscillation: Multiscale Variability and Global and Regional Impacts*, edited by Henry F. Diaz and Vera Barkgraf, 207–295. Cambridge: Cambridge University Press.

Ouimet, William B., Kelin X. Whipple, Leigh H. Royden, et al. 2007. "The Influence of Large Landslides on River Incision in a Transient Landscape: Eastern Margin of the Tibetan Plateau (Sichuan, China)." *GSA Bulletin* 119 (11–12): 1462–1476.

Owen, Lewis A., Marc W. Caffee, Robert C. Finkel, et al. 2008. "Quaternary Glaciation of the Himalayan–Tibetan Orogen." *Journal of Quaternary Science* 23 (6–7): 513–531.

Owen, Lewis A., and Jason M. Dortch. 2014. "Nature and timing of Quaternary glaciation in the Himalayan–Tibetan orogen." *Quaternary Science Reviews* 88: 14–54.

Owen, Lewis A., Robert C. Finkel, and Marc W. Caffee. 2002. "A Note on the Extent of Glaciation throughout the Himalaya during the Global Last Glacial Maximum." *Quaternary Science Reviews* 21 (1): 147–157.

Palumbo, Luigi, Ralf Hetzel, Mingxin Tao, et al. 2009. "Deciphering the Rate of Mountain Growth during Topographic Presteady State: An Example from the NE Margin of the Tibetan Plateau." *Tectonics* 28 (4): TC4017.

Pan, Xingliang, L Xu, Z Yang, et al. 2017. "Payments for Ecosystem Services in China: Policy, Practice, and Progress." *Journal of Cleaner Production* 158 (1): 200–208.

Pandit, Chetan, and Asit K. Biswas. 2019. "India's National Water Policy: 'Feel Good' Document, Nothing More." *International Journal of Water Resources Development* 35 (6): 1015–1028.

Parvin, Fahmida, and Shafi M. Tareq. 2021. "Impact of Landfill Leachate Contamination on Surface and Groundwater of Bangladesh: A Systematic Review and Possible Public Health Risks Assessment." *Applied Water Science* 11 (6): 100.

Parween, Fakeha, and Ajai Singh. 2021. "Implementation of National Water Policy by Eastern States of India—A Review and Analysis." *World Water Policy* 7 (1): 63–73.

Patton, Annette I., Sara L. Rathburn, and Denny M. Capps. 2019. "Landslide Response to Climate Change in Permafrost Regions." *Geomorphology* 340: 116–128.

Payne, Ashley E., Marie-Estelle Demory, L. Ruby Leung, et al. 2020. "Responses and Impacts of Atmospheric Rivers to Climate Change." *Nature Reviews Earth & Environment* 1 (3): 143–157.

Pederson, Neil, Amy E. Hessl, Nachin Baatarbileg, et al. 2014. "Pluvials, Droughts, the Mongol Empire, and Modern Mongolia." *Proceedings of the National Academy of Sciences* 111 (12): 4375–4379.

People's Government of the Xizang Autonomous Region. 2013. *Nature Reserves in Xizang Autonomous Region (2012)*. Lhasa: People's Government of the Xizang Autonomous Region.

People's Government of the Xizang Autonomous Region. 2021. *Fourteenth Five-Year Plan of Social and Economical Development*. Lhasa: People's Government of the Xizang Autonomous Region.

Perdue, Peter C. 1987. *Exhausting the Earth: State and Peasant in Hunan, 1500–1850*. Cambridge, MA: Harvard University Press.

Perdue, Peter C. 1998. "Comparing Empires: Manchu Colonialism." *The International History Review* 20 (2): 255–262.

Pereira, Winin. 1992. "The Sustainable Lifestyle of the Warlis." *India International Centre Quarterly* 19 (1/2): 188–204.

Pfeffer, W. Tad, Anthony A. Arendt, Andrew Bliss, et al. 2014. "The Randolph Glacier Inventory: A Globally Complete Inventory of Glaciers." *Journal of Glaciology* 60 (221): 537–552.

Phuntsho, Tshering. 2016. "History of Black-Necked Crane Festival in Bhutan." *Royal Society for Protection of Nature*, March 15. https://www.rspnbhutan.org/history-of-black-necked-crane-festival-in-bhutan/.

Phuntsho, Tshering, and Jigme Tshering. 2014. "Black-Necked Crane (*Grus nigricollis*) in Bhutan: Current Population Status and Conservation Initiatives." *Zoological Research* 35 (S1): 10–19.

Pietz, David A. 2015. *The Yellow River: The Problem of Water in Modern China*. Cambridge, MA: Harvard University Press.

Ping, Yan, Dong Zhibao, Dong Guangrong, et al. 2001. "Preliminary Results of Using137cs to Study Wind Erosion in the Qinghai-Tibet Plateau." *Journal of Arid Environments* 47 (4): 443–452.

Pirie, Fernanda. 2013. "Law and Religion in Historic Tibet." In *Religion in Disputes: Pervasiveness of Religious Normativity in Disputing Processes*, edited by Franz von Benda-Beckmann, Keebet von Benda-Beckmann, Martin Ramstedt, et al., 231–247. New York: Palgrave Macmillan.

Pitt, Mark M., Mark R. Rosenzweig, and M. Nazmul Hassan. 2021. "Identifying the Costs of a Public Health Success: Arsenic Well Water Contamination and Productivity in Bangladesh." *The Review of Economic Studies* 88 (5): 2479–2526.

Pittock, J., M. Thieme, E. Blom, et al. 2018. "Freshwater Protected Area Corridors." In *Freshwater Ecosystems in Protected Areas: Conservation and Management*, edited by C. Max Finlayson, Angela H. Arthington, and Jamie Pittock, 177–189. London: Routledge.

Pittock, Jamie. 2015. "Biodiversity and the Climate, Energy and Water Nexus." In *Climate, Energy and Water*, edited by Jamie Pittock, Karen Hussey, and Stephen Dovers, 283–302. Cambridge: Cambridge University Press.

Pittock, Jamie. 2019. "Pumped-Storage Hydropower: Trading off Environmental Values?" *Australian Environment Review* 33 (10): 195–200.

Pittock, Jamie. 2021. "Hydropower." In *Transitioning to a Prosperous, Resilient and Carbon-Free Economy: A Guide for Decision-Makers*, edited by Kenneth G. H. Baldwin, Mark Howden, Michael H. Smith, et al., 125–138. Cambridge: Cambridge University Press.

Pittock, Jamie, David Dumaresq, and Stuart Orr. 2017. "The Mekong River: Trading Off Hydropower, Fish, and Food." *Regional Environmental Change* 17 (8): 2443–2453.

Pittock, Jamie, Max Finlayson, Angela H. Arthington, et al. 2015b. "Managing Freshwater, River, Wetland and Estuarine Protected Areas." In *Protected Area Governance and Management*, edited by G. L. Worboys, M. Lockwood, A. Kothari, et al., 569–608. Canberra: ANU Press.

Pittock, Jamie, Colin Finlayson, Alex Gardner, et al. 2010. "Changing Character: The Ramsar Convention on Wetlands and Climate Change in the Murray-Darling Basin, Australia." *Environmental and Planning Law Journal* 27 (6): 401–425.

Plumwood, Val. 1993. *Feminism and the Mastery of Nature*. London: Routledge.

Pomeranz, Kenneth. 2009. "The Great Himalayan Watershed: Water Shortages, Mega-Projects and Environmental Politics in China, India, and Southeast Asia." *Asia–Pacific Journal: Japan Focus* 7 (30): 1–29.

Pomeranz, Kenneth. 2017. "Water, Energy, and Politics: Chinese Industrial Revolutions in Global Environmental Perspective." In *Economic Development and Environmental History in the Anthropocene: Perspectives on Asia and Africa*, edited by Gareth Austin, 271–290. London: Bloomsbury.

Powers, John. 2007. *Introduction to Tibetan Buddhism*. Boulder, CO: Shambhala Publications.

Pradhan, Neera Shrestha, Partha Jyoti Das, Nishikant Gupta, et al. 2021. "Sustainable Management Options for Healthy Rivers in South Asia: The Case of Brahmaputra." *Sustainability* 13 (3): 1087.

Prajapati, Shakti. 2021. "Stranding Cases of Endangered Ganges River Dolphins in the Ghaghara–Sharada Irrigation Canals, Ganges River Basin, India: Conservation Implications." *Mammalia* 85 (1): 39–46.

Prosser, Diann J., Peng Cui, John Y. Takekawa, et al. 2011. "Wild Bird Migration across the Qinghai-Tibetan Plateau: A Transmission Route for Highly Pathogenic H5N1." *PLOS ONE* 6 (3): e17622.

Qadir, Lieutenant Colonel Hammad. 2008. "Water—A Source of Conflict in South Asia." *NDU Journal* 2: 141–192.

Qi, Jianfeng Jeffrey, and Peter Dauvergne. 2022. "China and the Global Politics of Nature-Based Solutions." *Environmental Science & Policy* 137: 1–11.

Qiu, J. 2008. "China: The Third Pole." *Nature* 454 (7203): 393–396.

Qu, B., J. Ming, S. C. Kang, et al. 2014. "The Decreasing Albedo of the Zhadang Glacier on Western Nyainqentanglha and the Role of Light-Absorbing Impurities." *Atmospheric Chemistry and Physics* 14 (20): 11117–11128.

Ralph, F. M., and M. D. Dettinger. 2012. "Historical and National Perspectives on Extreme West Coast Precipitation Associated with Atmospheric Rivers during December 2010." *Bulletin of the American Meteorological Society* 93 (6): 783–790.

Ramble, Charles. 2008. *The Navel of the Demoness: Tibetan Buddhism and Civil Religion in Highland Nepal*. New York: Oxford University Press.

Ramble, Charles. 2013. "Both Fish and Fowl? Preliminary Reflections on Some Representations of a Tibetan Mirror-World." In *Nepalica-Tibetica: Festgabe for Christoph Cüppers*, edited by Franz-Karl Ehrhard, and Petra Maurer, 75–90. Halle: Prof. Schuch Securities GmbH.

RAMSAR. 2018. Strategic framework and guidelines for the future development of the List of Wetlands of International Importance of the Convention on Wetlands (Ramsar, Iran, 1971)—2018 update. Gland: Ramsar Convention on Wetlands.

RAMSAR. 2023. "The Convention on Wetlands." *RAMSAR*. Accessed 19 November 2023. https://www.ramsar.org/.

RAMSAR CONVENTION. 2021. *Ramsar Sites Information Service*. Gland: Ramsar Convention on Wetlands.

Rang byung rdo rje. 2006. "Rang byung rdo rje'i tshigs bcad ma." In *Rang byung rdo rje'i gsung 'bum*, Vol. Nga, 377–414. Mtshur phu: Mkhan po lo yag bkra shis.

Rao, Mukund P., Edward R. Cook, Benjamin I. Cook, et al. 2020. "Seven Centuries of Reconstructed Brahmaputra River Discharge Demonstrate Underestimated High Discharge and Flood Hazard Frequency." *Nature Communications* 11 (1): 6017.

Rasmusson, Eugene M., and John M. Wallace. 1983. "Meteorological Aspects of the El Nino/Southern Oscillation." *Science* 222 (4629): 1195–1202.

Raymo, M. E., and W. F. Ruddiman. 1992. "Tectonic Forcing of Late Cenozoic Climate." *Nature* 359 (6391): 117–122.

Rees, H. Gwyn, and David N. Collins. 2006. "Regional Differences in Response of Flow in Glacier-Fed Himalayan Rivers to Climatic Warming." *Hydrological Processes* 20 (10): 2157–2169.

Reinhardt, Anne. 2018. *Navigating Semi-Colonialism: Shipping, Sovereignty, and Nation-Building in China, 1860–1937*. Cambridge, MA: Harvard University Press.

Reinfeld, Marti Ann. 2003. "Tourism and the Politics of Cultural Preservation: A Case Study of Bhutan." *Journal of Public and International Affairs* 14: 1–26.

Relyea, Scott. 2015. "Yokes of Gold and Threads of Silk: Sino-Tibetan Competition for Authority in Early Twentieth Century Kham." *Modern Asian Studies* 49 (4): 963–1009.

Relyea, Scott. 2019. "Settling Authority: Sichuanese Farmers in Early Twentieth-Century Eastern Tibet." In *Frontier Tibet: Patterns of Change in the Sino-Tibetan Borderlands*, edited by Stéphane Gros, 179–216. Amsterdam: Amsterdam University Press.

Relyea, Scott. 2022. “Toward Simla: Proposing Provinces and Claiming Territory on the Tibetan Plateau.” In *Boundaries and Borderlands: A Century after the 1914 Simla Convention*, edited by Alka Acharya, 74–105. Abingdon: Routledge.

Ren, Guoyu. 2000. “Decline of the mid- to Late Holocene Forests in China: Climatic Change or Human Impact?” *Journal of Quaternary Science* 15 (3): 273–281.

Renwick, William H. 1992. “Equilibrium, Disequilibrium, and Nonequilibrium Landforms in the Landscape.” *Geomorphology* 5 (3): 265–276.

Rey, Patrice F., Tristan Salles, Sabin Zahirovic, et al. 2022. “A Glimpse into a Possible Geomorphic Future of Tibet.” *Nature Reviews Earth & Environment* 3 (10): 613–615.

Rhode, David, David B. Madsen, P. Jeffrey Brantingham, et al. 2007. “Yaks, Yak Dung, and Prehistoric Human Habitation of the Tibetan Plateau.” *Developments in Quaternary Sciences* 9: 205–224.

Richards, Alan, and Nirvikar Singh. 2002. “Inter-State Water Disputes in India: Institutions and Policies.” *International Journal of Water Resources Development* 18 (4): 611–625.

Richardson, H. E. 1985. *A Corpus of Early Tibetan Inscriptions.* London: Royal Asiatic Society.

Rikey, Thupten K. 2011. “The Nature-Deities of Tibet: A Discussion on the Tale ‘The Subduing and Putting under Oath of Tibet’s Malignant lha’ dre’ in Padma bka’ thang.” *Studia Orientalia Electronica* 109: 119–140.

Rikiishi, Kunio, and Haruka Nakasato. 2006. “Height Dependence of the Tendency for Reduction in Seasonal Snow Cover in the Himalaya and the Tibetan Plateau Region, 1966–2001.” *Annals of Glaciology* 43: 369–377.

Rnam par dag pa’i ’bum bzhi las rin po che gtsang ma’i klu ’bum bzhugs so. 1999. In *Bon Canon*, edited by Mongyal Lhasey, Vol. 139. Chengdu: Xizang zangwen guji chubanshe.

Robinson, Ruth A. J., Cynthia A. Brezina, Randall R. Parrish, et al. 2014. “Large Rivers and Orogens: The Evolution of the Yarlung Tsangpo–Irrawaddy System and the Eastern Himalayan Syntaxis.” *Gondwana Research* 26 (1): 112–121.

Rockhill, William Woodville. 1891. *The Land of the Lamas: Notes of a Journey through China, Mongolia and Tibet.* New York: Century.

Romain, William F. 2021. “Subduing the Demons of Tibet: Geomantic Magic during the Yarlung Dynasty: A Landscape Archaeology Assessment.” *Time and Mind* 14 (1): 33–71.

Rowan, Ann V. 2017. “The ‘Little Ice Age’ in the Himalaya: A Review of Glacier Advance Driven by Northern Hemisphere Temperature Change.” *The Holocene* 27 (2): 292–308.

Royal Government of Bhutan. 1986. *6th Five Year Plan.* Thimphu: Gross National Happiness Commission, Royal Government of Bhutan.

Royal Government of Bhutan. 2018a. *12th Five Year Plan.* Thimphu: Gross National Happiness Commission, Royal Government of Bhutan.

Royal Government of Bhutan. 2018b. *National Report to Ramsar COP13 Royal Government of Bhutan.* Gland: Ramsar Convention on Wetlands.

Ruffo, Susan, and Peter M. Kareiva. 2009. “Using Science to Assign Value to Nature.” *Frontiers in Ecology and the Environment* 7 (1): 3–3.

Rugenstein, Jeremy K. Caves, Daniel E. Ibarra, and Friedhelm von Blanckenburg. 2019. “Neogene Cooling Driven by Land Surface Reactivity Rather Than Increased Weathering Fluxes.” *Nature* 571 (7763): 99–102.

Rytwinski, Trina, Lisa A. Kelly, Lisa A. Donaldson, et al. 2021. "What Evidence Exists for Evaluating the Effectiveness of Conservation-Oriented Captive Breeding and Release Programs for Imperilled Freshwater Fishes and Mussels?" *Canadian Journal of Fisheries and Aquatic Sciences* 78 (9): 1332–1346.

Sa skya bsod nams rgyal mtshan gyis brtsamsb. 2002. *rGyal rabs gsal ba'i me long*. Pe cin: Mi rigs dpe skrun khang.

Saikia, Arupjyoti. 2009. "Kaziranga National Park: Dynamics of Social and Political History." *Conservation & Society* 7 (2): 113–129.

Saikia, Arupjyoti. 2019. *The Unquiet River: A Biography of the Brahmaputra*. New Delhi: Oxford University Press.

Saito, Kohei. 2021. "Primitive Accumulation as the Cause of Economic and Ecological Disaster." In *Rethinking Alternatives with Marx: Economy, Ecology and Migration*, edited by Marcello Musto, 93–112. Cham: Springer International Publishing.

Saji, N. H., B. N. Goswami, P. N. Vinayachandran, et al. 1999. "A Dipole Mode in the Tropical Indian Ocean." *Nature* 401 (6751): 360–363.

Salles, Tristan, Laurent Husson, Patrice Rey, et al. 2023. "Hundred Million Years of Landscape Dynamics from Catchment to Global Scale." *Science* 379 (6635): 918–923.

Samaranayake, Nilanthi, Satu Limaye, and Joel Wuthnow. 2016. *Water Resource Competition in the Brahmaputra River Basin: China, India, and Bangladesh*. Arlington: CNA Analysis & Solutions.

Samuel, Geoffrey. 1993. *Civilized Shamans: Buddhism in Tibetan Societies*. Washington, DC: Smithsonian Institution Press.

Samuel, Geoffrey. 2014. "Body and Mind in Tibetan Medicine and Tantric Buddhism." In *Bodies in Balance: The Art of Tibetan Medicine*, edited by Theresia Hofer, 32–45. Seattle: University of Washington Press.

Sangs rgyas rgya mtsho. 1997. *Phugs lugs rtsis kyi legs bshad bai ḍūra dkar po, stod cha and smad cha*. Sining: Krung go'i bod kyi shes rigs dpe skrun khang.

Sarkar, Soma. 2022. "Drought and Flood Dynamics of Godavari Basin, India: A Geospatial Perspective." *Arabian Journal of Geosciences* 15 (8): 772.

Sattar, Ashim, Ajanta Goswami, Anil V. Kulkarni, et al. 2021. "Future Glacial Lake Outburst Flood (GLOF) Hazard of the South Lhonak Lake, Sikkim Himalaya." *Geomorphology* 388: 107783.

Schellert, W. P., Z. Chen, V. Strak, et al. 2019. "Pacific Subduction Control on Asian Continental Deformation Including Tibetan Extension and Eastward Extrusion Tectonics." *Nature Communications* 10 (1): 4480.

Schmidt, Jennifer L., Peter K. Zeitler, Frank J. Pazzaglia, et al. 2015. "Knickpoint Evolution on the Yarlung River: Evidence for Late Cenozoic Uplift of the Southeastern Tibetan Plateau Margin." *Earth and Planetary Science Letters* 430: 448–457.

Schwanghart, Wolfgang, Raphael Worni, Christian Huggel, et al. 2016. "Uncertainty in the Himalayan Energy–Water Nexus: Estimating Regional Exposure to Glacial Lake Outburst Floods." *Environmental Research Letters* 11 (7): 074005.

Schwieger, Peter. 2023. "The Tax System in Central and Far East Tibet towards the End of the Ganden Phodrang Reign: An Outline of Its Structure and Terminology." In *Taxation in Tibetan Societies: Rules, Practices and Discourses*, edited by Alice Travers, Peter Schwieger, and Charles Ramble, 11–53. Leiden: Brill.

Scott, James C. 1998. *Seeing Like a State: How Certain Schemes to Improve the Human Condition Have Failed*. New Haven: Yale University Press.

Scott, James C. 2009. *The Art of Not Being Governed: An Anarchist History of Upland Southeast Asia*. New Haven, CT: Yale University Press.

Scott, James C. 2017. *Against the Grain: A Deep History of the Earliest States*. New Haven, CT: Yale University Press.

Searle, Mike. 2017. *Colliding Continents: A Geological Exploration of the Himalaya, Karakoram, and Tibet*. Oxford: Oxford University Press.

Seeber, L., and V. Gornitz. 1983. "River Profiles along the Himalayan Arc as Indicators of Active Tectonics." *Tectonophysics* 92 (4): 335–367.

Sen, Sudipta. 2019. *Ganges: The Many Pasts of an Indian River*. New Haven: Yale University Press.

Seong, Yeong Bae, Lewis A. Owen, Chaolu Yi, et al. 2009. "Geomorphology of Anomalously High Glaciated Mountains at the Northwestern End of Tibet: Muztag Ata and Kongur Shan." *Geomorphology* 103 (2): 227–250.

Shah, Tushaar. 2011. "Past, Present, and the Future of Canal Irrigation in India." In *India Infrastructure Report*, edited by Infrastructure Development Finance Company, 70–87. New Delhi: Oxford University Press.

Shakabpa, Tsepon Wangchuck. 2009. *One Hundred Thousand Moons: An Advanced Political History of Tibet*. Leiden: Brill.

Shakya, Tsering. 1986. "The Making of the Great Game Players: Tibetan Students in England 1912–1916." *Tibetan Review* XXI (1): 12–17.

Shalehin, Md. Shoyeb, Md. Taskin Parvez, Martyn C. Lucas, et al. 2022. "A Case Study of Illegal Fishing Causes during Seasonal Fishery Closure in Kaptai Lake, Bangladesh." *Fisheries Management and Ecology* 29 (5): 542–551.

Shan, Jie, and Xiaoyi Lin. 2020. "China to Build Historic Yarlung Zangbo River Hydropower Project in Tibet." *Global Times*, November 29.

Shang, Yanjun, Zhifa Yang, Lihui Li, et al. 2003. "A Super-Large Landslide in Tibet in 2000: Background, Occurrence, Disaster, and Origin." *Geomorphology* 54 (3): 225–243.

Sharma, Aashna, Vineet Kumar Dubey, Jeyaraj Antony Johnson, et al. 2021. "Is There Always Space at the Top? Ensemble Modeling Reveals Climate-Driven High-Altitude Squeeze for the Vulnerable Snow Trout Schizothorax Richardsonii in Himalaya."*Ecological Indicators* 120: 106900.

Sharma, Kishor, Bhoj Kumar Acharya, Ghanashyam Sharma, et al. 2020. "Land Use Effect on Butterfly Alpha and Beta Diversity in the Eastern Himalaya, India." *Ecological Indicators* 110: 105605. doi: 10.1016/j.ecolind.2019.105605.

Sharma, Milap Chand, and Lewis A. Owen. 1996. "Quaternary Glacial History of NW Garhwal, Central Himalayas." *Quaternary Science Reviews* 15 (4): 335–365.

Sharma, Narpati, Mukunda Dev Behera, Abhaya Prasad Das, et al. 2019. "Plant Richness Pattern in an Elevation Gradient in the Eastern Himalaya." *Biodiversity and Conservation* 28 (8): 2085–2104.

Shen, Wei, and Dong Jiang. 2021. "Making Authoritarian Environmentalism Accountable? Understanding China's New Reforms on Environmental Governance." *The Journal of Environment & Development* 30 (1): 41–67.

Shen, Xiaoming, Yuntao Tian, Ying Wang, et al. 2021. "Enhanced Quaternary Exhumation in the Central Three Rivers Region, Southeastern Tibet." *Frontiers in Earth Science* 9: 741491.

Shi, Feng, Keyan Fang, Chenxi Xu, et al. 2017. "Interannual to Centennial Variability of the South Asian Summer Monsoon over the Past Millennium." *Climate Dynamics* 49 (7): 2803–2814.

Shi, Feng, Anmin Duan, Qiuzhen Yin, et al. 2021. "Modulation of the Relationship between Summer Temperatures in the Qinghai–Tibetan Plateau and Arctic over the Past Millennium by External Forcings." *Quaternary Research* 103: 130–138.

Shi, Feng, Honglin He, Yiduo Liu, et al. 2022a. "Active Tectonics of the Nantinghe Fault in Southeastern Tibetan Plateau and Its Implications for Continental Collision." *Frontiers in Earth Science* 9: 818225.

Shi, Jiangfeng, Huayu Lu, Jinbao Li, et al. 2015. "Tree-Ring Based February–April Precipitation Reconstruction for the Lower Reaches of the Yangtze River, Southeastern China." *Global and Planetary Change* 131: 82–88.

Shi, Tong, Jinyun Guo, Haoming Yan, et al. 2022b. "Assessing Height Variations in Qinghai–Tibet Plateau from Time-Varying Gravity Data and Hydrological Model." *Remote Sensing* 14 (19): 4707.

Shi, Yafeng, Shiyin Liu, Donghui Shangguan, et al. 2006. "Peculiar Phenomena Regarding Climatic and Glacial Variations on the Tibetan Plateau." *Annals of Glaciology* 43: 106–110.

Shrestha, Arun Bhakta, Nand Kishor Agrawal, Björn Alfthan, et al. 2015. *The Himalayan Climate and Water Atlas: Impact of Climate Change on Water Resources in Five of Asia's Major River Basins*. Kathmandu: International Centre for Integrated Mountain Development (ICIMOD).

Shukla, Tanuj, and Indra S. Sen. 2021. "Preparing for Floods on the Third Pole." *Science* 372 (6539): 232–234.

Sima, Qian. 2009. *Xin bai hua Shi ji*. Beijing: Zhonghua shu ju.

Simmonds, Ian, Daohua Bi, and Pandora Hope. 1999. "Atmospheric Water Vapor Flux and Its Association with Rainfall over China in Summer." *Journal of Climate* 12 (5): 1353–1367.

Simpson, Thomas. 2021. *The Frontier in British India: Space, Science, and Power in the Nineteenth Century*. Cambridge: Cambridge University Press.

Simpson, Thomas. 2023. "Find the River: Discovering the Tsangpo-Brahmaputra in the Age of Empire." *Modern Asian Studies* 58 (1): 127–162.

Sinha, Ashish, Max Berkelhammer, Lowell Stott, et al. 2011. "The Leading Mode of Indian Summer Monsoon Precipitation Variability during the Last Millennium." *Geophysical Research Letters* 38 (15): L15703.

Singh, Riddhi, and Rohini Kumar. 2021. "The Suitability of Water Scarcity Indicators to the Indian Context." *Water Security* 14: 100097.

Smith, Brian D., A. K. M. Aminul Haque, M. Shakhawat Hossain, et al. 1998. "River Dolphins in Bangladesh: Conservation and the Effects of Water Development." *Environmental Management* 22 (3): 323–335.

Smith, Brian D., Benazir Ahmed, Muhammad Edrise Ali, et al. 2001. "Status of the Ganges River Dolphin or Shushuk Platanista Gangetica in Kaptai Lake and the Southern Rivers of Bangladesh." *Oryx* 35 (1): 61–72.

Song, Chunqiao, Yongwei Sheng, Jida Wang, et al. 2017. "Heterogeneous Glacial Lake Changes and Links of Lake Expansions to the Rapid Thinning of Adjacent Glacier Termini in the Himalayas." *Geomorphology* 280: 30–38.

Song, Chunqiao, Yongwei Sheng, Linghong Ke, et al. 2016. "Glacial Lake Evolution in the Southeastern Tibetan Plateau and the Cause of Rapid Expansion of Proglacial Lakes Linked to Glacial-Hydrogeomorphic Processes." *Journal of Hydrology* 540: 504–514.

Song, Ling-Ling, Qing Tian, and Zong-Jie Li. 2020. "Has Wind Erosion in the Source Region of the Yangtze River Been Strengthened?" *Polish Journal of Environmental Studies* 29 (2): 1351–1359.

Sørensen, Per K. 1994. *The Mirror Illuminating the Royal Genealogies: An Annotated Translation of the XIVth Century Tibetan Chronicle, Rgyal rabs gsal ba'i me long*. Berlin: Verlag Asiatische Forschungen.

Sørensen, Per K., and Guntram Hazod. 2007. *Rulers on the Celestial Plain. Ecclesiastic and Hegemonic Rule in Central Tibet*. Verlag der Österreichischen Akademie der Wissenschaften.

Spate, Michael, Guilin Zhang, Mumtaz Yatoo, et al. 2017. "New Evidence for Early 4th Millennium BP Agriculture in the Western Himalayas: Qasim Bagh, Kashmir." *Journal of Archaeological Science: Reports* 11: 568–577.

Sperling, Elliot. 1987. "Lama to the King of Hsia." *The Journal of the Tibet Society* 7: 31–50.

Sperling, Elliot. 1994. "Rtsa-mi lo-tsā-ba Sangs-rgyas Grags-pa and the Tangut Background to Early Mongol-Tibetan Relations." In *Tibetan Studies: Proceedings of the 6th Seminar of the International Association for Tibetan Studies*, edited by Per Kværne, 801–824. Oslo: The Institute for Comparative Research in Human Culture.

Spicer, R. A., Tao Su, Paul J. Valdes, et al. 2021. "Why 'the Uplift of the Tibetan Plateau' Is a Myth." *National Science Review* 8 (1): nwaa091.

Srivathsa, A., Vasudev, D., Nair, T. et al. 2023. "Prioritizing India's Landscapes for Biodiversity, Ecosystem Services and Human Well-Being." *Nature Sustainability* 6: 568–577.

Srivastava, P., A. Kumar, S. Chaudhary, et al. 2017. "Paleofloods Records in Himalaya." *Geomorphology* 284: 17–30.

Staisch, Lydia M., Nathan A. Niemi, Marin K. Clark, et al. 2020. "The Cenozoic Evolution of Crustal Shortening and Left-Lateral Shear in the Central East Kunlun Shan: Implications for the Uplift History of the Tibetan Plateau." *Tectonics* 39 (9): e2020TC006065.

Stearns, Cyrus. 2002. The Buddha from Dolpo: *A Study of the Life and Thought of the Tibetan Master Dolpopa Sherab Gyaltsen*. Delhi: Motilal Banarsidass Publishers.

Steffen, Will, Wendy Broadgate, Lisa Deutsch, et al. 2015. "The Trajectory of the Anthropocene: The Great Acceleration." *The Anthropocene Review* 2 (1): 81–98.

Stengers, Isabelle. 2018. *Another Science is Possible: A Manifesto for Slow Science*. Cambridge: Polity Press.

Stokes, Maya F., and J. Taylor Perron. 2020. "Modeling the Evolution of Aquatic Organisms in Dynamic River Basins." *Journal of Geophysical Research: Earth Surface* 125 (9): e2020JF005652.

Strang, Veronica. 2005. "Common Senses: Water, Sensory Experience and the Generation of Meaning." *Journal of Material Culture* 10 (1): 92–120.

Stroeven, A. P., C. Hättestrand, J. Heyman, et al. 2009. "Landscape Analysis of the Huang He Headwaters, NE Tibetan Plateau—Patterns of Glacial and Fluvial Erosion." *Geomorphology* 103 (2): 212–226.

Su, Qi, Xianyan Wang, Huayu Lu, et al. 2022. "River Piracy and Its Geomorphic Effects in the Northern Qilian Shan, Northeastern Qinghai–Tibet Plateau." *Palaeogeography, Palaeoclimatology, Palaeoecology* 601: 111147.

Sullivan, Sian. 2011. "Banking Nature? The Financialisation of Environmental Conservation." *Open Anthropology Cooperative Press*, March 11.

Sun, Jiaopeng, Yunpeng Dong, Licheng Ma, et al. 2021. "Devonian to Triassic Tectonic Evolution and Basin Transition in the East Kunlun–Qaidam Area, Northern Tibetan Plateau: Constraints from Stratigraphy and Detrital Zircon U–Pb Geochronology." *GSA Bulletin* 134 (7–8): 1967–1993.

Talley, L. D., G. L. Pickard, W. J. Emery, et al. 2011. *Descriptive Physical Oceanography: An Introduction*. London: Academic Press.

Tan, Gillian G. 2014. "An Ecology of Religiosity: Re-Emphasizing Relationships between Humans and Nonhumans." *Journal for the Study of Religion, Nature and Culture* 8 (3): 307–328.

Tan, Gillian G. 2016. "'Life' and 'freeing life' (*tshe thar*) among Pastoralists of Kham: Intersecting Religion and Environment." Études *mongoles et sibériennes, centrasiatiques et tibétaines [En ligne]* 47.

Tan, Gillian G. 2020. "Smoky Relations: Beyond Dichotomies of Substance on the Tibetan Plateau." In *Exploring Materiality and Connectivity in Anthropology and beyond*, edited by Philipp Schorch, Martin Saxer, and Marlen Elders, 145–161. London: University College London Press.

Tan, Liangcheng, Yanjun Cai, Zhisheng An, et al. 2017. "Decreasing Monsoon Precipitation in Southwest China during the Last 240 Years Associated with the Warming of Tropical Ocean." *Climate Dynamics* 48 (5): 1769–1778.

Tan, Ziyuan, Yuzhi Liu, Qingzhe Zhu, et al. 2021. "Impact of Massive Topography on the Dust Cycle Surrounding the Tibetan Plateau." *Atmospheric Environment* 264: 118703.

Tao, Yaling, Jianguo Xiong, Huiping Zhang, et al. 2020. "Climate-Driven Formation of Fluvial Terraces across the Tibetan Plateau since 200 ka: A Review." *Quaternary Science Reviews* 237: 106303.

Tapponnier, P., G. Peltzer, A. Y. Le Dain, et al. 1982. "Propagating Extrusion Tectonics in Asia: New Insights from Simple Experiments with Plasticine." *Geology* 10 (12): 611–616.

Tapponnier, Paul, Xu Zhiqin, Françoise Roger, et al. 2001. "Oblique Stepwise Rise and Growth of the Tibet Plateau." *Science* 294 (5547): 1671–1677.

Tariq, Muhammad Atiq Ur Rehman, Kelden Wangchuk, and Nitin Muttil. 2021. "A Critical Review of Water Resources and Their Management in Bhutan." *Hydrology* 8 (1): 31.

Tashi, Kelzang T. 2021. "The (un)Changing Karma: Pollution Beliefs, Social Stratification and Reincarnisation in Bhutan." *The Asia Pacific Journal of Anthropology* 22 (1): 41–57.

Tashi, Kelzang T. 2023. *World of Worldly Gods: The Persistence and Transformation of Shamanic Bon in Buddhist Bhutan*. New York: Oxford University Press.

Taylor, Caroline, Tom R. Robinson, Stuart Dunning, et al. 2023. "Glacial Lake Outburst Floods Threaten Millions Globally." *Nature Communications* 14 (1): 487.

Taylor, Michael, Adam Forte, Andrew Laskowski, et al. 2021a. "Active Uplift of Southern Tibet Revealed." *GSA Today* 31: 4–10.

Taylor, William T. T., Mélanie Pruvost, Cosimo Posth, et al. 2021b. "Evidence for Early Dispersal of Domestic Sheep into Central Asia." *Nature Human Behaviour* 5 (9): 1169–1179.

Teng, Yanmin, Jinyan Zhan, Wei Liu, et al. 2021. "Spatiotemporal Dynamics and Drivers of Wind Erosion on the Qinghai–Tibet Plateau, China." *Ecological Indicators* 123: 107340.

Tenpa, Lobsang. 2018. *An Early History of the Mon Region (India) and Its Relationship with Tibet and Bhutan*. Dharamshala: Library of Tibetan Works and Archives.

Tenzin, Jigme, and Phub Dhendup. 2023. "Habitat Characteristics, Relative Abundance and Conservation Threats of Himalayan Bull Frogs (*Nanorana leibigii*

Günther, 1860) in Primary Tributaries of Simkhar River, Bhutan." *Bhutan Journal of Natural Resources and Development* 4 (2): 29–38.

Thapa, Kritika, Theodore A. Endreny, and Craig R. Ferguson. 2018. "Atmospheric Rivers Carry Nonmonsoon Extreme Precipitation into Nepal." *Journal of Geophysical Research: Atmospheres* 123 (11): 5901–5912.

Thompson, David W. J., and John M. Wallace. 1998. "The Arctic Oscillation Signature in the Wintertime Geopotential Height and Temperature Fields." *Geophysical Research Letters* 25 (9): 1297–1300.

Thompson, L. G., E. Mosley-Thompson, M. E. Davis, et al. 1990. "Glacial Stage Ice-Core Records from the Subtropical Dunde Ice Cap, China." *Annals of Glaciology* 14: 288–297.

Thompson, L. G., T. Yao, M. E. Davis, et al. 1997. "Tropical Climate Instability: The Last Glacial Cycle from a Qinghai–Tibetan Ice Core." *Science* 276 (5320): 1821–1825.

Thompson, L. G., T. Yao, E. Mosley-Thompson, et al. 2011. "Data from: NOAA/WDS Paleoclimatology—Dasuopu Ice Core 1000 Year d18O, Dust, Anion and Accumulation Data." (dataset). NOAA National Centers for Environmental Information.

Thompson, L. G., T. Yao, E. Mosley-Thompson, et al. 2000. "A High-Resolution Millennial Record of the South Asian Monsoon from Himalayan Ice Cores." *Science* 289 (5486): 1916–1919.

Thompson, Lonnie G., Ellen Mosley-Thompson, Mary E. Davis, et al. 2018. "Global-Scale Abrupt Climate Events and Black Swans: An Ice-Core-Derived Palaeoclimate Perspective from Earth's Highest Mountains." In *The Himalayan Cryosphere: Past and Present*, edited by N. C. Pant, R. Ravindra, D. Srivastava, et al., Vol. 462, 7–22. London: Geological Society.

Thompson, Lonnie G., Yao Tandong, Mary E. Davis, et al. 2006. "Holocene Climate Variability Archived in the Puruogangri Ice Cap on the Central Tibetan Plateau." *Annals of Glaciology* 43: 61–69.

Tian, Lide, Tandong Yao, K. MacClune, et al. 2007. "Stable Isotopic Variations in West China: A Consideration of Moisture Sources." *Journal of Geophysical Research: Atmospheres* 112 (D10): D10112.

Tremblay, Marissa M., Matthew Fox, Jennifer L. Schmidt, et al. 2015. "Erosion in Southern Tibet Shut down at ~10 Ma due to Enhanced Rock Uplift within the Himalaya." *Proceedings of the National Academy of Sciences* 112 (39): 12030–12035.

Tsomu, Yudru. 2018. "Women as Chieftains in Modern Kham History." *Inner Asia* 20 (1): 107–131.

Turner, Andrew G., and H. Annamalai. 2012. "Climate Change and the South Asian Summer Monsoon." *Nature Climate Change* 2 (8): 587–595.

Tvedt, Terje. 2019. "Rivers of History." In *Mediterranean Rivers in Global Perspective*, edited by Johannes C. Bernhardt, Markus Koller, and Achim Lichtenberger, 49–68. Paderborn: Brill Schöningh.

UNESCO. 1994. Convention on Wetlands of International Importance especially as Waterfowl Habitat (Ramsar, 2.2.1971) as amended by the Protocol of 3.12.1982 (Paris) and the Amendments of 28.5.1987 (Regina). Paris: UNESCO.

UNESCO World Heritage Centre. 2024. World Heritage List, UNESCO World Heritage Centre. Accessed 10 June 2024. https://whc.unesco.org/en/list/

Vaidya, Ramesh Ananda, David James Molden, Arun Bhakta Shrestha, et al. 2021. "The Role of Hydropower in South Asia's Energy Future." *International Journal of Water Resources Development* 37 (3): 367–391.

Van Schaik, Sam. 2008. "The Golden Turtle: A Sino-Tibetan Divination Manuscript." *EarlyTibet.com*. Accessed 2 June 2024. https://earlytibet.com/2008/11/28/the-golden-turtle/.

Van Schaik, Sam. 2011. "A New Look at the Source of the Tibetan Script." In *New Studies of the Old Tibetan Documents: Philology, History, and Religion*, edited by Yoshiro Imaeda, Matthew Kapstein, and Tsuguhito Takeuchi, 45–96. Tokyo: Research Institute for Languages and Cultures of Asia and Africa, Tokyo University of Foreign Studies.

Van Spengen, Wim. 1995. "The Geo-History of Long-Distance Trade in Tibet 1850–1950." *The Tibet Journal* 20 (2): 18–63.

Van Spengen, Wim. 2000. *Tibetan Border Worlds: A Geo-Historical Analysis of Trade and Traders*. London: Routledge.

Veh, Georg, Oliver Korup, and Ariane Walz. 2020. "Hazard from Himalayan Glacier Lake Outburst Floods." *Proceedings of the National Academy of Sciences* 117 (2): 907–912.

Visbeck, Martin H., James W. Hurrell, Lorenzo Polvani, et al. 2001. "The North Atlantic Oscillation: Past, Present, and Future." *Proceedings of the National Academy of Sciences* 98 (23): 12876–12877.

Viste, E., and A. Sorteberg. 2015. "Snowfall in the Himalayas: An Uncertain Future from a Little-Known Past." *The Cryosphere* 9 (3): 1147–1167.

Viveiros de Castro, Eduardo. 2004. "Perspectival Anthropology and the Method of Controlled Equivocation." *Tipití* 2 (1): 3–22.

Walker, G. T. 1924. *Correlation in Seasonal Variations of Weather, IX: Further Study of World Weather*. Poona: Meteorological Office.

Walker, G. T., and E. W. Bliss. 1932. "World Weather V." *Memoirs of the Royal Meteorological Society* 4: 53–84.

Wallace, John M. 2000. "North Atlantic Oscillatiodannular Mode: Two Paradigms—One Phenomenon." *Quarterly Journal of the Royal Meteorological Society* 126 (564): 791–805.

Wallace, John M., and David S. Gutzler. 1981. "Teleconnections in the Geopotential Height Field during the Northern Hemisphere Winter." *Monthly Weather Review* 109 (4): 784–812.

Walvoord, Michelle A., and Barret L. Kurylyk. 2016. "Hydrologic Impacts of Thawing Permafrost—A Review." *Vadose Zone Journal* 15 (6): 1–20.

Wang, Bai-Qiu, Wei Wang, and Mei-Fu Zhou. 2013a. "Provenance and Tectonic Setting of the Triassic Yidun Group, the Yidun Terrane, Tibet." *Geoscience Frontiers* 4 (6): 765–777.

Wang, Bin, Qing Bao, Brian Hoskins, et al. 2008. "Tibetan Plateau Warming and Precipitation Changes in East Asia." *Geophysical Research Letters* 35 (14): L14702.

Wang, Bin, Jian Liu, Hyung-Jin Kim, et al. 2013b. "Northern Hemisphere Summer Monsoon Intensified by Mega-El Niño/southern Oscillation and Atlantic Multidecadal Oscillation." *Proceedings of the National Academy of Sciences* 110 (14): 5347–5352.

Wang, Gaochun, Hans Thybo, and Irina M. Artemieva. 2021a. "No Mafic Layer in 80 km Thick Tibetan Crust." *Nature Communications* 12 (1): 1069.

Wang, Hongru, Melinda A. Yang, Shargan Wangdue, et al. 2023a. "Human Genetic History on the Tibetan Plateau in the Past 5100 Years." *Science Advances* 9 (11): eadd5582.

Wang, Jianglin, Bao Yang, and Fredrik Charpentier Ljungqvist. 2020a. "Moisture and Temperature Covariability over the Southeastern Tibetan Plateau during the Past Nine Centuries." *Journal of Climate* 33 (15): 6583–6598.

Wang, Juan, Wei Zhou, and Yanjun Guan. 2022a. "Optimization of Management by Analyzing Ecosystem Service Value Variations in Different Watersheds in the Three-River Headwaters Basin." *Journal of Environmental Management* 321: 115956.

Wang, Ninglian, L. G. Thompson, M. E. Davis, et al. 2003. "Influence of Variations in NAO and SO on Air Temperature over the Northern Tibetan Plateau as Recorded by $\delta^{18}O$ in the Malan Ice Core." *Geophysical Research Letters* 30 (22): 2167.

Wang, Ninglian, Tandong Yao, L. G. Thompson, et al. 2002a. "Evidence for Cold Events in the Early Holocene from the Guliya Ice Core, Tibetan Plateau, China." *Chinese Science Bulletin* 47 (17): 1422–1427.

Wang, Ninglian, Tandong Yao, Lonnie G. Thompson, et al. 2002b. "Indian Monsoon and North Atlantic Oscillation Signals Reflected by Cl^- and Na^+ in a Shallow Ice Core from Dasuopu Glacier, Xixabangma, Himalaya." *Annals of Glaciology* 35: 273–277.

Wang, Ninglian, Xi Jiang, Lonnie G. Thompson, et al. 2007. "Accumulation Rates over the Past 500 Years Recorded in Ice Cores from the Northern and Southern Tibetan Plateau, China." *Arctic, Antarctic, and Alpine Research* 39 (4): 671–677.

Wang, Lei, Tandong Yao, Chenhao Chai, et al. 2021b. "TP-River: Monitoring and Quantifying Total River Runoff from the Third Pole." *Bulletin of the American Meteorological Society* 102 (5): E948–E965.

Wang, Li, Fan Zhang, Suhua Fu, et al. 2020b. "Assessment of Soil Erosion Risk and Its Response to Climate Change in the Mid-Yarlung Tsangpo River Region." *Environmental Science and Pollution Research* 27 (1): 607–621.

Wang, Licheng, Lin Ding, Eduardo Garzanti, et al. 2022b. "Mid-Cretaceous Drainage Reorganization and Exorheic to Endorheic Transition in Southeast Tibet." *Sedimentary Geology* 439: 106221.

Wang, Ping, Dirk Scherler, Jing Liu-Zeng, et al. 2014. "Tectonic Control of Yarlung Tsangpo Gorge Revealed by a Buried Canyon in Southern Tibet." *Science* 346 (6212): 978–981.

Wang, Xin, Shiyin Liu, Hongwei Mo, et al. 2011. "Expansion of Glacial Lakes and Its Implication for Climate Changes in the Chinese Himalaya." *Acta Geographica Sinica* 66 (7): 895–904.

Wang, S. 2022. *Effects of Ecological Restoration on Sustainable Output of Fishery Resources in the Yangtze Estuary*. Shanghai: East China Sea Fisheries Researcfh Institute, Chinese Academy of Fishery Sciences.

Wang, S., J. Liu, H. D. Pritchard, et al. 2023b. "Characterizing 4 Decades of Accelerated Glacial Mass Loss in the West Nyainqentanglha Range of the Tibetan Plateau." *Hydrology and Earth System Sciences* 27 (4): 933–952.

Wang, S., L. Zhao, and S. Li. 2002c. "Interaction between Permafrost and Desertification on the Qinghai–Tibet Plateau."*Journal of Desert Research* 22 (1): 33–39.

Wang, Xiaokui, Olusola Bamisile, Shuheng Chen, et al. 2022c. "Decarbonization of China's Electricity Systems with Hydropower Penetration and Pumped-Hydro

Storage: Comparing the Policies with a Techno-Economic Analysis." *Renewable Energy* 196: 65–83.

Wang, Xin, Tenglong Shi, Xueying Zhang, et al. 2020c. "An Overview of Snow Albedo Sensitivity to Black Carbon Contamination and Snow Grain Properties Based on Experimental Datasets across the Northern Hemisphere." *Current Pollution Reports* 6 (4): 368–379.

Wang, Xu, Florian Siegert, Ai-guo Zhou, et al. 2013c. "Glacier and Glacial Lake Changes and Their Relationship in the Context of Climate Change, Central Tibetan Plateau 1972–2010." *Global and Planetary Change* 111: 246–257.

Wang, Xun, Peifang Wang, Chao Wang, et al. 2020d. "Do Bacterioplankton Respond Equally to Different River Regulations? A Quantitative Study in the Single-Dammed Yarlung Tsangpo River and the Cascade-Dammed Lancang River." *Environmental Research* 191: 110194.

Wang, Yanren, Yu Gao, Jishuai Yang, et al. 2021c. "New Evidence for Early Human Habitation in the Nyingchi Region, Southeast Tibetan Plateau." *The Holocene* 31 (2): 240–246.

Wang, Ye, Chunrong Mi, and Yumin Guo. 2020e. "Satellite Tracking Reveals a New Migration Route of Black-Necked Cranes (*Grus nigricollis*) in Qinghai–Tibet Plateau." *PeerJ* 8: e9715.

Wang, Yiru, Yu Gao, Zhengwei Zhang, et al. 2023c. "Human–animal–environment Dynamics and Formation of Pastoralism in the Southern Tibetan Plateau during the Middle–Late Holocene." *Quaternary Research* 114: 30–51.

Wangchuk, Jigme, Dawa Yoezer, Norbu Wangdi, et al. 2017. "Macroinvertebrate and Fish Diversity in Mangdechhu Hydropower Plant, Trongsa Bhutan." *NeBIO* 8 (4): 335–342.

Wangchuk, Pema. 2018. *The Birds Have Lost Their Way: Essays on Hydropower and Climate Change Issues in Sikkim*. Gangtok: Lokta Books.

Wangdu, Pasang, and Hildegard Diemberger. 2000. *Dba' bzhad: The Royal Narrative Concerning the Bringing of the Buddha's Doctrine to Tibet*. Wien: Verlag der Österreichischen Akademie der Wissenshaften.

Wanner, Heinz, Stefan Brönnimann, Carlo Casty, et al. 2001. "North Atlantic Oscillation – Concepts and Studies." *Surveys in Geophysics* 22 (4): 321–381.

Warde, Paul, Libby Robin, and Sverker Sörlin. 2018. *The Environment: A History of the Idea*. Baltimore: Johns Hopkins University Press.

Wasson, R. J. 2003. "A Sediment Budget for the Ganga–Brahmaputra Catchment." *Current Science* 84 (8): 1041–1047.

Wasson, R. J., Y. P. Sundriyal, Shipra Chaudhary, et al. 2013. "A 1000-Year History of Large Floods in the Upper Ganga Catchment, Central Himalaya, India." *Quaternary Science Reviews* 77: 156–166.

Wasson, Robert, Arupjyoti Saikia, Priya Bansal, et al. 2020. "Flood Mitigation, Climate Change Adaptation and Technological Lock-In in Assam." *Ecology, Economy and Society–the INSEE Journal* 3 (2): 83–104.

Webber, Michael, Britt Crow-Miller, and Sarah Rogers. 2017. "The South–North Water Transfer Project: remaking the geography of China." *Regional Studies* 51 (3): 370–382.

Webster, Peter J., Andrew M. Moore, Johannes P. Loschnigg, et al. 1999. "Coupled Ocean–atmosphere Dynamics in the Indian Ocean during 1997–98." *Nature* 401 (6751): 356–360.

Wegerich, Kai, Daniel Van Rooijen, Ilkhom Soliev, et al. 2015. "Water Security in the Syr Darya Basin." *Water* 7 (9): 4657–4684.

Wei, Da, Hui Zhao, Lin Huang, et al. 2020. "Feedbacks of Alpine Wetlands on the Tibetan Plateau to the Atmosphere." *Wetlands* 40 (4): 787–797.

Wei, Feixiang, Julie Prytulak, Jiandong Xu, et al. 2017. "The Cause and Source of Melting for the Most Recent Volcanism in Tibet: A Combined Geochemical and Geophysical Perspective." *Lithos* 288–289: 175–190.

Wei, Ting, Janice Brahney, Zhiwen Dong, et al. 2021. "Hf–Nd–Sr Isotopic Composition of the Tibetan Plateau Dust as a Fingerprint for Regional to Hemispherical Transport." *Environmental Science & Technology* 55 (14): 10121–10132.

Weil, Benjamin. 2006. "Conservation, Exploitation, and Cultural Change in the Indian Forest Service, 1875–1927." *Environmental History* 11 (2): 319–343.

Weng, M. 2012. "The Cause and Evolution of Tibetan Fish-Eating Avoidance" (In Chinese.) M.Sc. Thesis, Lanzhou University.

Wesnousky, Steven G. 2020. "Great Pending Himalaya Earthquakes." *Seismological Research Letters* 91 (6): 3334–3342.

Wester, Philippus, Arabinda Mishra, Aditi Mukherji, et al. 2019. *The Hindu Kush Himalaya Assessment: Mountains, Climate Change, Sustainability and People.* Switzerland: Springer Open.

Whyte, Kyle Powys. 2013. "On the Role of Traditional Ecological Knowledge as a Collaborative Concept: A Philosophical Study." *Ecological Processes* 2 (1): 1–12.

Williams, Mark W., Alana Wilson, Dendup Tshering, et al. 2016. "Using Geochemical and Isotopic Chemistry to Evaluate Glacier Melt Contributions to the Chamkar Chhu (river), Bhutan." *Annals of Glaciology* 57 (71): 339–348.

Wilmsen, Brooke, Michael Webber, and Yuefang Duan. 2011. "Involuntary Rural Resettlement: Resources, Strategies, and Outcomes at the Three Gorges Dam, China." *The Journal of Environment & Development* 20 (4): 355–380.

Winichakul, Thongchai. 1997. *Siam Mapped: A History of the Geo-Body of a Nation.* Honolulu: University of Hawaii Press.

Wittfogel, Karl August. 1957. *Oriental Despotism: A Comparative Study of Total Power.* New Haven: Yale University Press.

Woerd, Jerome Van Der, Paul Tapponnier, Frederick J. Ryerson, et al. 2002. "Uniform Postglacial Slip-Rate along the Central 600 km of the Kunlun Fault (Tibet), from ^{26}Al, ^{10}Be, and ^{14}C Dating of Riser Offsets, and Climatic Origin of the Regional Morphology." *Geophysical Journal International* 148 (3): 356–388.

Woeser. 2009. "Tibet's Water Pollution and China's 'Global Warming'." *High Peaks Pure Earth*, December 21. Accessed 15 July 2021. https://highpeakspureearth.com/tibets-water-pollution-and-chinas-global-warming-by-woeser/.

Wohl, Ellen E. 2020. *Rivers in the Landscape.* Hoboken: John Wiley & Sons.

Wolf, Sebastian G., Ritske S. Huismans, Jean Braun, et al. 2022. "Topography of Mountain Belts Controlled by Rheology and Surface Processes." *Nature* 606 (7914): 516–521.

Wong, Soo Mun Theresa. 2010. "Making the Mekong: Nature, Region, Postcoloniality." PhD Thesis, Ohio State University.

Woodhouse, E., Mills, M.A., McGowan, et al., 2015. "Religious Relationships with the Environment in a Tibetan Rural Community: Interactions and Contrasts with Popular Notions of Indigenous Environmentalism." *Human Ecology* 43: 295–307.

Woodward, Guy, Daniel M. Perkins, and Lee E. Brown. 2010. "Climate Change and Freshwater Ecosystems: Impacts across Multiple Levels of Organization." *Philosophical Transactions of the Royal Society B: Biological Sciences* 365 (1549): 2093–2106.

Worboys, Graeme L., Michael Lockwood, Ashish Kothari, et al., eds. 2015. *Protected Area Governance and Management.* Canberra: ANU Press.

World Bank Group. 2020. *Bangladesh Water Sector Diagnostic: Priorities for the New Decade (English)*. Washington, DC: World Bank Group.

World Commission on Dams. 2000. *Dams and Development: A New Framework for Decision-Making*. London: Earthscan Publications.

World Heritage Convention. 2022. *World Heritage List*. Paris: UNESCO.

World Wildlife Fund (WWF). 2023a. "Black-Necked Crane." *World Wildlife Fund*. Accessed 17 August 2023. https://en.wwfchina.org/en/what_we_do/species/fs/black_necked_crane/.

World Wildlife Fund (WWF). 2023b. "Ganges River Dolphin." *World Wildlife Fund*. Accessed 17 August 2023. https://www.worldwildlife.org/species/ganges-river-dolphin.

World Wildlife Fund (WWF) Bhutan. 2011. *The Wetlands Sky High: Mapping Wetlands in Bhutan*. Thimphu: WWF Bhutan.

World Wildlife Fund (WWF) / The Nature Conservancy (TNC). 2019. "Freshwater Ecoregions of the World: A Global Biogeographical Regionalization of the Earth's Freshwater Biodiversity." *WWF/TNC*. Accessed 19 November 2023. https://www.feow.org/.

Xiao, Weiye, Liyang Xiong, Guoan Tang, et al. 2014. "Geomorphological Divisions of the Tibet Plateau Based on Topographical Feature Point Groups from DEMs." *Annals of GIS* 20 (4): 245–253.

Xin, Xiaoge, Tianjun Zhou, and Rucong Yu. 2010. "Increased Tibetan Plateau Snow Depth: An Indicator of the Connection between Enhanced Winter NAO and Late-Spring Tropospheric Cooling over East Asia." *Advances in Atmospheric Sciences* 27 (4): 788–794.

Xin, Xiaoge, Wei Xue, Minghua Zhang, et al. 2015. "How Much of the NAO Monthly Variability Is from Ocean–atmospheric Coupling: Results from an Interactive Ensemble Climate Model." *Climate Dynamics* 44: 781–790.

Xinhua. 2021. "Three More Chinese Irrigation Projects Designated World Heritage Structures." *Xinhua*, November 27. Accessed 11 November 2023. http://www.news.cn/english/2021-11/27/c_1310337170.htm.

Xu, Hongzhang, and Jamie Pittock. 2018. "Giant Hydropower Dams and Freshwater Biodiversity Protection on the Mekong River." *Global Water Forum*, November 27. https://www.globalwaterforum.org/2018/11/27/giant-hydropower-dams-and-freshwater-biodiversity-protection-on-the-mekong-river/.

Xu, Hongzhang, and Jamie Pittock. 2019. "Limiting the Effects of Hydropower Dams on Freshwater Biodiversity: Options on the Lancang River, China." *Marine and Freshwater Research* 70 (2): 169–194.

Xu, Hongzhang, and Jamie Pittock. 2020a. "Policy Changes in Dam Construction and Biodiversity Conservation in the Yangtze River Basin, China." *Marine and Freshwater Research* 72 (2): 228–243.

Xu, Hongzhang, and Jamie Pittock. 2020b. "Saving Fish by Removing Dams on Small Rivers – The Case of the Upper Yangtze River Basin." *Global Water Forum*, August 21. https://www.globalwaterforum.org/2020/08/21/saving-fish-by-removing-dams-on-small-rivers-the-case-of-the-upper-yangtze-river-basin/.

Xu, Hongzhang, Jamie Pittock, and Katherine A. Daniell. 2021a. "China: A New Trajectory Prioritizing Rural Rather Than Urban Development?" *Land* 10 (5): 514.

Xu, Xiangke, and Chaolu Yi. 2014. "Little Ice Age on the Tibetan Plateau and Its Bordering Mountains: Evidence from Moraine Chronologies." *Global and Planetary Change* 116: 41–53.

Xu, Hongzhang, Aaron Tang, and Jamie Pittock. 2021b. "The Dark Side of Ambition: Side-Effects of China's Climate Policy." *Environmental Research Letters* 16 (11): 111001.

Xu, Hongzhang, Jamie Pittock, and Katherine Daniell. 2022. "Sustainability of What, for Whom? A Critical Analysis of Chinese Development Induced Displacement and Resettlement (DIDR) Programs." *Land Use Policy* 115: 106043.

Xu, Qinqin, Jianqing Ji, Dalai Zhong, et al. 2020c. "Post-Glacial Entrenchment and Knickpoint Migration of the Yarlung Tsangpo Gorge, Southeastern Tibetan Plateau." *Journal of Asian Earth Sciences* 195: 104337.

Xu, Weihua, Xinyue Fan, Jungai Ma, et al. 2019. "Hidden Loss of Wetlands in China." *Current Biology* 29 (18): 3065–3071.e2.

Yan, G., F. B. Wang, G. R. Shi, et al. 1999. "Palynological and Stable Isotopic Study of Palaeoenvironmental Changes on the Northeastern Tibetan Plateau in the last 30,000 Years." *Palaeogeography, Palaeoclimatology, Palaeoecology* 153 (1): 147–159.

Yan, Taiming, Jiayang He, Deying Yang, et al. 2022. "Fish Community Structure and Biomass Particle-Size Spectrum in the Upper Reaches of the Jinsha River (China)." *Animals* 12 (23): 3412.

Yang, Anna, Hao Wang, Weiming Liu, et al. 2022a. "Two Megafloods in the Middle Reach of Yarlung Tsangpo River since Last-Glacial Period: Evidence from Giant Bars." *Global and Planetary Change* 208: 103726.

Yang, Hai-Jun, and Liang-Jun Hu. 2013. "Reconciling the Regulations and Conservation of the Lhasa River, Tibet." *Journal of Soil and Water Conservation* 68 (2): 55A–57A.

Yang, Meixue, Tandong Yao, Yuanqing He, et al. 2000. "ENSO Events Recorded in the Guliya Ice Core." *Climatic Change* 47 (4): 401–409.

Yang, Nyimatashi Gongwei. 2017. "Water Politics and Religious Practices in Kangding." *Himalaya* 37 (1): 117–121.

Yang, Peng, Shengqing Zhang, Jun Xia, et al. 2022b. "Analysis of Drought and Flood Alternation and Its Driving Factors in the Yangtze River Basin under Climate Change." *Atmospheric Research* 270: 106087.

Yang, Rong, Hakro Ahmed Suhail, Loraine Gourbet, et al. 2020. "Early Pleistocene Drainage Pattern Changes in Eastern Tibet: Constraints from Provenance Analysis, Thermochronometry, and Numerical Modeling." *Earth and Planetary Science Letters* 531: 115955.

Yang, Rong, Sean D. Willett, and Liran Goren. 2015. "In Situ Low-Relief Landscape Formation as a Result of River Network Disruption." *Nature* 520 (7548): 526–529.

Yang, Y. C. Ethan, Sungwook Wi, Patrick A. Ray, et al. 2016a. "The Future Nexus of the Brahmaputra River Basin: Climate, Water, Energy and Food Trajectories." *Global Environmental Change* 37: 16–30.

Yang, Youqing, and Mian Liu. 2002. "Cenozoic Deformation of the Tarim Plate and the Implications for Mountain Building in the Tibetan Plateau and the Tian Shan." *Tectonics* 21 (6): 1059.

Yang, Yuzhong, Qingbai Wu, Hanbo Yun, et al. 2016b. "Evaluation of the Hydrological Contributions of Permafrost to the Thermokarst Lakes on the Qinghai–Tibet Plateau Using Stable Isotopes." *Global and Planetary Change* 140: 1–8.

Yang, Z., and K. Duan. 2022. "Intelligent Detection UAV for Illegal Fishing Based on YOLOv5 Algorithm." *2022 IEEE International Conference on Artificial Intelligence and Computer Applications (ICAICA)*: 950–957.

Yangka, Dorji, Vanessa Rauland, and Peter Newman. 2019. "Carbon Neutral Policy in Action: The Case of Bhutan." *Climate Policy* 19 (6): 672–687.

Yao, Tandong, Jianchen Pu, Anxin Lu, et al. 2007. "Recent Glacial Retreat and Its Impact on Hydrological Processes on the Tibetan Plateau, China, and Surrounding Regions." *Arctic, Antarctic, and Alpine Research* 39 (4): 642–650.

Yao, Tandong, Lonnie G. Thompson, Volker Mosbrugger, et al. 2012a. "Third Pole Environment (TPE)." *Environmental Development* 3: 52–64.

Yao, Tandong, Lonnie Thompson, Wei Yang, et al. 2012b. "Different Glacier Status with Atmospheric Circulations in Tibetan Plateau and Surroundings." *Nature Climate Change* 2 (9): 663–667.

Yao, Tandong, Valerie Masson-Delmotte, Jing Gao, et al. 2013. "A Review of Climatic Controls on $\delta^{18}O$ in Precipitation over the Tibetan Plateau: Observations and Simulations." *Reviews of Geophysics* 51 (4): 525–548.

Yasuda, Yumiko, Dipankar Aich, Douglas Hill, et al. 2017. *Transboundary Water Cooperation over the Brahmaputra River: Legal Political Economy Analysis of Current and Future Potential Cooperation.* The Hague: Hague Institute for Global Justice.

Yeakley, J. Alan, David Ervin, Heejun Chang, et al. 2016. "Ecosystem Services of Streams and Rivers." In *River Science: Research and Management for the 21st Century*, edited by David J. Gilvear, Malcolm T. Greenwood, Martin C. Thoms, et al., 335–352. Chichester: John Wiley & Sons.

Yeh, Emily T. 2005. "Green Governmentality and Pastoralism in Western China: 'Converting Pastures to Grasslands'." *Nomadic Peoples* 9 (1–2): 9–30.

Yeh, Emily T. 2009. "From Wasteland to Wetland? Nature and Nation in China's Tibet." *Environmental History* 14 (1): 103–137.

Yeh, Emily T. 2013. *Taming Tibet: Landscape Transformation and the Gift of Chinese Development.* New York: Cornell University Press.

Yin, Guoan, Fujun Niu, Zhanju Lin, et al. 2017. "Effects of Local Factors and Climate on Permafrost Conditions and Distribution in Beiluhe Basin, Qinghai-Tibet Plateau, China." *Science of the Total Environment* 581–582: 472–485.

Yoshikawa, Saeko. 2020. *William Wordsworth and Modern Travel: Railways, Motorcars and the Lake District, 1830–1940.* Liverpool: Liverpool University Press.

You, Qinglong, Fangying Wu, Liucheng Shen, et al. 2020. "Tibetan Plateau Amplification of Climate Extremes under Global Warming of 1.5°C, 2°C and 3°C." *Global and Planetary Change* 192: 103261.

You, Qinglong, Jinzhong Min, and Shichang Kang. 2016. "Rapid Warming in the Tibetan Plateau from Observations and CMIP5 Models in Recent Decades." *International Journal of Climatology* 36 (6): 2660–2670.

You, Q., S. Kang, G. Ren, et al. 2011. "Observed Changes in Snow Depth and Number of Snow Days in the Eastern and Central Tibetan Plateau." *Climate Research* 46 (2): 171–183.

Yuan, Y., Yang, S. and Zhang, Z., 2012. "Different Evolutions of the Philippine Sea Anticyclone between the Eastern and Central Pacific El Niño: Possible Effects of Indian Ocean SST." *Journal of Climate* 25(22): 7867–7883.

Yuan, X. P., R. Jiao, J. Liu-Zeng, et al. 2023. "Downstream Propagation of Fluvial Erosion in Eastern Tibet." *Earth and Planetary Science Letters* 605: 118017.

Yue, Han, Zheng-Kang Shen, Zeyan Zhao, et al. 2022. "Rupture Process of the 2021 M7.4 Maduo Earthquake and Implication for Deformation Mode of the

Songpan-Ganzi Terrane in Tibetan Plateau." *Proceedings of the National Academy of Sciences* 119 (23): e2116445119.

Zeitler, Peter K., Anne S. Meltzer, Peter O. Koons, et al. 2001. "Erosion, Himalayan Geodynamics, and the Geomorphology of Metamorphism." *GSA Today* 11 (1): 4–9.

Zeitoun, Mark, and Jeroen Warner. 2006. "Hydro-Hegemony – A Framework for Analysis of Trans-Boundary Water Conflicts." *Water Policy* 8 (5): 435–460.

Zeren, Bashen. 2021. "A Comparative Study of the Klu 'bum: Tibetan Bonpo Sources for an Understanding of the Cult of the Klu (Serpent Spirits)." PhD thesis (Paris).

Zhang, Ling. 2016. *The River, the Plain, and the State: An Environmental Drama in Northern Song China, 1048–1128*. Cambridge: Cambridge University Press.

Zhang, Maoliang, Zhengfu Guo, Sheng Xu, et al. 2021. "Linking Deeply-Sourced Volatile Emissions to Plateau Growth Dynamics in Southeastern Tibetan Plateau." *Nature Communications* 12 (1): 4157.

Zhang, Peiqi, Xinjun Zhang, Xiaoling Zhang, et al. 2022a. "Denisovans and Homo Sapiens on the Tibetan Plateau: Dispersals and Adaptations." *Trends in Ecology & Evolution* 37 (3): 257–267.

Zhang, Qi-Bin, Michael N. Evans, and Lixin Lyu. 2015. "Moisture Dipole over the Tibetan Plateau during the Past Five and a Half Centuries." *Nature Communications* 6 (1): 8062.

Zhang, Xiaolong, Sheng Xu, Lifeng Cui, et al. 2022b. "Erosions on the southern Tibetan Plateau: Evidence from In-Situ Cosmogenic Nuclides ^{10}Be and ^{26}Al in Fluvial Sediments." *Journal of Geographical Sciences* 32 (2): 333–357.

Zhang, Xuelei, Lijing Zhao, Daniel Q. Tong, et al. 2016. "A Systematic Review of Global Desert Dust and Associated Human Health Effects." *Atmosphere* 7 (12): 158.

Zhang, Y., S. Kang, M. Sprenger, et al. 2018. "Black Carbon and Mineral Dust in Snow Cover on the Tibetan Plateau." *The Cryosphere* 12 (2): 413–431.

Zhang, Yongqiang, Hongxing Zheng, Natasha Herron, et al. 2019. "A Framework Estimating Cumulative Impact of Damming on Downstream Water Availability." *Journal of Hydrology* 575: 612–627.

Zhang, Zhi, and Yu-lian Jia. 2019. "Different Provenance of Separate Loess Sites in Yangtze River Basin and Its Paleoenvironmental Implications." *Journal of Mountain Science* 16 (7): 1616–1628.

Zhao, H. L., X. Y. Zhao, et al. 2005. "Desertification Processes Due to Heavy Grazing in Sandy Rangeland, Inner Mongolia." *Journal of Arid Environments* 62 (2): 309–319.

Zhao, Shuangfeng, Wen Chen, Bin Zhang, et al. 2020. "Mesozoic and Cenozoic Exhumation History and Magmatic-Hydrothermal Events of the Central Tianshan Mt. Range, NW China: Evidence from (U–Th)/He and ^{40}Ar/^{39}Ar Dating." *Journal of Asian Earth Sciences* 194: 104057.

Zhao, Suisheng. 2004. *A Nation-State by Construction: Dynamics of Modern Chinese Nationalism*. Stanford: Stanford University Press.

Zhao, Xudong, Huiping Zhang, Ralf Hetzel, et al. 2021. "Existence of a Continental-Scale River System in Eastern Tibet during the Late Cretaceous–Early Palaeogene." *Nature Communications* 12 (1): 7231.

Zhao, Yahui, Jie Zhang, and Chunguang Zhang. 2008. "Fish Diversity on the Qinghai–Tibet Plateau." *Bulletin of Biology* 43 (7): 8–10.

Zhao, Z. L., Y. L. Zhang, L. S. Liu, et al. 2014. "Advances in Research on Wetlands of the Tibetan Plateau." *Progress in Geography* 33 (9): 1218–1230.

Zhao, Zhigang, and Xiaoming Shi. 2020. "Ecosystem Evolution of Alpine Wetland in Tibetan Plateau and Consideration for Ecological Restoration and Conservation." *Science and Technology Review* 38: 33–41.

Zhenhan, W., P. J. Barosh, W. Zhonghai, H. Daogong, Z. Xun, and Y. Peisheng (2008). "Vast Early Miocene Lakes of the Central Tibetan Plateau." *GSA Bulletin*, 120 (9/10), 1326–1337.

Zheng, Guoxiong, Simon Keith Allen, Anming Bao, et al. 2021. "Increasing Risk of Glacial Lake Outburst Floods from Future Third Pole Deglaciation." *Nature Climate Change* 11 (5): 411–417.

Zheng, Hongbo, Peter D. Clift, Mengying He, et al. 2020. "Formation of the First Bend in the Late Eocene Gave Birth to the Modern Yangtze River, China." *Geology* 49 (1): 35–39.

Zheng, Jingyun, Lingbo Xiao, Xiuqi Fang, et al. 2014. "How Climate Change Impacted the Collapse of the Ming Dynasty." *Climatic Change* 127 (2): 169–182.

Zhou, Tuyu, Hongzhang Xu, Jorrit Gosens, et al. 2022. *China's Net-Zero Plans: Policy Brief on Near-Term Policy Challenges & Australia-China Links in Decarbonization*. Canberra: Australian National University.

Zhou, Xiongdong, Zhaoyin Wang, Mengzhen Xu, et al. 2017. "Stream Power as a Predictor of Aquatic Macroinvertebrate Assemblages in the Yarlung Tsangpo River Basin (Tibetan Plateau)." *Hydrobiologia* 797 (1): 215–230.

Zhou, Zhen-Qiang, Shang-Ping Xie, and Renhe Zhang. 2021. "Historic Yangtze Flooding of 2020 Tied to Extreme Indian Ocean Conditions." *Proceedings of the National Academy of Sciences* 118 (12): e2022255118.

Zhu, Di-Cheng, Qing Wang, Peter A. Cawood, et al. 2017. "Raising the Gangdese Mountains in Southern Tibet." *Journal of Geophysical Research: Solid Earth* 122 (1): 214–223.

Zhu, Di-Cheng, Zhi-Dan Zhao, Yaoling Niu, et al. 2011. "The Lhasa Terrane: Record of a Microcontinent and Its Histories of Drift and Growth." *Earth and Planetary Science Letters* 301 (1): 241–255.

Zhu, Qingzhe, Yuzhi Liu, Tianbin Shao, et al. 2021a. "Role of the Tibetan Plateau in Northern Drought Induced by Changes in the Subtropical Westerly Jet." *Journal of Climate* 34 (12): 4955–4969.

Zhu, Ren, Dekui He, Xiu Feng, et al. 2021b. "The New Record of the Highest Distribution Altitude of Cyprinid Fishes in the World." *Journal of Applied Ichthyology* 37 (3): 474–478.

Zhu, Song, Zhen-han Wu, Xi-tao Zhao, et al. 2013. "Glacial Dammed Lakes in the Tsangpo River during Late Pleistocene, Southeastern Tibet." *Quaternary International* 298: 114–122.

Zhu, Tingbing, Xuemei Li, Xingbing Wu, et al. 2019. "Growth and Thermal Tolerance of a Tibet Fish Schizopygopsis Younghusbandi Juveniles Acclimated to Three Temperature Levels." *Journal of Applied Ichthyology* 35 (6): 1281–1285.

INDEX

Note: **Bold** page numbers refer to tables; *italic* page numbers refer to figures and page numbers followed by "n" denote endnotes.